SCHAUM'S OUTLINE OF

THEORY AND PROBLEMS

OF

PHYSICS FOR PRE-MED, BIOLOGY, AND ALLIED HEALTH STUDENTS

•

GEORGE J. HADEMENOS, Ph.D.
Visiting Assistant Professor
University of Dallas

•

SCHAUM'S OUTLINE SERIES

McGRAW-HILL

New York   Chicago   San Francisco   Lisbon
London   Madrid   Mexico City   Milan   New Delhi
San Juan   Seoul   Singapore   Sydney   Toronto

Dr GEORGE HADEMENOS, originally from San Angelo, Texas, earned his B.S. in Physics from Angelo State University, and both his M.S. and Ph.D. in Physics from the University of Texas at Dallas. While in graduate school, he developed a strong interest in biophysics and relevant applications of physics to the medical/biological sciences. Following receipt of his doctorate, he completed a postdoctoral fellowship in nuclear medicine at the University of Massachusetts Medical Center in Worcester, Massachusetts, followed by another postdoctoral fellowship in radiological sciences at UCLA Medical Center in Los Angeles. He is currently a Visiting Assistant Professor of Physics at the University of Dallas, where, among other courses, he is teaching General Physics geared toward biology and pre-med students. He is also in the process of developing an active research program for students interested in problems of a biophysical nature.

Schaum's Outline of Theory and Problems of
PHYSICS FOR PRE-MED, BIOLOGY, AND ALLIED HEALTH STUDENTS

4 5 6 7 8 9 10 11 12 13 14 15 16 17 18 19 20   VFM   VFM   0 9 8 7 6 5

ISBN 0-07-025474-5

Sponsoring Editor: Barbara Gilson
Production Supervisor: Sherri Souffrance
Editing Supervisor: Maureen B. Walker

Library of Congress Cataloging-in-Publication Data

Hademenos, George J.
    Schaum's outline of physics for pre-med, biology, and allied
health students / George J. Hademenos.
        p.   cm.— (Schaum's outline series)
    Includes index.
    ISBN 0-07-025474-5
    1. Biophysics—Outlines, syllabi, etc. 2. Physics—Outlines,
syllabi, etc.   I. Title   II. series.
QH505.H23   1997
612'.014—dc21   97-31139
                                                 CIP

McGraw-Hill
A Division of The McGraw-Hill Companies

This book is dedicated to:
Kelly, Alexandra, and Dr. George Kattawar,
my three greatest motivators

Preface

   Schaum's Outline of Physics for Pre-Med, Biology, and Allied Health Students is a text which complements the standard physics textbook in a fundamental physics course. The book is structured in a format similar to that of most physics textbooks used for such courses, except that the presented topics—including kinematics, statics, dynamics, thermal physics, electricity and magnetism, optics, and radioactivity—are supplemented with illustrative examples from the biological and medical sciences. The primary goal of this book is to provide the physics student with unique examples and applications to the life sciences as well as to simplify and enlighten the biology student in these sometimes difficult physical concepts and related principles.

   Although a tremendous amount of painstaking effort was devoted to this project, there will, most probably, be errors in this publication that were overlooked throughout the editorial process and I assume full responsibility for these errors. I would greatly appreciate if these errors could be brought to my attention either by direct correspondence with me or with the editorial staff at McGraw-Hill.

GEORGE J. HADEMENOS

Contents

Chapter 1

Mathematics Fundamentals

Physical principles and interactions encountered in physics are expressed typically in terms of qualitative reasoning and observations and rely on mathematics to translate these laws into quantitative or substantive results. Mathematics allows the student to intuitively and intellectually grasp physical concepts and to understand the effects of various factors on the particular physics problem. This chapter provides a brief description of the mathematics integral in the description and exemplification of physics as it relates to biological and medical applications.

1.1 FUNCTIONAL ANALYSIS

A function is defined as a mathematical one-to-one relationship between elements of a given set $X = \{x\}$, referred to as the *domain*, and another set $Y = \{y\}$, referred to as the *range*. A function represents a mathematical relationship between a dependent variable y and an independent variable x, which can represent any number within a specified range or interval. A function, denoted typically as $y = f(x)$, allows one to mathematically characterize a physical system in terms of variables that directly influence the system's behavior and stability.

ILLUSTRATIVE EXAMPLE 1.1.   Poiseuille's Law and Blood Flow

Blood flow through a blood vessel in the human circulation can be approximated by the elementary case of fluid flow through a rigid tube. Suppose, for example, we want to determine the influence of tube radius R on the rate of fluid flow Q through the tube. Although there exist variables such as fluid density and viscosity which are embedded in the system function as constants, theory and experimental data reveal a relationship between the variable R and the system variable Q according to

$$Q = \frac{\pi \Delta P R^4}{8L\eta}$$

where Q is the volumetric flow rate, ΔP is the pressure gradient, L is tube length, and η is the fluid viscosity. The relationship is known as *Poiseuille's law*. The rate of fluid flow Q is a dependent variable since its value depends upon the value of R, which is an independent variable and can be expressed as a function

$$Q = f(R)$$

When displayed graphically on a rectangular coordinate system with the two axes representing the two variables Q and R, the function defined by $Q = f(R) = \text{constant} \times R^4$ is described by a continuous curve in two dimensions, as shown in Figure 1-1. This example illustrates the strong "power" dependence of flow on the tube radius.

Solved Problem 1.1.   Identify the following mathematical statements as either functions or relations:

(a) $A = \pi R^2$

(b) $x + 5y = 15$

(c) $F = \{\text{human fingerprints}\}, P = \{\text{humans}\}$

(d) $x^2 + y^2 = 16$

(e) $f(x) = 7(-\infty < x < \infty)$

(f) $b(r) = 3r^2$

(g) $f(x) = \sqrt{x}$

(h) $x = \frac{1}{2}at^2$

1

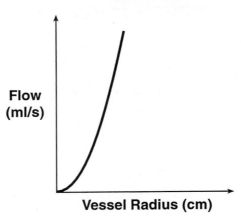

Poiseuille's Law and Blood Flow

Fig. 1-1

Solution

(a)   Function, $A = A(r)$ where π is a fixed constant.

(b)   Function, $y = 3 - x/5$.

(c)   Function, there is a one-to-one relationship between humans and their fingerprints.

(d)   Relation, $y^2 = 25 - x^2$ is not a function since for a single value of x, y can be either positive or negative.

(e)   Relation, not a function since different x values are one-to-one related to the same element, 7, of the range.

(f)   Function, since one value of r corresponds to a single value of $b(r)$.

(g)   Relation, depending on the domain. This is a function only for positive real x, and the square root means the positive square root.

(h)   Function, for reasons similar to (f) in that one value of t corresponds to a unique value of x.

1.1.1   Slope and y Intercept

Consider the function describing a line:

$$y = mx + b$$

Two important features of a function presented by the above equation are the y intercept b and the slope of the function or curve m. These features are illustrated in Figure 1-2. Ideal functions originate at the origin $(0, 0)$ of the coordinate system. Most functions, however, either originate from or pass through a point along the y axis other than the origin $(0, 0)$. The value along the y axis, known as the y intercept, is given by the parameter b. The *slope* defines the rate of change of the curve in the y axis with respect to the corresponding change in the x direction and is given mathematically as

$$\text{Slope} = m = \frac{\Delta y}{\Delta x}$$

The above relation is applicable only to linear functions.

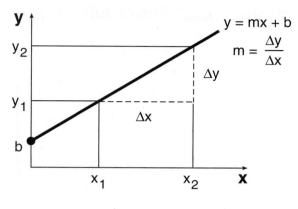

Fig. 1-2

ILLUSTRATIVE EXAMPLE 1.2.   Poiseuille's Law and Blood Flow Revisited

As mentioned previously, Poiseuille's law for fluid flow is a power function described by

$$y = cx^n \Rightarrow Q = \frac{\pi \Delta P}{8\eta L} R^4$$

where c is a constant and n is any real number. The power function presented by Poiseuille's law serves as a unique variation of the above discussion. As can be seen from Figure 1-1, the curve describing fluid flow is not linear, and thus the above expression for slope cannot be applied directly. It is possible, however, to transform the power function to a linear one by taking the logarithm of both sides:

$$\log y = n \log x + \log c$$
$$y = mx + b$$

By comparison, the slope for a power function is given by

$$n = \frac{\Delta(\log y)}{\Delta(\log x)}$$

Application of Poiseuille's law for fluid flow yields

$$\log Q = 4 \log R + \log \frac{\pi \Delta P}{8\eta L}$$

Thus, the slope of Poiseuille's law is 4, implying a 4-fold increase in flow with respect to a change in tube radius. To put this in perspective, when applied to the human circulation, doubling the radius of a blood vessel increases blood flow by a factor of 16 whereas halving the radius reduces blood flow by a factor of $\frac{1}{16}$.

1.1.2 Coordinate Systems

Almost as important as the function itself are the geometric boundaries within which the function is defined. This can be understood by the following illustrative example. Consider a point in space $p(x, y, z)$ bounded by a rectangular coordinate system given by $\mathbf{x} = x$, $\mathbf{y} = y$, and $\mathbf{z} = z$. However, if that same point is now defined within a different geometry such as a cylinder or sphere, it is a cumbersome, although possible, task to define the point in rectangular coordinates. It thus becomes a matter of convenience to denote any point within a sphere or cylinder in terms of the variables that more precisely define its geometry in space through coordinate transformations between different geometries.

Cylindrical and Spherical Coordinates

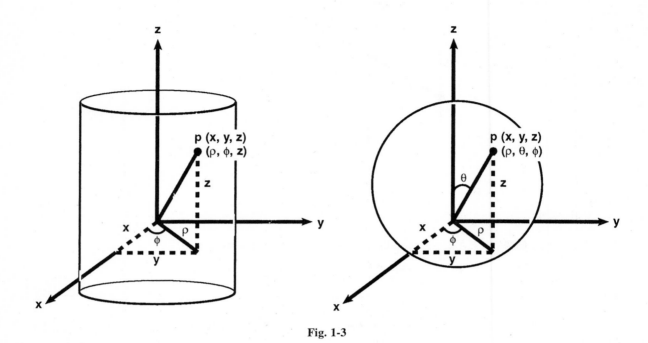

Fig. 1-3

Given the geometry of the cylinder and sphere depicted in Figure 1-3, the three-dimensional coordinates of a point in space which were defined originally as x, y, and z, in the rectangular coordinate system, are now redefined for the cylinder in cylindrical coordinates (ρ, ϕ, z) as

$$x = \rho \cos \phi$$
$$y = \rho \sin \phi$$
$$z = z$$

where $\rho \geq 0$, $0 \leq \phi < 2\pi$, $-\infty < z < \infty$; and for the sphere in spherical coordinates (ρ, θ, ϕ) as

$$x = \rho \sin \theta \cos \phi$$
$$y = \rho \sin \theta \sin \phi$$
$$z = \rho \cos \theta$$

where $\rho \geq 0$, $0 \leq \theta \leq \pi$, $0 \leq \phi < 2\pi$.

1.2 TYPES OF FUNCTIONS

Although numerous mathematical statements meet the criteria of a function, there are five general classifications of functions, described below.

1.2.1 Polynomial Functions

Polynomial functions are the most general of the five types of functions and of the form

$$y = f(x) = a_0 + a_1 x + a_2 x^2 + a_3 x^3 + \cdots + a_n x^n$$

Polynomial Functions

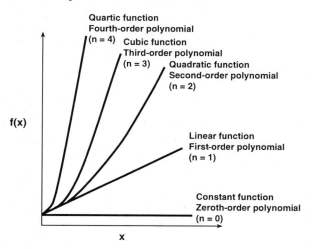

Fig. 1-4

where a_0, a_1, a_3, a_n are coefficients (constants) and n is an integer which describes the degree or order of the polynomial. The most general polynomial function is the first-order polynomial function given by

$$y = f(x) = a_0 + a_1 x \qquad a_1 \neq 0$$

which is a linear function (compare with Section 1.1.1). In the linear function, a_1 corresponds to the slope m. Figure 1-4 displays graphically the more common polynomial functions.

ILLUSTRATIVE EXAMPLE 1.3.   Applications of Polynomial Functions

Polynomial functions represent a wide range of mathematical functions and can be found in numerous physical applications as they pertain to biological sciences, including:

(1) *Ohm's law of electricity.* The voltage V applied across a wire is related to the current flowing through the wire I, multiplied by the resistance R of the wire, or

$$V = IR$$

(2) *Temperature conversion.* The Celsius temperature scale (°C) and the Fahrenheit scale (°F) are linearly related by

$$T_{\,^\circ\mathrm{C}} = \frac{5}{9}(T_{\,^\circ\mathrm{F}} - 32)$$

(3) *Lineweaver-Burk plot of enzyme kinetics.* The influence of an enzyme on the speed of a chemical reaction can be observed through the relationship between the substrate concentration $[S]$ of the enzyme and the velocity of the reaction v by

$$\frac{1}{v} = \frac{K_m}{V_{\max}} \frac{1}{[S]} + \frac{1}{V_{\max}}$$

where $V_{\max}$ is the maximum velocity of the chemical reaction and K_m is a constant.

Solved Problem 1.2.   Normal human body temperature is approximately 98.6°F. Calculate this temperature in the Celsius scale.

Solution

The mathematical relation of importance is

$$T_{^\circ\text{C}} = \frac{5}{9}(T_{^\circ\text{F}} - 32)$$

Substituting the body temperature 98.6°F yields

$$T_{^\circ\text{C}} = \frac{5}{9}(98.6 - 32) = 37^\circ\text{C}$$

Solved Problem 1.3. For the Lineweaver-Burk equation of enzyme kinetics, identify the slope and y intercept.

Solution

The Lineweaver-Burk equation of enzyme kinetics is

$$\frac{1}{v} = \frac{K_m}{V_{\max}}\frac{1}{[S]} + \frac{1}{V_{\max}}$$

Since it is a linear function, the above equation follows the general form of

$$y = mx + b$$

By an elementary comparison, the slope m is

$$\text{Slope} = \frac{K_m}{V_{\max}}$$

and the y intercept b is

$$y \text{ intercept} = \frac{1}{V_{\max}}$$

Solved Problem 1.4. The absorption of ethyl alcohol into the bloodstream in the human body is one of the few known kinetic processes that is zeroth-order. Assume that the process may be represented as

Ethyl alcohol → bloodstream

Derive the integrated rate law for the biochemical process, and show that this is a linear function.

Solution

Assuming that ethyl alcohol can be represented by $[B]$, the reaction for the process can be mathematically characterized by

$$-\frac{dB}{dt} = k$$
$$-dB = k\,dt$$

Integrating both sides of the equation yields

$$\int_{B_0}^{B} dB = -k\int_{0}^{t} dt$$
$$B - B_0 = -kt$$
$$B = B_0 - kt$$

Integration is covered in greater detail in Section 1.4. By comparison, it can be seen that this rate law is a linear function, where $-k$ is the slope and B_0 is the y intercept.

Power Functions

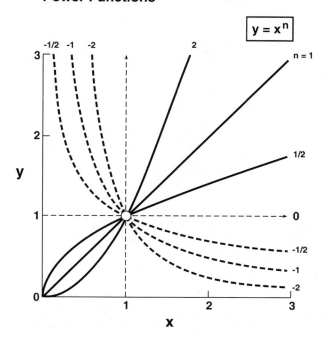

Fig. 1-5

1.2.2 Power Functions

Power functions are defined by

$$y = cx^n$$

where c is a constant and n is a real number, representing the mathematical power of the function. Typical power functions are displayed in Figure 1-5. Mathematical rules involving powers are summarized below: Let p and q be any numbers and $a > 0$ but $a \neq 1$:

$$a^p a^q = a^{p+q} \qquad a^p + a^q \neq a^{p+q} \qquad \frac{a^p}{a^q} = a^{p-q} \qquad (a^p)^q = a^{pq}$$

$$a^{m/n} = \sqrt[n]{a^m} \qquad a^{-n} = \frac{1}{a^n} \qquad a^{1/n} = \sqrt[n]{a} \qquad a^0 = 1$$

Several forms of the power function include

$$n = -2: \qquad y = x^{-2} = \frac{1}{x^2}$$

$$n = -1: \qquad y = x^{-1} = \frac{1}{x}$$

$$n = -\tfrac{1}{2}: \qquad y = x^{-1/2} = \frac{1}{\sqrt{x}}$$

$$n = 0: \qquad y = x^0 = 1$$

$$n = \tfrac{1}{2}: \qquad y = x^{1/2} = \sqrt{x}$$

$$n = 1: \qquad y = x^1 = x$$

$$n = 2: \qquad y = x^2$$

ILLUSTRATIVE EXAMPLE 1.4. Applications of Power Functions

Examples of power functions in the physical and biological sciences include these:

(1) *Blood flow in the circulation.* Blood flow is typically described by Poiseuille's law for fluid flow in which the rate of fluid flow Q through a tube is dependent on the tube radius R, tube length L, and pressure gradient ΔP according to the relation

$$Q = \frac{\pi \Delta P}{8\eta L} R^4$$

(2) *Basal metabolic rate.* Basal metabolic rate (BMR) represents the rate of metabolic energy production for an average person at rest and, for mammals, is approximately related to the body mass M by

$$\text{BMR} = aM^b$$

where a and b are constants.

(3) *Sedimentation coefficient of linear DNA molecules.* The sedimentation coefficient s reflects the rate at which a macromolecule sediments or separates from a solution in a centrifugal field created by rapid rotational motion and is related to the molecular weight M of the macromolecule. For linear DNA molecules, the sedimentation coefficient is defined empirically by

$$s = 2.8 + 8.34 \times 10^{-3} M^{0.49}$$

Solved Problem 1.5. By how much will blood flow through an artery be reduced for a sudden (*a*) one-quarter, (*b*) one-half, and (*c*) three-quarters reduction in blood vessel radius?

Solution

Since we are concerned only with the vessel radius, all other flow parameters are assumed constant, and Poiseuille's law can be written as

$$Q = (\text{constant})R^4$$

(*a*) A vessel reduced by one-quarter implies a vessel radius of three-quarters. The flow reduction is

$$Q_{\text{new}} = (\text{constant})\left(\frac{3}{4}\right)^4 = 0.32 Q_{\text{original}}$$

(*b*) A vessel reduced by one-half implies a vessel radius of one-half. The flow reduction is

$$Q_{\text{new}} = (\text{constant})\left(\frac{1}{2}\right)^4 = 0.063 Q_{\text{original}}$$

(*c*) A vessel reduced by three-quarters implies a vessel radius of one-quarter. The flow reduction is

$$Q_{\text{new}} = (\text{constant})\left(\frac{1}{4}\right)^4 = 0.039 Q_{\text{original}}$$

1.2.3 Trigonometric Functions

With the periodic nature of many biological processes including cardiac physiology and blood flow, one can deduce easily the potential importance of trigonometric functions and their role in the biological sciences. Trigonometric functions are employed typically in the description of periodic or oscillatory phenomena. The family of trigonometric functions includes

$$\sin\theta, \cos\theta, \tan\theta, \cot\theta, \sec\theta, \csc\theta$$

where the argument θ is a given angle in units of radians ($1\,\text{rad} = 57.3°$). Figure 1-6 defines analytically the trigonometric identities with respect to the components of a triangle with a corresponding graph of the variation of angle θ for each of the trigonometric functions, illustrating their periodic nature.

Trigonometric Identities

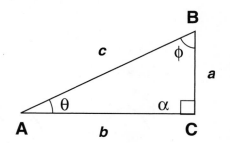

$$\mathbf{sin}\ \theta = \frac{a}{c} = \frac{\text{opposite}}{\text{hypotenuse}} \qquad \mathbf{cos}\ \theta = \frac{b}{c} = \frac{\text{adjacent}}{\text{hypotenuse}}$$

$$\mathbf{tan}\ \theta = \frac{a}{b} = \frac{\text{opposite}}{\text{adjacent}} \qquad \mathbf{cot}\ \theta = \frac{b}{a} = \frac{\text{adjacent}}{\text{opposite}}$$

$$\mathbf{sec}\ \theta = \frac{c}{b} = \frac{\text{hypotenuse}}{\text{adjacent}} \qquad \mathbf{csc}\ \theta = \frac{c}{a} = \frac{\text{hypotenuse}}{\text{opposite}}$$

Fig. 1-6a

These are some of the more important relations between the trigonometric functions:

$$\sin(-\theta) = -\sin\theta \qquad \cos(-\theta) = \cos\theta \qquad \tan(-\theta) = -\tan\theta$$

$$\sec\theta = \frac{1}{\cos\theta} \qquad \csc\theta = \frac{1}{\sin\theta} \qquad \cot\theta = \frac{1}{\tan\theta}$$

$$\tan\theta = \frac{\sin\theta}{\cos\theta} \qquad \sin^2\theta + \cos^2\theta = 1 \qquad \sec^2\theta - \tan^2\theta = 1$$

ILLUSTRATIVE EXAMPLE 1.5.   Applications of Trigonometric Functions

Examples of trigonometric functions in the physical and biological sciences include these:

(1) *Resolution of vectors.* The most important application of trigonometric functions to physics is the resolution of vectors into their respective components. Given a vector $\mathbf{V} = V_x\mathbf{i} + V_y\mathbf{j}$, each vector component is determined by projecting the vector along its respective coordinate axis according to the following:

$$V_x = \mathbf{V}\cos\theta \qquad \text{and} \qquad V_y = \mathbf{V}\sin\theta$$

(2) *Simple harmonic motion.* An object exhibiting simple harmonic or periodic motion can be described by the wave equation

$$x = A\cos(\omega t + \phi)$$

where x is the position or instantaneous amplitude of the object at a given time t, A is the maximum amplitude, ω is the angular frequency, and ϕ is the phase constant.

(3) *Snell's law of refraction.* Light rays traveling from one medium, described by n_1 and θ_1, will refract upon entrance at an angle θ_1 into a second medium, described by n_2, at an angle θ_2, according to the relation

$$n_1 \sin\theta_1 = n_2 \sin\theta_2$$

where the angles θ_1 and θ_2 are defined with respect to a line perpendicular to the interface of the two media.

Solved Problem 1.6.   In moving a wagon, Max exerted a 5-N force at an angle of 40°. How much of the force was exerted along the x direction, and how much of the force was exerted along the y direction?

Solution

This problem can be solved by resolving the force vector into its components:

$$V_x = 5\,\text{N}\cos 40° = (5\,\text{N})(0.766) = 3.8\,\text{N}$$
$$V_y = 5\,\text{N}\sin 40° = (5\,\text{N})(0.642) = 3.2\,\text{N}$$

Trigonometric Functions

y = sin x

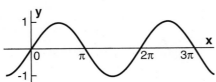

y = cos x

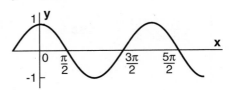

y = tan x

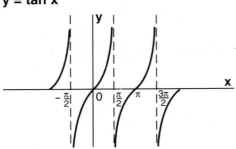

y = cot x

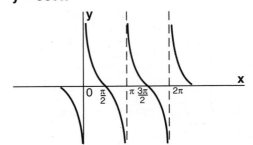

y = sec x

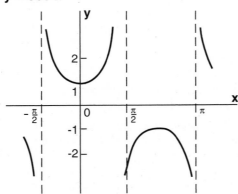

y = csc x

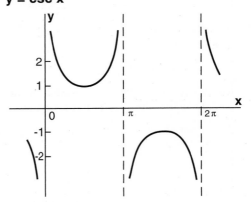

Fig. 1-6b

Exponential Functions

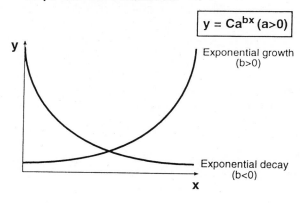

Fig. 1-7

1.2.4 Exponential Functions

Exponential functions are defined mathematically as $f(x) = Ca^{bx}$ for $a > 0$, where a, b, and C are constants. Exponential functions describe many biological processes, particularly when $b > 0$ (growth processes) and when $b < 0$ (decay processes). They are especially useful when a is equal to $e = 2.718$. In this case, exponential functions are expressed mathematically as $f(x) = Ce^{bx}$. Figure 1-7 shows exponential functions representing growth and decay processes.

ILLUSTRATIVE EXAMPLE 1.6.   Applications of Exponential Functions

Examples of exponential functions in the biological and physical sciences include these:

(1) *Radioactive decay.* Decay of a radioactive nucleus proceeds according to the following exponential relation:

$$N = N_0 e^{-\lambda t}$$

where N is the number of radioactive atoms in a sample at a given time t, N_0 is the original number of radioactive atoms, and λ is a decay constant or a constant describing the rate of decay unique to the radioactive nucleus.

(2) *Beer-Lambert law.* As light is transmitted through an absorber of thickness x, the change in its intensity can be mathematically described by

$$I(x) = I_0 e^{-kx}$$

where $I(x)$ is the intensity at a given thickness of the absorber, I_0 is the initial light intensity, and k is a proportionality constant, specific for the particular absorber.

(3) *Bacterial growth.* The growth of *Escherichia coli* in a sufficient nutrient medium can be mathematically described by

$$N(t) = N_0 2^{t/T_0}$$

where N is the number of bacteria at a given time t, N_0 is the original number of bacteria, and T_0 is the generation or doubling time.

Solved Problem 1.7.   Carbon dating or biochronology is based on the fact that all living organisms incorporate the carbon necessary to maintain metabolic functions from CO_2 in the atmosphere either directly or indirectly. And ^{14}C is produced in the atmosphere from ^{14}N according to the reaction

$$^{14}N + neutron \longrightarrow \ ^{14}C + \ ^1H$$
(from cosmic radiation)

As a result, living tissue contains ^{14}C sufficient to produce a specific radioactivity of 15.3 ^{14}C disintegrations per minute per gram of total carbon in the tissue (dpm/g_C). After death, the organism stops the incorporation of ^{14}C, which does, however, continue to disintegrate with a half-life of $t_{1/2} = 5730$ years. If $N(t)$ represents the number of ^{14}C atoms at present (at time t), then at $t = 0$ (at time of death)

$$N(t) = N(0)e^{-(\ln 2/t_{1/2})}$$

A skull found at an archeological dig exhibited a ^{14}C activity of 10 dpm/g_C. What is the age of the skull? (*Hint*: The specific radioactivity is proportional to the number of ^{14}C atoms.)

Solution

From the hint,

$$\frac{a(t)}{a(0)} = \frac{N(t)}{N(0)}$$

From the above equation, we must solve for t:

$$
\begin{aligned}
t &= \frac{t_{1/2}}{\ln 2} \ln \frac{N(0)}{N(t)} = \frac{t_{1/2}}{\ln 2} \ln \frac{a(0)}{a(t)} \\
&= \frac{5730 \text{ years}}{0.693} \ln \frac{15.3 \text{ dpm}/g_C}{a(t) \text{ dpm}/g_C} \\
&= 8.27 \times 10^3 \text{ years} \times \ln \frac{15.3 \text{ dpm}/g_C}{10.0 \text{ dpm}/g_C} \\
&= 3.52 \times 10^3 \text{ years}
\end{aligned}
$$

Solved Problem 1.8. Bacterial growth can be described mathematically by

$$N(t) = N_0 2^{t/T_0}$$

where $N(t)$ is the number of bacteria at time t, N_0 is the original number of bacteria, and T_0 is the generation or doubling time. Calculate the number of bacteria generated overnight ($t \approx 15\,h$) from a single bacterium, assuming a generation time of 30 min.

Solution

$$N(t = 15\,h) = (1 \text{ bacterium}) \times 2^{15\,h/0.5\,h} = 2^{30} \text{ bacteria} = 10^9 \text{ bacteria}$$

1.2.5 Logarithmic Functions

The logarithm of a number is the exponent to which a given base must be raised to give you the number. For example, if 10 is the base, then $\log_{10} 100 = 2$. Logarithmic functions, shown in Figure 1-8, are defined by $y = \log_a x$ for $a > 0$, $a \neq 1$. Several mathematical relations involving logarithmic functions are expressed below:

$$\log(y_1 y_2) = \log y_1 + \log y_2 \qquad y_1 y_2 > 0$$
$$\log \frac{y_1}{y_2} = \log y_1 - \log y_2 \qquad y_2 \neq 0, \frac{y_1}{y_2} > 0$$
$$\log(y_1)^{y_2} = y_2 \log y_1 \qquad y_1 > 0$$

The argument of a logarithmic function, that is, x, can be determined easily by

$$x = a^{\log_a x} = a^y$$

Given the exponential function $f(x) = e^x$, the logarithm of $f(x) = \log_e e^x$ and can also be stated as $f(x) = \ln e^x$.

Logarithmic Functions

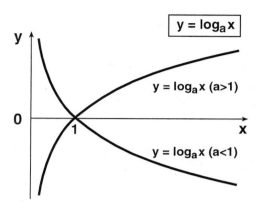

Fig. 1-8

ILLUSTRATIVE EXAMPLE 1.7.   Applications of Logarithmic Functions

Examples of logarithmic functions in the biological and physical sciences include these:

(1) *The pH of a solution.* The *pH* of a solution represents the hydrogen ion activity H^+ of a solution and is defined by

$$pH = -\log[H^+] = \log\frac{1}{[H^+]}$$

Another related application is the Henderson-Hasselbalch equation. Buffer systems designed to resist changes in pH consist of a weak acid, HA, and a salt of the weak acid and can be developed according to the Henderson-Hasselbalch equation:

$$pH = pK_a + \log\frac{[A^-]}{[HA]}$$

where $[A^-]$ is the concentration of the dissociated anion (conjugate base) of the weak acid, $[HA]$ is the concentration of the weak acid, and pK_a is the negative logarithm of the equilibrium constant for the dissociation of the acid (K_a).

(2) *Decibel scale of sound intensity.* The intensity level or loudness of a sound is defined in terms of decibels (dB), expressed mathematically as

$$dB = 10\log_{10}\frac{I}{I_0}$$

where I is the intensity of a given sound and I_0 is a reference sound intensity ($=1 \times 10^{-12}$ W/m^2) corresponding to the lower limits of sound intensity detectable by the human ear.

(3) *Work by an ideal gas.* The work done by an ideal gas W undergoing an isothermal process is given by

$$W = nRT\ln\frac{V_2}{V_1}$$

where n is the number of moles of gas in a sample, R is the universal gas constant, T is the temperature, V_2 is the final volume upon expansion, and V_1 is the initial volume.

Solved Problem 1.9.   One mole of an ideal monoatomic gas initially at 5.0 L and 298 K experienced an isothermal compression to one-half the initial volume. Calculate the work done by the gas.

Solution

The work done by an ideal gas is given by

$$W = nRT \ln \frac{V_2}{V_1}$$

where n is the number of moles ($n = 1$), R is the universal gas constant

$$R = 1.987 \frac{\text{cal}}{(\text{mol} \cdot \text{K})}$$

T is the temperature ($T = 298\,\text{K}$), V_1 is the initial volume ($V_1 = 5\,\text{L}$), and V_2 is the final volume ($V_2 = V_1/2 = 2.5\,\text{L}$). Substituting into the above equation yields

$$W = (1\,\text{mol})\left[1.987 \frac{\text{cal}}{(\text{mol} \cdot \text{K})}\right](298\,\text{K}) \ln \frac{2.5}{5} = -410\,\text{cal}$$

1.3   DIFFERENTIAL CALCULUS

The basis for differential calculus is the derivative which is fundamental to the characterization of dynamic physical processes. The derivative of a function, depicted symbolically as $df(x)/dx (= dy/dx)$ or $f'(x)$, yields the tangent of a mathematical function or the rate of increase (or decrease) along the y axis with respect to the x axis (Figure 1-9). In the above expression, dy and dx are termed *differentials*. Although the differentials dx and dy are the same as Δx and Δy in meaning, they represent infinitesimal changes along their respective axes. The derivative dy/dx can be equated with the slope $\Delta y/\Delta x$ only in the limit as Δx becomes very small and approaches 0. This is expressed typically as

$$\frac{dy}{dx} = \lim_{\Delta x \to 0} \frac{\Delta y}{\Delta x}$$

Upon closer inspection, we see that this, in essence, is the definition of a slope for a linear function, and they are, in fact, one and the same. A tabulated list of the formulas for derivatives of the most common functions described in Section 1.2 is shown in Table 1.1.

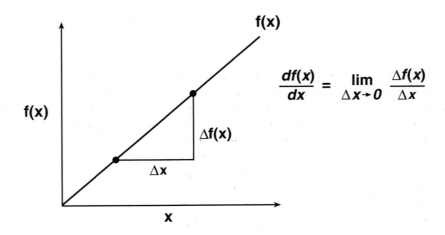

Fig. 1-9

ILLUSTRATIVE EXAMPLE 1.8.   Particle Motion

A physical application of the derivative can be illustrated with the problem of motion of a particle or object. Consider an object that moves in space according to the function

$$\mathbf{x} = f(t) = 4t^3 + 10t^2 + 6$$

The boldface type implies that the parameter represents a vector quantity (Chapter 3). Referring to Table 1.1, we see the derivative of a polynomial function is given as

$$\frac{df(x)}{dx} = \frac{d(x^n)}{dx} = nx^{n-1}$$

Then the velocity of the object is the derivative of $f(t)$ or $\mathbf{x}$,

$$\mathbf{v} = \frac{d\mathbf{x}}{dt} = 12t^2 + 20t$$

and the acceleration of the object is the derivative of the velocity $\mathbf{v}$, or the second derivative of $\mathbf{x}$, or

$$\mathbf{a} = \frac{d\mathbf{v}}{dt} = \frac{d^2\mathbf{x}}{dt^2} = 24t + 20$$

where the superscript 2 refers to the order of the derivative. The applications of a derivative include identification of maximum and minimum points of a function, description of the rate of change of a function $f(x)$ with respect to x, and determination of the velocity and acceleration of a particle given its position, as just demonstrated by the above example.

Solved Problem 1.10.   Consider a wave whose position x in space is defined by the relation

$$\mathbf{x} = A \cos \omega t$$

where A is the wave amplitude, ω is the angular frequency, and t is time. Determine the velocity and acceleration of the wave.

Solution

The velocity of the wave can be determined by the time derivative of the position:

$$\mathbf{v} = \frac{d\mathbf{x}}{dt} = -\omega A \sin \omega t$$

Similarly, the acceleration can be determined by the time derivative of the velocity:

$$\mathbf{a} = \frac{d\mathbf{v}}{dt} = \omega^2 A \cos \omega t$$

which can be simplified as

$$\mathbf{a} = \omega^2 \mathbf{x}$$

Table 1.1.   Differentiation Formulas for Common Mathematical Functions

Polynomial Functions

$$\frac{d}{dx}(a_0 x^n) = na_0 x^{n-1}$$

$$\frac{d}{dx}(a_0 u^n) = na_0 u^{n-1}\frac{du}{dx}$$

Trigonometric Functions

$$\frac{d}{dx}(\sin u) = \cos u \frac{du}{dx}$$

$$\frac{d}{dx}(\cos u) = -\sin u \frac{du}{dx}$$

$$\frac{d}{dx}(\tan u) = \sec^2 u \frac{du}{dx}$$

$$\frac{d}{dx}(\sec u) = -\sec u \tan u \frac{du}{dx}$$

$$\frac{d}{dx}(\csc u) = -\csc u \cot u \frac{du}{dx}$$

$$\frac{d}{dx}(\cot u) = -\csc^2 u \frac{du}{dx}$$

Exponential Functions

$$\frac{d}{dx}(a^u) = a^u \ln a$$

$$\frac{d}{dx}(e^u) = e^u \frac{du}{dx}$$

Logarithmic Functions

$$\frac{d}{dx}(\log_a u) = \frac{\log_a e}{u}\frac{du}{dx} \qquad a \neq 0, 1$$

$$\frac{d}{dx}(\ln u) = \frac{1}{u}\frac{du}{dx}$$

1.4   INTEGRAL CALCULUS

The integral is to integral calculus what the derivative is to differential calculus and serves just as important a role in the characterization of dynamic processes. Integration of a given function $f(x)$, in essence, is the inverse of differentiation and is defined mathematically by

$$\int f(x)\,dx$$

where dx is the differential along the x axis. The integral represents the area under the curve $y = f(x)$ bounded by limits of x_1 and x_2 along the x axis and predefined limits along the y axis. The area under the curve is subdivided into equal regions of size dx, as shown in Figure 1-10. By integrating the function, one is, in effect, summing all the dx regions bounded by the function curve.

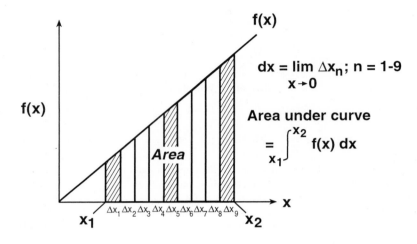

Fig. 1-10

ILLUSTRATIVE EXAMPLE 1.9.   Particle Motion Revisited

To show the relationship between the integral and derivative, we use the same illustrative example of particle motion presented in Section 1.3. However, to demonstrate integration, we must begin in a manner opposite to that presented earlier. Thus, the equation for the acceleration of the particle

$$\mathbf{a} = \frac{d\mathbf{v}}{dt} = \frac{d^2\mathbf{x}}{dt} = 24t + 20$$

is integrated to obtain the velocity of the particle. A tabulated list of the integral formulas of the most common mathematical functions is given in Table 1.2. The integral for a polynomial, given by

$$\int u^p\,du = \frac{u^{p+1}}{p+1} \qquad p \neq -1$$

is applied to the particle acceleration, and, upon solving, it yields the velocity

$$\mathbf{v} = \frac{d\mathbf{x}}{dt} = 12t^2 + 20t$$

Integrating once again, using the integral for a polynomial function, gives the position of the particle:

$$\mathbf{x} = f(t) = 4t^3 + 10t^2 + C$$

where C is an integration constant ($= 6$).

Solved Problem 1.11. For a particle whose velocity is described by

$$\mathbf{v} = \omega A \cos \omega t$$

determine its position.

Solution

The position of the particle can be determined by integrating the velocity:

$$x = \int \mathbf{v}\,dt = \int \omega A \cos \omega t\, dt = A \sin \omega t$$

Solved Problem 1.12. Given the function

$$y = x^3 + 2x^2 + 3x + 4$$

determine the (1) slope of the curve at $x = 5$ and (2) area under the curve from $x = 1$ to $x = 4$.

Solution

(1) The slope of the function is given by its derivative

$$\text{Slope} = \frac{dy}{dx} = \frac{d(x^3 + 2x^2 + 3x + 4)}{dx} = \frac{d(x^3)}{dx} + \frac{d(2x^2)}{dx} + \frac{d(3x)}{dx} + \frac{d(4)}{dx}$$
$$= 3x^2 + 4x + 3$$

At $x = 5$, the value of the slope is

Slope $(x = 5) = 3(5)^2 + 4(5) + 3 = 75 + 20 + 3 = 98$

(2) The area under the curve is simply the integral of the function over the limits $x = 1$ to $x = 4$:

$$\text{Area} = \int_{x=1}^{x=4} y\,dx = \int_{x=1}^{x=4} x^3\,dx + 2\int_{x=1}^{x=4} x\,dx + 3$$
$$\times \int_{x=1}^{x=4} x\,dx + 4\int_{x=1}^{x=4} dx$$
$$= \frac{x^4}{4}\Big|_{x=1}^{x=4} + 2\frac{x^3}{3}\Big|_{x=1}^{x=4} + 3\frac{x^2}{2}\Big|_{x=1}^{x=4} + 4x\Big|_{x=1}^{x=4}$$
$$= \left(\frac{4^4}{4} - \frac{1^4}{4}\right) + 2\left(\frac{4^3}{3} - \frac{1^3}{3}\right) + 3\left(\frac{4^2}{2} - \frac{1^2}{2}\right) + 4(4 - 1)$$
$$= 140.25$$

Table 1.2. Integration Formulas for Common Mathematical Functions

Polynomial Functions
$\int u^p\,du = \dfrac{u^{p+1}}{p+1} \qquad p \neq 1$
$\int u^{-1}\,du = \int \dfrac{du}{u} = \ln u$

Trigonometric Functions
$\int \sin u\,du = -\cos u$
$\int \cos u\,du = \sin u$
$\int \tan u\,du = -\ln \cos u$
$\int \cot u\,du = \ln \sin u$
$\int \sec u\,du = \ln(\sec u + \tan u)$
$\int \csc u\,du = \ln(cscu - \cot u)$

Exponential Functions
$\int a^u\,du = \dfrac{a^u}{\ln a} \qquad a > 0, a \neq 1$
$\int e^u\,du = e^u$

Logarithmic Functions
$\int \ln x\,dx = x \ln x - x$

1.5 ORDINARY DIFFERENTIAL EQUATIONS

As stated previously, a variety of physical and biological phenomena are dynamic and thus exhibit a time-dependent behaviour that can be described by using ordinary differential equations. Differential equations are defined simply as mathematical equations that contain derivatives. The solution of a differential equation is generally such that the derivative of a function either is identical or maintains a distinct similarity to the original function. Generally speaking, the two types of mathematical functions which satisfy this criterion in most cases are exponential functions and trigonometric functions. An example of a differential equation with numerous applications to physics and biology is the growth and decay of a given process.

ILLUSTRATIVE EXAMPLE 1.10.   Growth and Decay Biological Processes

Growth and decay processes describe the time rate of change of a physical entity in terms of its initial state and can be summarized mathematically as

$$\frac{dN(t)}{dt} = kN(t)$$

where $N(t)$ describes the entity at time t and k is a constant describing a growth process if $k > 0$ and a decay process is $k < 0$. The solution to this differential equation should be a function that is identical in form to its derivative. One type of function which satisfies this requirement is the exponential function, yielding a solution:

$$N(t) = N_o e^{kt}$$

where N_o describes the original value of the entity at $t = 0$.

ILLUSTRATIVE EXAMPLE 1.11.   Harmonic Oscillator

A harmonic oscillator describes any physical system capable of producing oscillatory or periodic phenomena such as a mass fastened to a spring. In this system, the mass is at rest and the system is in equilibrium. When pulled a finite distance x past the point of equilibrium and released, the object responds with oscillatory or periodic motion about the point of equilibrium until the energy of the spring ultimately dissipates and subsequently stops the imposed motion of the mass, as shown in Figure 1-11. Hooke's law relates the force exerted on the mass to the distance by the spring as

$$F = -kx$$

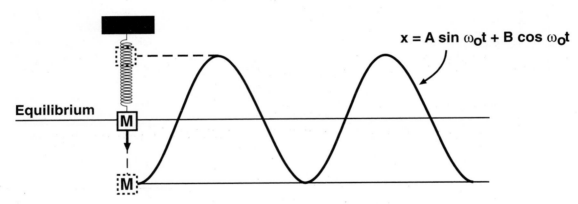

Fig. 1-11

The force applied to the attached mass is defined, according to Newton's second law, by the product of its mass and its acceleration, which in turn can be expressed as the second derivative of position versus time:

$$F = m\frac{d^2x}{dt^2}$$

Substituting into the above relation for Hooke's law reveals

$$m\frac{d^2x}{dt^2} + kx = 0$$

with the solution to this equation given by

$$x = A \sin \omega_0 t + B \cos \omega_0 t$$

where ω_0 is the angular frequency of the oscillating mass. This is an example of a differential equation solved in terms of trigonometric functions.

Supplementary Problems

1.1.   Classify the following functions in terms of their classification:

(a) $y = 3x + 4$

(b) $y = x^2 + x - 1$

(c) $y = e^{2x+6}$

(d) $y = \sin(x^2)$

(e) $y = \ln(\sin x)$

Answer

(a)   Polynomial function

(b)   Polynomial function

(c)   Exponential function

(d)   Trigonometric function

(e)   Logarithmic function

1.2.   Determine the slope of the function

$$y = \sin x \cos \pi$$

Answer

$$m = -\cos x$$

1.3.   Express the three directional coordinates of both (a) cylindrical (ρ, ϕ, z) and (b) spherical (ρ, θ, ϕ) coordinates in terms of rectangular coordinates.

Answer

(a)   Cylindrical coordinates:

$$\rho = \sqrt{x^2 + y^2}$$
$$\phi = \tan^{-1}\frac{y}{x}$$
$$z = z$$

(b)   Spherical coordinates:

$$\rho = \sqrt{x^2 + y^2 + z^2}$$
$$\theta = \cos^{-1}\frac{z}{\sqrt{x^2 + y^2 + z^2}}$$
$$\phi = \tan^{-1}\frac{y}{x}$$

1.4.   What is the derivative of the y intercept of a function?

Answer

0

1.5.   Given a periodic wave described by

$$P = ke^{i\omega t}$$

where i is a complex number ($i = \sqrt{-1}$), ω is angular frequency, and t is time, what is the derivative of this function with respect to time? With respect to frequency?

Answer

$$\frac{dP}{dt} = i\omega P$$

$$\frac{dP}{d\omega} = itP$$

1.6. The gaussian curve is used in many applications to the physical sciences ranging from radioactive decay to image processing to statistical physics and is given by

$$y = e^{-A^2 x^2}$$

where A is a constant with units inverse to those of x. the gaussian curve resembles a symmetric bell curve. An important feature of the gaussian curve or distribution is a quantitative measure of the *full width at half maximum* (FWHM), which reflects the spread or width of the curve in the central portion of the function. Derive an expression for the FWHM.

Answer

$$x^2 = 2.77/A^2$$

1.7. A function multiplied by its inverse is always equal to 1. Prove this for

$$f(x) = e^{ix}$$

where i is a complex number ($i = \sqrt{-1}$).

1.8. Given the equation

$$\frac{x}{p} + \frac{y}{q} = 1$$

define p and q.

Answer

Here p is the x intercept and q is the y intercept.

1.9. The effects of X-ray irradiation on a sample of bacteria in a nutrient suspension can be described by

$$N = N_0 e^{-kD} \tag{1}$$

where N is the number of surviving bacteria, N_0 is the initial number of bacteria, k is the cross-sectional or sample area irradiated (typically 5×10^{-11} cm^2), and D is the radiation dose.

(a) What are the units of the radiation dose?

(b) What dosage is required for survival of one-quarter of the original bacteria?

(c) What is the slope of the Eq. (1)?

Answer

(a) Particles per square centimeter

(b) $D = \dfrac{1.38}{k}$

(c) $-k$

1.10. Consider the kinetics of a zeroth-order chemical reaction

$$A \rightarrow B$$

The reaction rate can be expressed as $-d[A]/dt = k$, where k is a reacation rate constant. Derive an expression for the half-life of this process.

Solution

$$t_{1/2} = \frac{A_0}{2k}$$

1.11. Why is the mean age of cells in an exponentially growing bacterial culture always less than one-half the doubling time?

Solution

For each "dying" (dividing) cell, there are two cells born.

1.12. For the integral

$$\int_2^a \frac{dx}{x} = 1$$

what is the value of a?

Solution

5.435

1.13. Determine the area of the shaded star, shown in Figure 1-12, described by

$$\sqrt{x} + \sqrt{y} = 1$$

Solution

Total area $= \frac{2}{3} \text{ cm}^2$

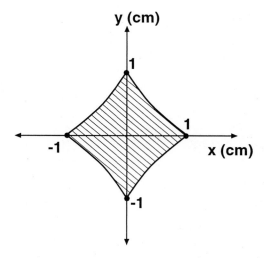

Fig. 1-12

1.14. Simplify the expression

$$1 + \frac{2}{x} \ln \sqrt{e^{-x}}$$

Answer

0

1.15. The ratio of absorbed to original light intensity given by

$$\frac{I}{I_0} = e^{-kl}$$

can also be expressed as

$$\frac{I}{I_0} = 10^{-\varepsilon Cl}$$

where I is the absorbed light intensity, I_0 is the original light intensity, k is a proportionality constant, l is the length of absorber, ε is the molar extinction coefficient (a characteristic of the absorber and the frequency of the radiation), and C is the molar concentration of the absorber. In comparing the two relations, determine an expression for k.

Answer

$2.303\varepsilon C$

1.16. A bacterial colony of *Escherichia coli* increased by 500,000 over 24 h. How many times did the bacteria divide?

Answer

19 divisions

1.17. The atmospheric pressure is related to the altitude by

$$p = 0.999^{h/8}$$

where p is atmospheric pressure, in atmospheres, and h is altitude, in meters. At what altitude would the atmospheric pressure be one-half the normal?

Answer

5.5 km

1.18. From the expression

$$N(t) = N_0(2^{t/D})$$

what is the slope of the curve?

Answer

$$\text{Slope} = \frac{0.30}{D}$$

1.19. For a sound wave interacting with a medium, the amplitude A at a depth x in the medium is related to the initial amplitude A_0 by

$$A = A_0 e^{-\alpha x}$$

where α is a coefficient related to the absorption of the sound wave by the medium. Find the intensity given that intensity is directly proportional to the square of the amplitude.

Answer

$$\left|\frac{I}{I_0}\right| = I_0 e^{-2\alpha x}$$

1.20. Differentiate the following functions with respect to x. All other variables are constants.

(1) $y = A \sin \omega x$

(2) $y = A e^{kt}$

(3) $y = B x^n$

(4) $y = C/x$

Answer

(1) $A\omega \cos \omega x$

(2) $k A e^{kt}$

(3) $n B x^{n-1}$

(4) $C \ln x$

1.21. Integrate the following functions.

(1) $\displaystyle \int A \sin \omega x \, dx$

(2) $\displaystyle \int A e^{kt} \, dt$

(3) $\displaystyle \int_0^{2\pi/(3\omega)} A \sin \omega x \, dx$

(4) $\displaystyle \int_4^7 x^2 \, dx$

Answer

(1) $-\dfrac{A}{\omega} \cos \omega x + C$

(2) $\dfrac{A}{k} e^{kt} + C$

(3) $1.5 \dfrac{A}{\omega}$

(4) 93

1.22. Breathing is a continual cyclical process required to maintain and regulate the oxygen demands of the circulatory system. Passive expiration or expiration in a normal rested state represents the outflow of air Q in the breathing cycle, which begins at a maximum value (tidal volume V_T) and steadily decreases at a rate proportional to decreasing lung volume V. Derive an expression for the variation in lung volume with respect to time.

Answer

$$V = V_T e^{-kt}$$

1.23. Consider light incident on a series of absorbers as arranged in Figure 1-13. How is the intensity I_2 related to the incident intensity I_0?

Solution

$$\frac{I_2}{I_0} = e^{-(k_1 x_1 + k_2 x_2)}$$

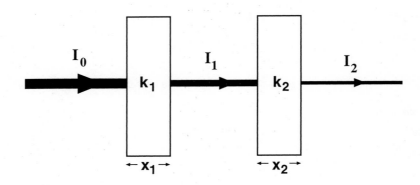

Fig. 1-13

Chapter 2

Units and Dimensions

As in any branch of science, physical concepts, principles, and interactions are explained and understood through various mathematical equations, relations, and, most importantly, quantities. A physical quantity reflects the physical basis of a given process and is described primarily by two factors: a fundamental unit and its corresponding magnitude or value, which are related by a dimension. A mathematical equation is valid and becomes applicable if and only if the quantities on both sides are equal in numerical value and in dimension. In this chapter we summarize the quantitative tools and techniques required for the solution of physics problems.

2.1 SIGNIFICANT FIGURES

Significant figures are a reflection of the experimental precision of a measurement or quantitative value. Every figure, with the exception of the last, in a numerical result should be significant. The significance of a figure corresponds to the accuracy of a measurement and implies a high probability of being correct. In physical measurements that contain zeros in the numerical result, one should not be misled by their appearance and interpret them as significant figures. The only purpose of zeros is to (1) convey magnitude and (2) identify the location of the decimal point.

Solved Problem 2.1. Determine the number of significant figures in the following numerical results:

(a) 0.1

(b) 0.00076

(c) 0.000009543

(d) 25,000

(e) 578,000

Solution

The number of significant figures in the numerical result is

(a) 1 significant figure

(b) 2 significant figures

(c) 4 significant figures

(d) 2 significant figures

(e) 3 significant figures

Solved Problem 2.2. The volumetric capacity of a particular pipette was measured as 0.02586 L.

(a) How many significant figures are included in this numerical result?

(b) What is the numerical result reflecting measurement accuracy to the nearest milliliter?

Solution

(a) There are 4 significant figures in the numerical result, 0.02586 L.

(b) Measurement of the volumetric capacity to the nearest milliliter is that numerical value to one-thousandth accuracy, or 0.025 L.

2.2 SCIENTIFIC NOTATION

In science, quantitative or numerical values of units can vary between wide extremes. For example, mass can range from 9.11×10^{-31} kg for the rest mass of the electron m_e to 5.98×10^{24} kg for the mass of the earth M_e. These mass values are presented in scientific notation. Scientific notation simplifies the presentation of numerical values by expressing the value as a number between 1 and 10, multiplied by 10 raised to a power determined by the number of zeros in the numerical result. Scientific notation is applicable to a wide quantitative range of numerical results, as illustrated by the table below.

Numerical Value		Prefix	Symbol	Magnitude
Expanded	**Scientific Notation**	**Prefix**	**Symbol**	**Magnitude**
0.000000000000001	1×10^{-15}	femto	f	one-quadrillionth
0.000000000001	1×10^{-12}	pico	p	one-trillionth
0.000000001	1×10^{-9}	nano	n	one-billionth
0.000001	1×10^{-6}	micro	μ	one-millionth
0.001	1×10^{-3}	milli	m	one-thousandth
0.01	1×10^{-2}	centi	c	one-hundredth
0.1	1×10^{-1}	deci	d	one-tenth
1.0	1×10^{0}	—	—	one
10.0	1×10^{1}	deca	da	ten
100.0	1×10^{2}	hecto	h	hundred
1000.0	1×10^{3}	kilo	k	thousand
1,000,000.0	1×10^{6}	mega	M	million
1,000,000,000.0	1×10^{9}	giga	G	billion
1,000,000,000,000.0	1×10^{12}	tera	T	trillion
1,000,000,000,000,000.0	1×10^{15}	peta	P	quadrillion

Solved Problem 2.3.   Express the following numerical values in scientific notation:

(a)   23

(b)   24,000

(c)   589,000,000

(d)   0.92

(e)   0.00067

(f)   0.0000005

Solution

The values expressed in scientific notation are

(a) 2.3×10^{1}

(b) 2.4×10^{4}

(c) 5.89×10^{8}

(d) 9.2×10^{-1}

(e) 6.7×10^{-4}

(f) 5×10^{-7}

Solved Problem 2.4. Appreciate the convenience of scientific notation by expressing the rest mass of the electron ($m_e = 9.11 \times 10^{-31}$ kg) and the mass of the earth ($M_e = 5.98 \times 10^{24}$ kg) in expanded numerical form.

Solution

In expanded numerical form, the rest mass of the electron and the mass of the earth are, respectively,

$$m_e = 9.11 \times 10^{-31} \text{ kg} = 0.000000000000000000000000000000911 \text{ kg}$$
$$M_e = 5.98 \times 10^{24} \text{ kg} = 5,980,000,000,000,000,000,000,000 \text{ kg}$$

Solved Problem 2.5. Express the following numerical results in terms of the appropriate numerical prefix:

(a) 10×10^9 bytes

(b) 0.635×10^7 V

(c) 1.4×10^3 m

(d) 15×10^2 L

(e) 78.9×10^{-2} g

(f) 0.0054 H

(g) 90×10^{-10} s

(h) 3.65×10^{-15} F

Solution

All these results can be expressed in terms of their appropriate numerical prefix by consulting the above table.

(a) 10 Gbytes   (b) 6.35 MV   (c) 1.4 km   (d) 15 hL

(e) 78.9 cg   (f) 5.4 mH   (g) 9.0 ns   (h) 3.65 fF

2.3 UNITS AND DIMENSIONAL ANALYSIS

In physics, the three most common dimensions used in the description of physical processes are mass M, length L, and time T. Consider, for example, the physical parameter of speed. The values of speed $v = 55$ mi h^{-1} = 88.5 km h^{-1} = 24.6 m s^{-1} = 80.7 ft s^{-1} = 1.48×10^5 cm min^{-1} = 5.81×10^4 in min^{-1} are identical. Although the fundamental unit (that is, mi h^{-1}, km h^{-1}, m s^{-1}, ft s^{-1}, cm min^{-1}, and in min^{-1}) and its corresponding magnitude (that is, 55, 88.5, 24.6, 80.7, 1.48×10^5, and 5.81×10^4, respectively) differ in value, each value for speed possesses the same dimension of $L T^{-1}$.

These are other examples of fundamental dimensions for physical quantities:

Length:	L
Area:	$L \times L = L^2$
Volume:	$L \times L \times L = L^3$
Position (= distance):	L
Velocity (= distance/time):	$L T^{-1}$
Acceleration (= velocity/time):	$(L T^{-1})/T = L T^{-2}$
Force (= mass × acceleration):	$M \times L T^{-2} = M L T^{-2}$
Pressure (= force/area):	$(M L T^{-2})/L^2 = M L^{-1} T^{-2}$
Energy (= force × distance):	$(M L T^{-2}) \times L = M L^2 T^{-2}$
Power (= energy/time):	$(M L^2 T^{-2})/T = M L^2 T^{-3}$

One should be aware of dimensions of variables embedded in exponential, power, logarithmic, and trigonometric functions. In each case, the argument must be dimensionless.

Solved Problem 2.6.   Determine the dimensions of the following variables: density and viscosity.

Solution

Solution of this problem requires fundamental knowledge of the definition of these variables. *Density* is defined as mass per unit volume, from which the dimensions can be determined:

$$\text{Density} = \frac{\text{mass}}{\text{volume}} = M L^{-3}$$

Viscosity is a physical property of a fluid describing its resistance to motion. Viscosity is measured in units of pressure multiplied by time. Thus, the dimensions of viscosity are

$$\text{Viscosity} = \text{pressure} \times \text{time} = (M L^{-1} T^{-2}) \cdot T = M L^{-1} T^{-1}$$

Solved Problem 2.7.   The pH is a quantitative measurement of the hydrogen ion activity of a solution and is defined as

$$\text{pH} = -\log [\text{H}^+]$$

where $[\text{H}^+]$ is the numerical value of the hydrogen ion activity. What are the units of pH?

Solution

Since pH is the negative logarithm of the hydrogen ion activity, pH is unitless and thus is expressed as a pure number.

Solved Problem 2.8.   The slope of a functional curve, defined as $\Delta y / \Delta x$, is sometimes calculated as $\tan \alpha$, where α is the angle of inclination. Why is this incorrect?

Solution

The slope of a curve, describing the rate of change of a function over predefined ranges along the x and y axes, generally has a dimension while $\tan \alpha$, a trigonometric function, is unitless and is a pure number.

2.4 CONVERSION FACTORS

Often, in the physical analysis of problems, the measurements of observed variables required for solution are recorded in different units. Since every physical equation must be dimensionally equal as well as numerically equal, measurements for all variables must conform to the same units. Conversion factors are used to convert units between equivalent terms through ratios of the same dimensions relating the desired unit to the unit expressed in the given quantity. Conversion factors ensure uniformity in all measurements and for all variables. Below is a list of conversion factors for representative variables found in the analysis of physical problems.

Mass	1 lb = 0.4536 kg
	1 g = 10^{-3} kg
	1 slug = 14.59 kg
Length	1 Å = 10^{-10} m
	1 in = 0.0254 m = 2.54 cm
	1 ft = 0.3048 m
	1 m = 1609 m

Area
$$1\,\text{in}^2 = 6.6452 \times 10^{-4}\,\text{m}^2$$
$$1\,\text{ft}^2 = 9.29 \times 10^{-2}\,\text{m}^2$$
$$1\,\text{mi}^2 = 2.59 \times 10^6\,\text{m}^2$$

Volume
$$1\,\text{in}^3 = 1.639 \times 10^{-5}\,\text{m}^3$$
$$1\,\text{ft}^3 = 2.832 \times 10^{-2}\,\text{m}^3$$

Time
$$1\,\text{yr} = 3.156 \times 10^7\,\text{s}$$
$$1\,\text{day} = 8.64 \times 10^4\,\text{s}$$

Speed
$$1\,\text{ft s}^{-1} = 0.3048\,\text{m s}^{-1}$$
$$1\,\text{mi h}^{-1} = 0.447\,\text{m s}^{-1}$$
$$1\,\text{km h}^{-1} = 0.2778\,\text{m s}^{-1}$$

Force
$$1\,\text{dyn} = 10^{-5}\,\text{N}$$
$$1\,\text{lb} = 4.448\,\text{N}$$

Pressure
$$1\,\text{dyn cm}^2 = 0.1\,\text{N m}^{-2}$$
$$1\,\text{mm Hg} = 1333\,\text{N m}^{-2}$$
$$1\,\text{N m}^{-2} = 1\,\text{pascal (Pa)}$$
$$1\,\text{torr} = 133.3\,\text{Pa}$$
$$1\,\text{atm} = 1.013 \times 10^5\,\text{Pa}$$

Energy
$$1\,\text{Btu} = 1055\,\text{J}$$
$$1\,\text{erg} = 10^{-7}\,\text{J}$$
$$1\,\text{cal} = 4.186\,\text{J}$$
$$1\,\text{kWh} = 3.6 \times 10^6\,\text{J}$$
$$1\,\text{eV} = 1.602 \times 10^{-19}\,\text{J}$$

Power
$$1\,\text{Btu h}^{-1} = 0.293\,\text{W}$$
$$1\,\text{hp} = 745.7\,\text{W}$$
$$1\,\text{cal s}^{-1} = 4.186\,\text{W}$$

Temperature
$$T_{\text{Fahrenheit}} = \frac{9}{5}(T_{\text{Celsius}} + 32)$$

$$T_{\text{Celsius}} = \frac{5}{9}T_{\text{Fahrenheit}} - 32$$

$$T_{\text{Kelvin}} = T_{\text{Celsius}} + 273.15$$

Viscosity
$$1\,\text{poise (P)} = 0.1\,\text{kg m}^{-1}\,\text{s}^{-1}$$

Solved Problem 2.9. Uncle Max has just celebrated his 55th birthday. Calculate his age in seconds.

Solution

Uncle Max's age, that is, 55 years, can be expressed in seconds through the use of appropriate conversion factors:

$$55\ \text{years}\left[\left(\frac{12\ \text{months}}{1\ \text{yr}}\right)\left(\frac{4\ \text{weeks}}{1\ \text{month}}\right)\left(\frac{7\ \text{days}}{1\ \text{week}}\right)\left(\frac{24\ \text{h}}{1\ \text{day}}\right)\left(\frac{60\ \text{min}}{1\ \text{h}}\right)\left(\frac{60\ \text{s}}{1\ \text{min}}\right)\right]$$

where 12 months/1 yr, 4 weeks/1 month, 7 days/1 week, 24 h/l day, 60 min/1 h, and 60 s/l min are conversion factors. Multiplication of the conversion factors negates all units of time in both the numerator and denominator with the exception of one—the desired unit of time, seconds. Therefore,

$$55\ \text{years} = 55(12 \times 4 \times 7 \times 24 \times 60 \times 60)\,\text{s} = 1.6 \times 10^9\,\text{s}$$

Solved Problem 2.10.   In the human circulation under normal conditions, blood is pumped from the heart under a systolic pressure of 120 mmHg. Express this value of pressure in terms of dynes per centimeter squared.

Solution

Solution of this problem requires the appropriate conversion factor relating pressure in millimeters of mercury to pressure in dynes per centimeter squared. Therefore, the systolic pressure is

$$\text{Systolic pressure} = 120 \text{ mmHg} \left(\frac{1333 \text{Nm}^{-2}}{1 \text{ mmHg}} \right) \left(\frac{1 \text{ dyn cm}^{-2}}{0.1 \text{ N m}^{-2}} \right) = 1.6 \times 10^6 \text{ dyn cm}^{-2}$$

2.5   SYSTEMS OF UNITS

The particular units corresponding to each dimension depend, for the most part, on the system of units. Currently, three systems of units exist: Système International d'Unités (SI), British, and gaussian (cgs). The units for the dimensions of mass M, length L, and time T in the three systems are given below:

Dimension	SI	British	Gaussian (cgs)
M	kg	lb	g
L	m	ft	cm
T	s	s	s

Although dimensional units from each of the aforementioned systems can be found in the physical and biological literature, a concerted effort has been made by the scientific community to adopt SI units for expression of all dimensional quantities.

2.6   PROBLEM-SOLVING TECHNIQUES

In a typical physics problem the student must assimilate a variety of numbers, values, equations, and relations to solve quantitatively and arrive at a numerical result. However, the successful calculation of a quantitative value for an answer to a problem does not necessarily constitute the completion of a problem and can, at times, mislead the student into a false sense of security.

- Read the problem carefully.
    What is the problem about, what is it describing, what is it asking me to solve, and what information is it giving me to solve the problem?
- Draw a rough sketch of the physical processes.
- Extract all pertinent descriptive information from the problem text.
    What known parameters and corresponding values are given in the problem? What unknown parameters are to be solved? What equations relate the known and unknown parameters?
- Upon substitution of all known values into the equations, are the units appropriate?
    If speed is the desired parameter, do all units within a given equation cancel to yield a final unit of speed (LT^{-1})? [Be sure to use the proper conversion formulas to ensure the uniformity of the units.]
- Is the final solution of appropriate magnitude and units?
    For example, if a certain speed is asked for in a problem, is the speed of a magnitude which makes sense, given the information of the problem, and is the value in units of LT^{-1}?
- Recheck problem to ensure accuracy of results.

Supplementary Problems

2.1. What are the dimensions of the following quantities?

(a) $d(m)$

(b) $d(v)/dt$

(c) d^2v/dt^2

Solution

(a) M

(b) LT^{-2}

(c) LT^{-3}

2.2. Given that entropy is defined as work/temperature where the dimension of temperature is Θ, what are the dimensions of entropy S?

Solution

$ML^2T^{-2}\Theta^{-1}$

2.3. If the displacement of a particle is described by

$$x = 4 + 9t - 2t^3$$

where x is in meters and t is time in seconds, what are the units of 4, 9, and 2?

Solution

4 m, 9 m/s, 2 m/s^3

2.4. Express $7\frac{3}{4}$ in in centimeters.

Solution

19.7 cm

2.5. Express the surface area (SA) of a spherical cell as a function of its volume.

Solution

$$SA = 4.84V^{2/3}$$

2.6. The angular speed ω of centrifuge rotors, defined as the time rate of change of an angle α, is often given in revolutions per minute (rpm).

(a) What are the dimensions of angular speed?

(b) Express 5.0×10^4 rpm in terms of s^{-1}.

Solution

(a) T^{-1}

(b) 5.24×10^3 s^{-1}

2.7. How many significant figures are included in the following numerical results?

(a) 0.0889

(b) 8.4

(c)   7576.77

(d)   12,400,000

(e)   12,400,000.8

(f)   0.0008808

(g)   60.006

(h)   204.0

Solution

(a)   3

(b)   2

(c)   6

(d)   3

(e)   9

(f)   4

(g)   5

(h)   3

2.8.   Express the following numbers in scientific notation:

(a) 9500, (b) 0.00045, (c) 15,600,000, (d) 0.8, (e) 0.00000074, (f) 27, (g) 178,000, (h) 0.00303

Solution

(a) 9.5×10^2, (b) 4.5×10^{-4}, (c) 1.56×10^7, (d) 8×10^{-1}, (e) 7.4×10^{-7}, (f) 2.7×10^1, (g) 1.78×10^5, (h) 3.03×10^{-3}

Chapter 3

Vectors

In physics, measurements of physical quantities, processes, and interactions that exhibit both magnitude and direction along a directed line segment or displacement can be represented by vectors. Several examples of vector quantities in the physical sciences include velocity, acceleration, and force. This is in contrast to scalar quantities or physical variables that exhibit only magnitude. Examples of scalar quantities include distance, speed, and time. A more complete list of examples of vector and scalar quantities encountered commonly in physical applications is given in Table 3.1.

Table 3.1. Scalar and Vector Quantities of Parameters in Selected Physical Topics

Scalar Quantities	Vector Quantities
Measurement	
Time	
Mass	
Area	
Volume	
Kinematics	
Distance	Displacement
Speed	Velocity
	Acceleration
Dynamics	
Energy	Force
Work	Momentum
Moment of inertia	Torque
Frequency	Electric current
Viscosity	Electrostatic field
	Electromagnetic field

3.1 DEFINITIONS OF VECTORS

A vector, symbolized by a boldface character, sometimes with an arrow on top, for example, $\vec{\mathbf{V}}$, represents a displacement or directed line segment between two points $P_1(x_1, y_1, z_1)$ and $P_2(x_2, y_2, z_2)$, whose distance in a given direction is determined by the difference in the respective coordinates, referred to as *vector components*:

$$V_x = x_2 - x_1 \qquad V_y = y_2 - y_1 \qquad V_z = z_2 - z_1$$

Vectors can be expressed in a coordinate system thus:

$$\text{One dimension:} \qquad \mathbf{V} = V_x \mathbf{i}$$
$$\text{Two dimensions:} \qquad \mathbf{V} = V_x \mathbf{i} + V_y \mathbf{j}$$
$$\text{Three dimensions:} \qquad \mathbf{V} = V_x \mathbf{i} + V_y \mathbf{j} + V_z \mathbf{k}$$

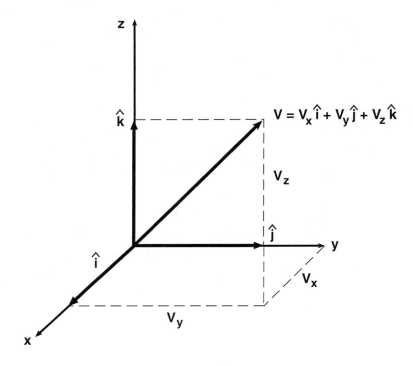

Fig. 3-1

where V_x, V_y, and V_z are coefficients representing scalar measurements and **i**, **j**, and **k** are unit vectors along the x, y, and z coordinates, respectively, as shown in Figure 3-1. Scalar quantities are positive or negative numbers and represent only the magnitude of a variable without any reference to direction.

The *magnitude* of a vector is the sum of the square of its components, or

$$|\mathbf{V}| = \sqrt{V_x^2 + V_y^2 + V_z^2}$$

and it is simply a number or scalar quantity. The above relation is known as *Pythagorean's theorem* for a right triangle from geometry.

3.2   RESOLUTION OF VECTORS

The vector is represented graphically by an arrow with the tail at the origin, pointing in a direction dictated by the problem with the size or length of the vector corresponding to its magnitude, as shown in Figure 3-2. The vector is resolved by projecting the vector components onto the x and y axes. Thus, the projection of the vector along the x axis, V_x, is determined by fundamental trigonometric relations:

$$V_x = V \cos \theta$$

and similarly, the projection of the vector along the y axis is

$$V_y = V \sin \theta$$

Given these components, one can work backward and determine the magnitude and direction of the vector by

$$\mathbf{V} = \sqrt{V_x^2 + V_y^2}$$

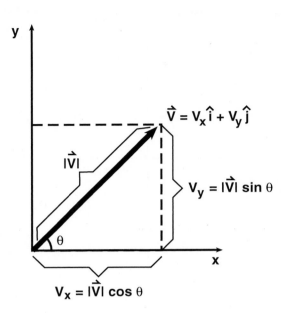

Fig. 3-2

and the direction of the vector by

$$\tan \theta = \frac{V_y}{V_x}$$

or

$$\theta = \tan^{-1} \frac{V_y}{V_x}$$

where θ is measured from the positive x axis in a counterclockwise direction.

Solved Problem 3.1. Given the vector $\mathbf{V} = 4\mathbf{i} + 2\mathbf{j}$, determine its magnitude and direction.

Solution

The magnitude of the vector $\mathbf{V}$ can be determined by

$$|\mathbf{V}| = \sqrt{V_x^2 + V_y^2} = \sqrt{4^2 + 2^2}$$
$$= \sqrt{16 + 4} = \sqrt{20} = 2\sqrt{5}$$

The direction or angle of orientation of the vector is given by

$$\theta = \tan^{-1} \frac{V_y}{V_x} = \tan^{-1} \frac{2}{4} = \tan^{-1} 0.5 = 26.6°$$

Both the magnitude and direction of this vector are schematically represented in Figure 3-3.

Solved Problem 3.2. On a typical bicycle ride to Sunset Mall, Alexandra travels 0.5 km to the east and 0.3 km to the north. Determine the vector (magnitude and direction) of her total displacement as a result of the bicycle trip.

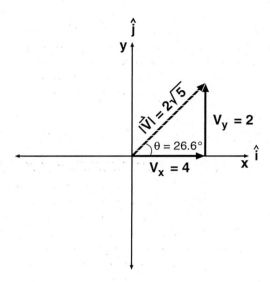

Fig. 3-3

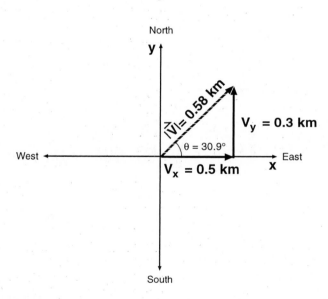

Fig. 3-4

Solution

As shown in Figure 3-4, the directions east and north can be represented on a two-dimensional coordinate system as x and y, respectively. The magnitude of the displacement vector is the square root of the sum of the squares of the directional components, or

$$|\vec{V}| = \sqrt{(0.5 \text{ km})^2 + (0.3 \text{ km})^2} = \sqrt{0.25 \text{ km}^2 + 0.09 \text{ km}^2}$$
$$= \sqrt{0.34 \text{ km}^2} = 0.58 \text{ km}$$

The direction of the resultant vector is determined by

$$\theta = \tan^{-1}\left(\frac{0.3 \text{ km}}{0.5 \text{ km}}\right) = \tan^{-1} 0.6 = 30.9°$$

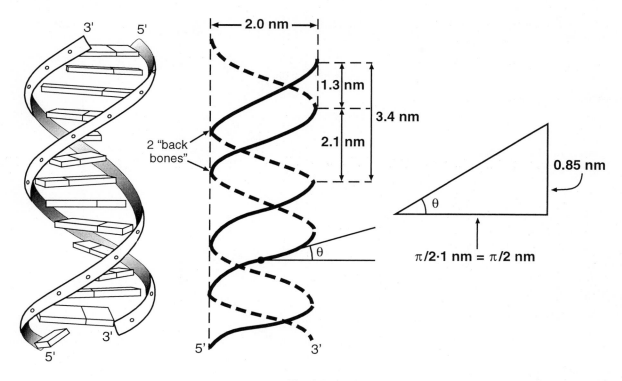

Fig. 3-5

ILLUSTRATIVE EXAMPLE 3.1.   Vector Representation of DNA

DNA (deoxyribonucleic acid) is a biomolecule that contains genetic information and is found in the nucleus of cells. DNA consists of two strands of polynucleotides running in opposite directions and twisted about a central axis to form a double helix, as shown in Figure 3-5. If one assumes that the two helices are smooth and circular, it is possible to approximate DNA geometry as a circular cylinder whose coordinates (ρ, θ, z) are related to cartesian or rectangular coordinates (x, y, z) by

$$x = \rho \cos \theta$$
$$y = \rho \sin \theta$$
$$z = z$$

As the variables related to the DNA geometry, ρ is the radius of the double helix (= 1.0 nm), θ is the rotation or screw angle of the DNA backbone as it winds about the central axis, and z is the distance along the central axis. Since there are two strands of nucleotides in the double helix, two vectors are required to represent the structure of the DNA molecule.

For the strand that is moving in the positive z direction (bottom to top), denoted by $\mathbf{S}_1$, the vector can be expressed as

$$\mathbf{S}_1 = S_{1x}\mathbf{i} + S_{1y}\mathbf{j} + S_{1z}\mathbf{k}$$
$$= [1.0 \cos \theta \text{ (nm)}]\mathbf{i} + [1.0 \sin \theta \text{ (nm)}]\mathbf{j} + [5.4 \, \theta \text{ (nm)}]\mathbf{k}$$

The other strand is similar to the previous one except that (1) it is moving in the negative z direction and (2) it is out of phase or offset by a small angle (approximately 2 rad). Keeping this in mind, we see that the vector for the other strand, denoted by $\mathbf{S}_2$, can be expressed as

$$\mathbf{S}_2 = S_{2x}\mathbf{i} + S_{2y}\mathbf{j} + S_{2z}\mathbf{k}$$
$$= [1.0 \cos (\theta - 2) \text{ (nm)}]\mathbf{i} + [1.0 \sin (\theta - 2) \text{ (nm)}]\mathbf{j} + [-5.4(\theta - 2) \text{ (nm)}]\mathbf{k}$$

It is possible to further refine these vectors by calculating the value for the rotation or screw angle θ and substituting into the above expressions for $\mathbf{S}_1$ and $\mathbf{S}_2$.

From the geometric considerations presented in Figure 3-5, the magnitude of the opposite side of the triangle must be calculated, followed by calculation of θ. Setting up ratios of distance, we have

$$\frac{2\pi}{3.4 \text{ nm}} = \frac{\pi/2}{y \text{ nm}}$$

Solving for y yields

$$y = \frac{(\pi/2)(3.4 \text{ nm})}{2\pi} = 0.85 \text{ nm}$$

Solving for the rotation angle θ gives

$$\tan \theta = \frac{y}{x} = \frac{0.85 \text{ nm}}{\pi/2 \cdot 1 \text{ nm} = \pi/2 \text{ nm}} = 0.54$$

$$\theta = \tan^{-1} 0.54 = 28°$$

Converting the rotation angle to radians gives

$$28° \cdot \frac{2\pi \text{ rad}}{360°} = 0.50 \text{ rad}$$

3.3   ADDITION OF VECTORS

In many cases, a physical system under consideration undergoes several changes in either magnitude or direction within any given process of time interval. This involves the incorporation of several vectors, one for each corresponding change in the system. The objective in such a problem is to add all vectors to determine the total or net effect contributed by each individual change or vector represented by a resultant vector. Addition of vectors can be performed according to two general methods: the graphical method and the analytic method.

3.3.1   Graphical Method of Vector Addition

Graphical methods of vector addition involve the proper arrangement of vectors on graph paper, where they can be drawn to scale and measured directly with a straightedge or ruler. There are two types of graphical methods for vector addition, each of which employs a different arrangement for the vectors: the parallelogram method and the polygon method.

3.3.1.1   Parallelogram Method of Vector Addition

The parallelogram method of vector addition is illustrated in Figure 3-6 with vectors **A** and **B**. In the parallelogram method of vector addition, the vectors are arranged so that their tails originate from the same point. The two vectors, as positioned, constitute one-half of a parallelogram. Imaginary lines are drawn on the opposing side of each vector so as to complete the parallelogram. The diagonal of the parallelogram is the resultant vector whose magnitude can be obtained by direct measurement with a ruler or straightedge. The angle that the resultant vector makes with the x axis can be measured with a protractor.

3.3.1.2   Polygon Method of Vector Addition

The polygon method of vector addition is illustrated in Figure 3-7 with the vectors **A** and **B**. In the polygon or "head-to-tail" method of vector addition, the two vectors are positioned so that the tail of one vector originates from the head of the other vector. The resultant vector is the vector drawn from the tail of the original vector to the head of the second vector. The magnitude and angle of the resultant vector can be determined through trigonometric relations of a triangle.

Polygon Method of Vector Addition

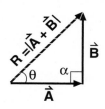

Fig. 3-7a

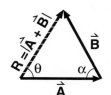

Fig. 3-7b

Parallelogram Method of Vector Addition

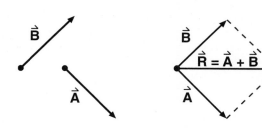

Fig. 3-6

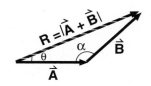

Fig. 3-7c

- If **A** is perpendicular to **B**, that is, $\alpha = 90°$ or it is a right angle, then the resultant vector $\mathbf{R} = |\mathbf{A} + \mathbf{B}|$ is [Figure 3-7(a)]

$$\mathbf{R} = \sqrt{A^2 + B^2}$$

$$\tan \theta = \frac{B}{A}$$

where A and B are the scalar magnitude of the vectors **A** and **B**.

- If **A** is not perpendicular to **B**, that is, $\alpha \neq 90°$ is either acute ($\alpha < 90°$) or obtuse ($\alpha > 90°$), then the resultant vector $\mathbf{R} = |\mathbf{A} + \mathbf{B}|$ is [Figure 3-7(b) and (c), respectively]

$$\mathbf{R} = \sqrt{A^2 + B^2 - 2AB \cos \alpha} \qquad \text{law of cosines}$$

$$\frac{R}{\sin \alpha} = \frac{B}{\sin \theta} = \frac{A}{\sin \gamma} \qquad \text{law of sines}$$

Solved Problem 3.3.   Given two vectors

$$\mathbf{A} : \text{magnitude} = 5\,\text{m}, \quad \text{direction along } x \text{ axis}$$

$$\mathbf{B} : \text{magnitude} = 8\,\text{m}, \quad \text{direction } 65° \text{ from } x \text{ axis}$$

use the polygon method of vector addition to determine the magnitude and direction of $\mathbf{A} + \mathbf{B}$.

Solution

The magnitude of the vector sum can be determined by resolving the vectors into their respective components:

A: $A_x = 5$ m $A_y = 0$
B: $B_x = 8 \cos 65°$ m $= 3.38$ m $B_y = 8 \sin 65°$ m $= 7.25$ m

In vector form,

$$\mathbf{A} = 3\mathbf{i} \text{ m} \qquad \mathbf{B} = 3.38\mathbf{i} \text{ m} + 7.25\mathbf{j} \text{ m}$$

Therefore, by drawing the vectors in a head-to-tail manner and making appropriate measurements, we can see from Figure 3-8 that

$$\mathbf{A} + \mathbf{B} = 6.38\mathbf{i} \text{ m} + 7.25\mathbf{j} \text{ m}$$
$$|\mathbf{A} + \mathbf{B}| = 9.66 \text{ m}$$
$$\theta = \tan^{-1}\left(\frac{7.25 \text{ m}}{6.38 \text{ m}}\right) = 48.7° \qquad \text{from positive } x \text{ axis}$$

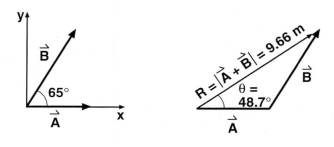

Fig. 3-8

Solved Problem 3.4.   Consider the two vectors **A** and **B**:

A: magnitude $= 3$ N, direction along x axis
B: magnitude $= 4$ N, direction $30°$ from positive x axis

Use the polygon method of vector addition to determine the magnitude and direction of the resultant vector **R**.

Solution

Graphical display of these two vectors, as shown in Figure 3-9, does not reveal a right triangle, and therefore the law of cosines must be used to determine the magnitude of **R**:

$$\mathbf{R} = \sqrt{A^2 + B^2 - 2AB \cos \gamma} = \sqrt{9 + 16 - 2 \cdot 3 \text{ m} \cdot 4 \text{ m} \cdot \cos 150°} = 6.7 \text{ m}$$

The angle α can be determined from the law of sines:

$$\frac{\sin \alpha}{B} = \frac{\sin \gamma}{R}$$

$$\frac{\sin \alpha}{4 \text{ m}} = \frac{\sin 150°}{6.7}$$

$$\sin \alpha = 0.298$$

$$\alpha = \sin^{-1} 0.298 = 17.4°$$

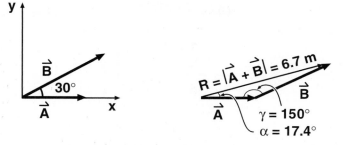

Fig. 3-9

3.3.2   *Analytic Method of Vector Addition*

In the analytic method of vector addition, each component of the individual vectors is added. The sum yields the component for the resultant vector. Consider two vectors defined below as

$$\mathbf{V}_1 = V_{1x}\mathbf{i} + V_{1y}\mathbf{j}$$
$$\mathbf{V}_2 = V_{2x}\mathbf{i} + V_{2y}\mathbf{j}$$

Their sum is

$$\mathbf{V}_{12} = \mathbf{V}_1 + \mathbf{V}_2 = (V_{1x} + V_{2x})\mathbf{i} + (V_{1y} + V_{2y})\mathbf{j}$$

Solved Problem 3.5.   Given two vectors

$$\mathbf{A} = \mathbf{i} - 4\mathbf{k} \qquad \mathbf{B} = \mathbf{j} + 4\mathbf{k}$$

determine the displacement vector $\mathbf{A} + \mathbf{B}$.

Solution

The sum of **A** and **B** is given by

$$\mathbf{A} + \mathbf{B} = (1 + 0)\mathbf{i} + (0 + 1)\mathbf{j} + (-4 + 4)\mathbf{k} = \mathbf{i} + \mathbf{j}$$

Solved Problem 3.6.   Determine the resultant sum vector of the three vectors **A**, **B**, and **C**, defined as

$$\mathbf{A} = 5\mathbf{i} + 3\mathbf{j} \qquad \mathbf{B} = 15\mathbf{i} + \mathbf{j} \qquad \mathbf{C} = 7\mathbf{i} + 10\mathbf{j}$$

Solution

The sum of the three vectors is the sum of the individual components from each of the three vectors:

$$\mathbf{R} = \mathbf{A} + \mathbf{B} + \mathbf{C} = (5 + 15 + 7)\mathbf{i} + (3 + 1 + 10)\mathbf{j} = 27\mathbf{i} + 14\mathbf{j}$$

3.4   SUBTRACTION OF VECTORS

Subtraction of two vectors can be done by reversing the direction of one vector and adding it to the remaining vector, as shown in Figure 3-10. Reversing the sign of the vector by multiplying by −1 simply changes the direction of the vector without affecting its magnitude.

Vector Subtraction

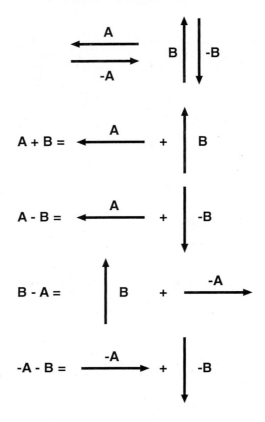

Fig. 3-10

Solved Problem 3.7.   Given the three vectors **A**, **B**, and **C**, defined as

$$\mathbf{A} = 5\mathbf{i} + 3\mathbf{j} \qquad \mathbf{B} = 15\mathbf{i} + \mathbf{j} \qquad \mathbf{C} = 7\mathbf{i} + 10\mathbf{j}$$

determine the resultant vector of $\mathbf{A} - \mathbf{B} - \mathbf{C}$.

Solution

This can be done by either subtracting the appropriate components

$$\mathbf{A} - \mathbf{B} - \mathbf{C} = (5 - 15 - 7)\mathbf{i} + (3 - 1 - 10)\mathbf{j} = -17\mathbf{i} - 8\mathbf{j}$$

or by multiplying **B** and **C** by -1 and adding the three vectors:

$$-\mathbf{B} = -(15\mathbf{i} + \mathbf{j}) \; = -15\mathbf{i} - \mathbf{j}$$
$$-\mathbf{C} = -(7\mathbf{i} + 10\mathbf{j}) = -7\mathbf{i} - 10\mathbf{j}$$

The difference can then be calculated according to the following:

$$\mathbf{A} - \mathbf{B} - \mathbf{C} = (5\mathbf{i} + 3\mathbf{j}) + (-15\mathbf{i} - \mathbf{j}) + (-7\mathbf{i} - 10\mathbf{j})$$
$$= (5 - 15 - 7)\mathbf{i} + (3 - 1 - 10)\mathbf{j} = -17\mathbf{i} - 8\mathbf{j}$$

Solved Problem 3.8.   For vectors **A** and **B** in Solved Problem 3.3, use the polygon method to determine the magnitude and direction of the difference $\mathbf{A} - \mathbf{B}$.

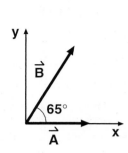
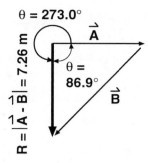

Fig. 3-11

Solution

The drawn vectors according to the polygon method of vector addition are displayed in Figure 3-11. Using the method of vector addition is acceptable because we are, in effect, adding a reversed or negative vector to another vector. As stated in the solution, the two vectors presented can be written in vector form as

$$\mathbf{A} = 3\mathbf{i} \text{ m} \qquad \mathbf{B} = 3.38\mathbf{i} \text{ m} + 7.25\mathbf{j} \text{ m}$$

Therefore, by drawing the vectors in a "head-to-tail" manner and making appropriate measurements, we see that

$$\mathbf{A} - \mathbf{B} = -0.38\mathbf{i} \text{ m} - 7.25\mathbf{j} \text{ m}$$
$$|\mathbf{A} - \mathbf{B}| = 7.26 \text{ m}$$
$$\theta = \tan^{-1}\left(\frac{-7.25 \text{ m}}{-0.38 \text{ m}}\right) = 86.9° \text{ or } 273.0° \text{ from } +x \text{ axis}$$

3.5 MULTIPLICATION OF VECTORS

Multiplication of vectors also serves an important role in the description of physical parameters and processes. There are, in effect, three different types of vector multiplication: multiplication of a vector by a scalar, dot product of vector multiplication, and cross product of vector multiplication.

3.5.1 Multiplication of a Vector by a Scalar

$$\lambda\mathbf{a} = \lambda(a_x\mathbf{i} + a_y\mathbf{j} + a_z\mathbf{k}) = \lambda a_x\mathbf{i} + \lambda a_y\mathbf{j} + \lambda a_z\mathbf{k} \qquad \text{where } \lambda \text{ is scalar quantity}$$

3.5.2 Dot Product of Vector Multiplication

$$\mathbf{A} \cdot \mathbf{B} = |A|\,|B| \cos\theta = A_xB_x + A_yB_y + A_zB_z$$

Note that

$$\mathbf{A} \cdot \mathbf{B} = \mathbf{B} \cdot \mathbf{A}$$
$$\hat{\mathbf{i}} \cdot \hat{\mathbf{i}} = \hat{\mathbf{j}} \cdot \hat{\mathbf{j}} = \hat{\mathbf{k}} \cdot \hat{\mathbf{k}} = 1$$
$$\hat{\mathbf{i}} \cdot \hat{\mathbf{j}} = \hat{\mathbf{j}} \cdot \hat{\mathbf{i}} = \hat{\mathbf{i}} \cdot \hat{\mathbf{k}} = \hat{\mathbf{k}} \cdot \hat{\mathbf{i}} = \hat{\mathbf{j}} \cdot \hat{\mathbf{k}} = \hat{\mathbf{k}} \cdot \hat{\mathbf{j}} = 0$$

The scalar product can be summarized by the multiplication table in Figure 3-12. The scalar product of two vectors is a scalar quantity. Physical quantities derived from scalar products include the work W done by a force $\mathbf{F}$ applied to a particle over a displacement $\mathbf{d}$ ($W = \mathbf{F} \cdot \mathbf{d}$) and the potential energy U of an electric dipole in an external electric field $\mathbf{E}$ given the electric dipole moment $\mathbf{p}$ ($U = -\mathbf{p} \cdot \mathbf{E}$).

Dot Products

	$\cdot\hat{\mathbf{i}}$	$\cdot\hat{\mathbf{j}}$	$\cdot\hat{\mathbf{k}}$
$\hat{\mathbf{i}}$	1	0	0
$\hat{\mathbf{j}}$	0	1	0
$\hat{\mathbf{k}}$	0	0	1

Fig. 3-12

3.5.3   Cross Product of Vector Multiplication

$$\mathbf{A} \times \mathbf{B} = |\mathbf{A}||\mathbf{B}|(\sin \theta)\mathbf{u} = \begin{vmatrix} \hat{\mathbf{i}} & \hat{\mathbf{j}} & \hat{\mathbf{k}} \\ A_x & A_y & A_z \\ B_x & B_y & B_z \end{vmatrix}$$

$$= \mathbf{i}(A_y B_z - A_z B_y) + \mathbf{j}(A_z B_x - A_x B_z) + \mathbf{k}(A_x B_y - A_y B_x)$$

where $\mathbf{u}$ is a unit vector indicating the direction of $\mathbf{A} \times \mathbf{B}$. The cross product can be summarized by the multiplication table in Figure 3-13.

The vector product of two vectors is a vector quantity. The magnitude of $\mathbf{A} \times \mathbf{B}$ is determined from the equation above; the direction of the resultant vector can be determined from the right-hand rule, as illustrated in Figure 3-14. When one extends the right hand with the fingers oriented along vector $\mathbf{A}$ and moving the fingers toward vector $\mathbf{B}$ as if clenching a fist, the extended thumb reveals the direction of the vector product. The product of two vectors defines the area of a parallelogram, as illustrated in Figure 3-15.

Cross Products

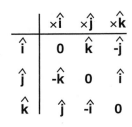

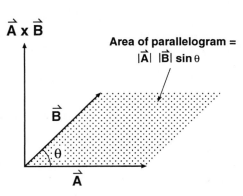

Fig. 3-13   Fig. 3-14   Fig. 3-15

Physical quantities derived from vector products are torque τ created by a force $\mathbf{F}$ exerted at a radial vector position $\mathbf{r}$ ($\tau = \mathbf{F} \times \mathbf{r}$) and the angular momentum $\mathbf{L}$ generated by linear momentum $\mathbf{p}$ exerted at a radial vector position $\mathbf{r}$ ($\mathbf{L} = \mathbf{r} \times \mathbf{p}$).

Solved Problem 3.9.   For the vectors

$$\mathbf{A} = 3\mathbf{i} + 4\mathbf{j} \qquad \text{and} \qquad \mathbf{B} = 2\mathbf{i} + 3\mathbf{j}$$

prove that the vectors are parallel.

Solution

This problem can be solved by taking the scalar or dot product:

$$\mathbf{A} \cdot \mathbf{B} = |\mathbf{A}||\mathbf{B}|\cos \theta = (3)(2) + (4)(3) = 6 + 12 = 18$$

The magnitudes of vectors $\mathbf{A}$ and $\mathbf{B}$ can easily be calculated as

$$|\mathbf{A}| = \sqrt{3^2 + 4^2} = \sqrt{9 + 16} = \sqrt{25} = 5$$
$$|\mathbf{B}| = \sqrt{2^2 + 3^2} = \sqrt{4 + 9} = \sqrt{13}$$

Making the appropriate substitutions, we get

$$18 = 5\overline{)13}\,\cos\theta$$

$$\cos\theta = \frac{18}{5\overline{)13}} = \frac{18}{18.03} \cong \frac{18}{18} = 1$$

$$\theta = \cos^{-1} 1 = 0$$

Therefore, **A** and **B** are parallel.

Solved Problem 3.10.   Determine the area of the parallelogram defined by the vectors

$$\mathbf{A} = \mathbf{i} + 3\mathbf{j} - \mathbf{k} \qquad\text{and}\qquad \mathbf{B} = 2\mathbf{i} - \mathbf{j} + 2\mathbf{k}$$

Solution

The area of the parallelogram can be determined by the cross product of vectors **A** and **B**:

$$\text{Area} = |\mathbf{A} \times \mathbf{B}|$$

$$\mathbf{A} \times \mathbf{B} = \begin{vmatrix} \mathbf{i} & \mathbf{j} & \mathbf{k} \\ 1 & 3 & -1 \\ 2 & -1 & 2 \end{vmatrix} = (6-1)\mathbf{i} + (-2-2)\mathbf{j} + (-1-6)\mathbf{k} = 5\mathbf{i} - 4\mathbf{j} - 7\mathbf{k}$$

$$|\mathbf{A} \times \mathbf{B}| = \overline{)5^2 + (-4)^2 + (-7)^2} = \overline{)90} = 3\overline{)10} = 9.48$$

3.6   PHYSICAL AND BIOLOGICAL APPLICATIONS OF VECTORS

Vectors are widely applicable in many areas of the biophysical sciences. Several examples are illustrated below.

ILLUSTRATIVE EXAMPLE 3.2.   Particle Motion and Dynamics

Since vectors are functions, all mathematical topics discussed previously, including differential and integral calculus, are readily applicable to vectors, as shown by the following:

Vector differentiation:

$$\frac{d\mathbf{V}}{dt} = \frac{dV_x}{dt}\mathbf{i} + \frac{dV_y}{dt}\mathbf{j} + \frac{dV_z}{dt}\mathbf{k}$$

Vector integration:

$$\int \mathbf{V}\,dt = \mathbf{i}\int V_x\,dt + \mathbf{j}\int V_y\,dt + \mathbf{k}\int V_z\,dt$$

Solved Problem 3.11.   The position of an oscillating mass, propelled into motion by an external force, is defined by the vector

$$\mathbf{x} = A\sin\omega t\,\mathbf{i} + A\cos\omega t\,\mathbf{j}$$

(a)   What is the particle's distance from the origin?

(b)   What is the velocity of this particle?

(c)   What are the speed and direction of the particle?

(d)   What is the acceleration of the particle?

(e)   What is the general direction of acceleration of the particle with respect to its velocity? Explain.

Solution

(a)   The particle's distance from the origin is given by

$$|\mathbf{x}| = \overline{)(A \sin \omega t)^2 + (A \cos \omega t)^2}$$
$$= \overline{)A^2 \sin^2 \omega t + A^2 \cos^2 \omega t}$$
$$= \overline{)A^2(\sin^2 \omega t + \cos^2 \omega t)} = A$$

(b)   The velocity of this particle is

$$\mathbf{v} = \frac{d\mathbf{x}}{dt} = \mathbf{i}A\omega \cos \omega t - \mathbf{j}A\omega \sin \omega t$$

(c)   The speed and direction of this particle are given by

$$v = |\mathbf{v}| = \overline{)A^2 \omega^2 \cos^2 \omega t + A^2 \omega^2 \sin^2 \omega t} = A\omega$$

The direction of the angle is defined by

$$\theta = \tan^{-1} \frac{V_y}{V_x}$$
$$= \tan^{-1} \left(\frac{-A\omega \sin \omega t}{A\omega \cos \omega t} \right) = \tan^{-1} \left(-\frac{\sin \omega t}{\cos \omega t} \right)$$

(d)   The acceleration of this particle is given by

$$\mathbf{a} = \frac{d\mathbf{v}}{dt} = -\mathbf{i}A\omega^2 \sin \omega t - \mathbf{j}A\omega^2 \cos \omega t$$

(e)   The acceleration is perpendicular to the velocity since the scalar or dot product of the vectors $\mathbf{v}$ and $\mathbf{a}$ is 0, or

$$\mathbf{v} \cdot \mathbf{a} = (A\omega \cos \omega t)(-A\omega^2 \sin \omega t) + (-A\omega \sin \omega t)(-A\omega^2 \cos \omega t) = 0$$

ILLUSTRATIVE EXAMPLE 3.3.   Skeletal Mechanics

The structure and function of the human skeleton, particularly in response to physical activity, can be understood and explained through representation by vectors. Applications include the following:

Standing

The leg of a standing person is exerting downward forces on the ground that are counteracted by an upward force from the ground. In particular, the upper leg and lower leg, representing the weight of the leg and the weight of the body supported by the leg, act downward while the ground is exerting a force equal in magnitude but opposite in direction to those forces of the leg. This is represented in Figure 3-16. All forces in a standing person can be represented by vectors which can easily be manipulated to understand the mechanics of the skeleton. Two additional examples are presented below.

Running

During running activities, the foot is subject to three forces: (1) the force exerted by the Achilles tendon, which connects the calf muscles to the back of the heel; (2) the force of the leg bones (both upper and lower legs) acting downward on the foot; and (3) the force of the ground acting upward to counteract the downward forces of the foot, which is equal in magnitude to the weight of the body. Again, each of the forces exerted on the foot can be represented by vectors, as shown in Figure 3-17.

Chewing

The process of chewing, or mastication, involves the exertion of forces generated by a group of muscles which control the position and motion of the upper (maxilla) and lower (mandible) jaws. The largest muscles of the group are the masseter muscle, whose action is to lower the mandible (opening the mouth), and the temporal muscle, which assists the masseter muscle in raising the mandible (closing the mouth). Vectors can be used to represent the magnitude and direction of the forces generated by these two muscles in the process of chewing. The resultant vector of the two vectors representing the masseter and temporal muscles is illustrated in Figure 3-18.

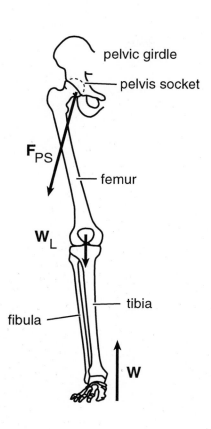

pelvic girdle

pelvis socket

$\mathbf{F}_{PS}$

femur

$\mathbf{W}_L$

tibia

fibula

$\mathbf{W}$

$\mathbf{F}_{PS}$ = force of the pelvis socket acting on the head of the femoral bone

$\mathbf{W}$ = upward force of the floor

$\mathbf{W}_L$ = weight of the leg acting downward at the center of gravity
$\mathbf{W}_L \cong W/7$

Fig. 3-16

Solved Problem 3.12. For a jaw of a particular mammal, the masseter muscle exerts a force of 275 N in a direction 125° from the +x axis, and the temporal muscle exerts a force of 220 N in a direction of 20° from the +x axis. Determine the magnitude and direction of the resultant force acting on the lower jaw.

Solution

To determine the resultant force, we must first determine and add the components of force in the x and y directions:

$$\Sigma F_x = -275 \text{ N} \cos 55° + 220 \text{ N} \cos 20° = -157.7 \text{ N} + 206.7 \text{ N} = 49.0 \text{ N}$$
$$\Sigma F_y = 275 \text{ N} \sin 55° + 220 \text{ N} \sin 20° = 225.3 \text{ N} + 75.2 \text{ N} = 300.5 \text{ N}$$

The magnitude and direction of the resultant vector can be calculated accordingly:

$$R = \sqrt{(\Sigma F_x)^2 + (\Sigma F_y)^2} = \sqrt{(49.0 \text{ N})^2 + (300.5 \text{ N})^2} = \sqrt{(2401 \text{ N}^2) + (90,300 \text{ N}^2)} = 304.5 \text{ N}$$

$$\theta = \tan^{-1}\left(\frac{\Sigma F_y}{\Sigma F_x}\right) = \tan^{-1}\frac{300.5 \text{ N}}{49.0 \text{ N}} = \tan^{-1} 6.1 = 80.7°$$

ILLUSTRATIVE EXAMPLE 3.4. Vessel Bifurcation

The human circulation consists of a complex distribution of blood vessels that connect the heart to all organs and tissues in the human body. As they distribute blood about the circulatory system, blood vessels gradually become smaller and tend to bifurcate or branch into yet smaller arteries. This arterial bifurcation or "fork in the road" can be the source of major abnormal developments, leading to stroke and heart disease. The point of importance in vessel bifurcations occurs where the longitudinal stresses of the parent (main stem) and daughter (branches) arteries are exerted on the apex. Each of the vessels involved in the birfurcation can be represented by vectors with a magnitude of force and a certain direction or angle with respect to the apex in the form of a freebody diagram (Figure 3-19). The vectors are then solved to reveal the total or resultant force acting in both the x and y directions. In static equilibrium, the only forces acting on the bifurcation apex are

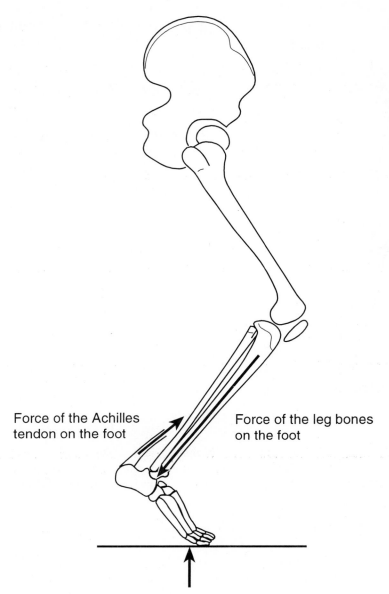

Force of the Achilles
tendon on the foot

Force of the leg bones
on the foot

Force of the ground on the foot

Fig. 3-17

the forces of the arteries F_p, F_{d1}, F_{d2} which can be written in equation form as

$$F_p + F_{d1} + F_{d2} = 0$$

These vectors can be resolved into their geometric components along each axis:

x axis:
$$(F_{d1})_x + (F_{d2})_x = 0$$
$$F_{d1} \cos \theta_1 + F_{d2} \cos \theta_2 = 0$$

y axis:
$$(F_p)_y + (F_{d1})_y + (F_{d2})_y = 0$$
$$F_{d1} \sin \theta_1 + F_{d2} \sin \theta_2 - F_p = 0$$

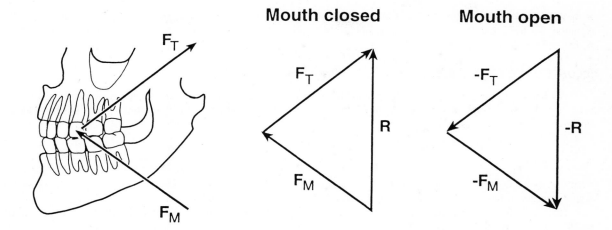

F_T = force of temporal muscle

F_M = force of masseter muscle

Fig. 3-18

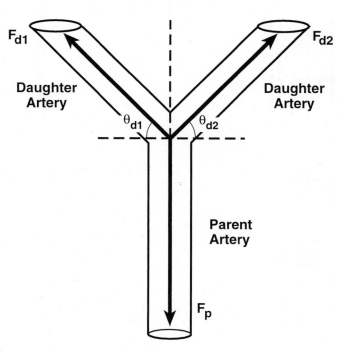

Fig. 3-19

If the magnitude of one of the forces is known, the two other forces can be determined by solving simultaneously the equations for the x and y axis. This is an example of free-body diagrams used in the analysis of physics problems, which is discussed in detail in Chapter 5.

ILLUSTRATIVE EXAMPLE 3.5.   Velocity Profile of Blood Flow

The final application of vectors in this context is the vector representation of the velocity profile of blood flow. Under elementary approximations of steady blood flow through a rigid vessel, blood flow velocity exhibits a parabolic profile along the cross section of the vessel according to

$$v = \frac{\Delta P}{4L\eta}(R^2 - r^2)$$

where ΔP is the pressure gradient applied to both ends of the vessel to initiate blood flow, L is the length of the vessel, η is the viscosity of blood, R is the vessel radius, and r is the radial component of the vessel.

The profile is physically significant in that it resembles the distribution of flow velocity and wall shear stress, each of which can be represented by vectors (see Figure 3-20). At the center of the profile where the flow velocity is the largest,

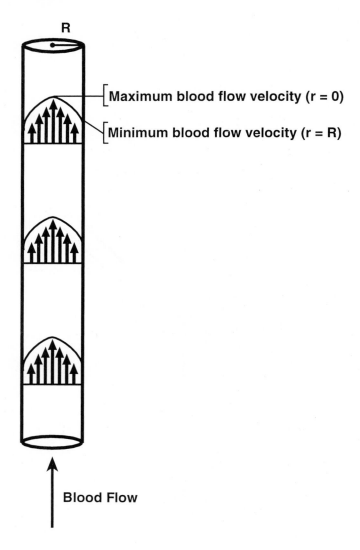

Fig. 3-20

the corresponding vector is longest. The flow velocity is minimal at the ends of the profile where the vectors are the smallest. The exact opposite is true for wall shear stress. The maximum wall shear stress occurs at the end of the profile or at the wall of the tube, while it is at a minimum in the center of the profile.

Supplementary Problems

3.1. Answer the following statements True or False.

(a) The components of a vector depend on the choice of the coordinate system. *True*

(b) Multiplying each component of a vector by -1 does not change the vector. *False*

(c) $\mathbf{A} \times \mathbf{B} = 0$ means that either $\mathbf{A}$ or $\mathbf{B}$ must be zero. *False*

(d) $\mathbf{A}/|\mathbf{A}|$ is a unit vector. *True*

(e) $\mathbf{A} \times \mathbf{B} = \mathbf{B} \times \mathbf{A}$. *False*

(f) The vectors with components $(-3, 0, 4)$ and $(8, 6, 6)$ are

equal	*False*
parallel	*False*
scalar quantities	*False*
perpendicular	*True*
opposite	*False*

3.2. Find the magnitude and direction of the following vectors:

(a) $\mathbf{V} = 4\mathbf{i} + 6\mathbf{j}$ m

(b) $\mathbf{V} = -2\mathbf{i} - \mathbf{j}\,\mathrm{s}^{-1}$

(c) $\mathbf{V} = 5\mathbf{i} + 10\mathbf{j}\,\mathrm{m\,s}^{-2}$

(d) $\mathbf{V} = 3\mathbf{i} + 7\mathbf{j}\,\mathrm{N}$

Answer

(a) $|V| = 7.2$ m, $\theta = 56.3°$

(b) $|V| = 2.2\,\mathrm{m\,s}^{-1}$, $\theta = 206.6°$

(c) $|V| = 11.2\,\mathrm{m\,s}^{-2}$, $\theta = 63.4°$

(d) $|V| = 7.6\,\mathrm{N}$, $\theta = 66.8°$

3.3. Given the following vector components, determine the magnitude and direction of the resultant vector.

(a) $V_x = 4$, $V_y = 3$

(b) $V_x = 40\,\mathrm{cm}$, $V_y = 0.0005\,\mathrm{km}$

(c) $V_x = 2\,\mathrm{N}$, $V_y = -5\,\mathrm{N}$

(d) $V_x = 0.6\,\mathrm{m\,s}^{-1}$, $V_y = 80\,\mathrm{cm\,s}^{-1}$

(e) $V_x = 2\,\mathrm{m\,s}^{-1}$, $V_y = 10\,\mathrm{m\,s}^{-1}$

Answer

(a) $R = 5$, $\theta = 36.8°$

(b) $R = 64$ cm, $\theta = 51.3°$

(c) $R = \sqrt{29}$, $\theta = -68.2°$ or $291.8°$ from $+x$ axis

(d) $R = 100\,\mathrm{cm\,s}^{-1}$, $\theta = 53.1°$

(e) $R = 10.2\,\mathrm{m\,s}^{-1}$, $\theta = 78.7°$

3.4.   A 6-N force is applied to a block at a 45° angle. Resolve the vector into its x and y components.

Solution

$$V_x = 4.2\,\text{N} \qquad V_y = 4.2\,\text{N}$$

3.5.   Resolve the vector 7, 65° into components.

Solution

$$V_x = 2.9\,\text{N} \qquad V_y = 6.3\,\text{N}$$

3.6.   Given that $\mathbf{A} = 2\mathbf{i} + 3\mathbf{j}$, find the magnitude and direction of the resultant vector.

Solution

$$|\mathbf{A}| = \sqrt{13} \qquad \theta = 56.3°$$

3.7.   For $\mathbf{A} = \mathbf{i} + \mathbf{j}$ and $\mathbf{B} = -2\mathbf{i} + 2\mathbf{j}$, determine (1) $\mathbf{A} + \mathbf{B}$, (2) $\mathbf{A} \cdot \mathbf{B}$, and (3) $\mathbf{A} \times \mathbf{B}$.

Solution

(1) $\mathbf{A} + \mathbf{B} = -\mathbf{i} + 3\mathbf{j}$

(2) $\mathbf{A} \cdot \mathbf{B} = 0$

(3) $\mathbf{A} \times \mathbf{B} = 4\mathbf{k}$

3.8.   Given two vectors $\mathbf{A} = 1\mathbf{i} + 3\mathbf{j} + 2\mathbf{k}$ and $\mathbf{B} = 4\mathbf{i} + 3\mathbf{j} + 9\mathbf{k}$, determine (1) the displacement vector $\mathbf{B} - \mathbf{A}$, (2) $\mathbf{A} \cdot \mathbf{B}$ and the angle between $\mathbf{A}$ and $\mathbf{B}$, and (3) the length of the vector $\mathbf{A} \times \mathbf{B}$.

Solution

(1) $3\mathbf{i} + 7\mathbf{k}$

(2)   31, 36.4°

(3)   22.87

3.9.   Given two forces $\mathbf{F}_1 = 2\mathbf{i} - 4\mathbf{j} + 3\mathbf{k}$ and $\mathbf{F}_2 = -3\mathbf{i} - 2\mathbf{j} + 1\mathbf{k}$, determine the force $\mathbf{F}_3$ that would balance $\mathbf{F}_1$ and $\mathbf{F}_2$.

Solution

$\mathbf{F}_3 = \mathbf{i} + 6\mathbf{j} - 4\mathbf{k}$

3.10.   What is the area of the parallelogram defined by these vectors?

$$\mathbf{A} = \mathbf{i} + 3\mathbf{j} - \mathbf{k} \qquad \text{and} \qquad \mathbf{B} = 2\mathbf{i} - \mathbf{j} + 2\mathbf{k}$$

Solution

$|\mathbf{A} \times \mathbf{B}| = 3\sqrt{10}$

3.11.   Find the velocity and acceleration of a particle whose position is defined by

(a) $R(t) = (2t - t^2)\mathbf{i}$

(b) $R(t) = 3\cos 2t\,\mathbf{i} + 2\sin 3t\,\mathbf{j}$.

Solution

(a) $\mathbf{v} = (2 - 2t)\mathbf{i}$
       $\mathbf{a} = -2\mathbf{i}$

(b) $\mathbf{v} = -6\sin 2t\,\mathbf{i} + 6\cos 3t\,\mathbf{j}$
       $\mathbf{a} = -12\cos 2t\,\mathbf{i} - 18\sin 3t\,\mathbf{j}$

3.12.   On a baseball field, the bases are separated by 90 ft, positioned in the form of a diamond, as shown in Figure 3-21. If J. P. hits a home run and rounds the bases at a constant speed, what is the vector displacement (magnitude and direction) of his total trip around the bases?

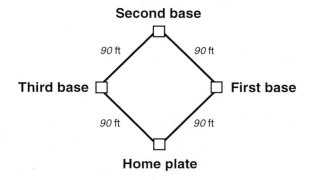

Fig. 3-21

Solution

Zero displacement

3.13.   Consider two points defined in the *xy* plane: $A = (3, 5)$ m and $B = (2, 4)$ m. (1) Express points A and B as vectors. (2) Determine the magnitude and direction of the two vectors $\mathbf{A}$ and $\mathbf{B}$. (3) Use either graphical method to calculate the resultant vectors $\mathbf{A} + \mathbf{B}$ and $\mathbf{A} - \mathbf{B}$. (4) Determine the magnitude and direction of $\mathbf{A} + \mathbf{B}$ and $\mathbf{A} - \mathbf{B}$.

Solution

(1) $\mathbf{A} = 3\,\text{m}\,\mathbf{i} + 5\,\text{m}\,\mathbf{j}$ and $\mathbf{B} = 2\,\text{m}\,\mathbf{i} + 4\,\text{m}\,\mathbf{j}$
(2) $|\mathbf{A}| = 5.8$ m, $|\mathbf{B}| = 4.5$ m
(3) $|\mathbf{A} + \mathbf{B}| = 5\,\text{m}\,\mathbf{i} + 9\,\text{m}\,\mathbf{j}$, $|\mathbf{A} - \mathbf{B}| = 1\,\text{m}\,\mathbf{i} + 1\,\text{m}\,\mathbf{j}$
(4) $|\mathbf{A} + \mathbf{B}| = 10.3$ m, $\theta = 60.9°$; $|\mathbf{A} - \mathbf{B}| = 1.4$ m, $\theta = 45.0°$

3.14.   You are given the following two vectors:

$$\mathbf{A} = \mathbf{i} + 3\mathbf{j} \qquad \text{and} \qquad \mathbf{B} = -2\mathbf{i} + \mathbf{j}$$

(1) What are the vector components of $\mathbf{A}$ and $\mathbf{B}$? (2) Determine the magnitudes and angles of $\mathbf{A}$ and $\mathbf{B}$. (3) Determine a vector $\mathbf{C}$ which, when added to $\mathbf{A}$ and $\mathbf{B}$, makes the resultant sum zero.

Solution

(1) $A_x = 1, A_y = 3; B_x = -2, B_y = 1$
(2)   3.2, 71.6°; 2.2, 206.6°
(3) $\mathbf{i} - 4\mathbf{j}$

3.15. Given the vectors

$$\mathbf{A} = \mathbf{i} + 4\mathbf{j} - 3\mathbf{k} \qquad \text{and} \qquad \mathbf{B} = 4\mathbf{i} + 2\mathbf{j} + 4\mathbf{k}$$

determine the dot product and cross product.

Solution

$$\mathbf{A} \cdot \mathbf{B} = 0 \qquad \mathbf{A} \times \mathbf{B} = 22\mathbf{i} - 16\mathbf{j} - 14\mathbf{k}$$

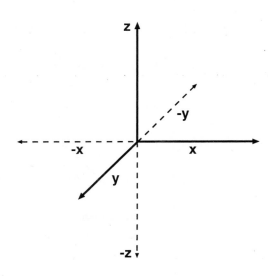

Fig. 3-22

3.16. Assume that each unit vector along a given coordinate axis in Figure 3-22 is 4 units in length. Determine (1) the magnitude of the dot product and (2) the magnitude and direction of the cross product for (*a*) **X** and **Y**, (*b*) **X** and −**X**, (*c*) **Y** and **X**, (*d*) **Z** and −**Z**, and (*e*) **Z** and **Y**.

Solution

(*a*) $\mathbf{X} \cdot \mathbf{Y} = 0$; $\mathbf{X} \times \mathbf{Y} = 16$; direction of $\mathbf{X} \times \mathbf{Y}$ is −**Z**

(*b*) $\mathbf{X} \cdot (-\mathbf{X}) = 16$; $\mathbf{X} \times (-\mathbf{X}) = 0$; direction of $\mathbf{X} \times (-\mathbf{X})$ is undefined

(*c*) $\mathbf{Y} \cdot \mathbf{X} = 0$; $\mathbf{Y} \times \mathbf{X} = 16$; direction of $\mathbf{Y} \times \mathbf{X}$ is +**Z**

(*d*) $\mathbf{Z} \cdot (-\mathbf{Z}) = 16$; $\mathbf{Z} \times (-\mathbf{Z}) = 0$; direction of $\mathbf{Z} \times (-\mathbf{Z})$ is undefined

(*e*) $\mathbf{Z} \cdot \mathbf{Y} = 0$; $\mathbf{Z} \times \mathbf{Y} = 16$; direction of $\mathbf{Y} \times \mathbf{X}$ is +**X**

3.17. The position vector of an object is given as a function of time:

$$\mathbf{r}(t) = 4t^4\mathbf{i} + 3t^3\mathbf{j} + 2t^2\mathbf{k}$$

For the object, determine the (1) force ($\mathbf{F} = m\mathbf{a}$), (2) angular momentum ($\mathbf{L} = \mathbf{r} \times m\mathbf{v}$), and (3) torque ($\boldsymbol{\tau} = \mathbf{r} \times \mathbf{F}$).

Solution

(1) $\mathbf{F} = m(48t^2\mathbf{i} + 18t\mathbf{j} + 4\mathbf{k})$

(2) $\mathbf{L} = m(-6t^4\mathbf{i} + 16t^5\mathbf{j} - 8t^6\mathbf{k})$

(3) $\boldsymbol{\tau} = m(-24t^3\mathbf{i} + 80t^4\mathbf{j} - 72t^5\mathbf{k})$

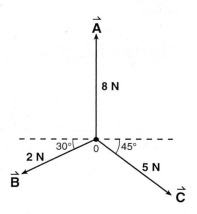

Fig. 3-23

3.18.   Consider a physical system in which an 8-N force is acting north, a 5-N force is acting southeast at an angle of 45°, and a 2-N force is acting southwest at an angle of 30°, as shown in Figure 3-23. Determine the magnitude and direction of the resultant vector.

Solution

$$\mathbf{R} = 3.89\,\text{N} \qquad \theta = 62.4°$$

Chapter 4

Planar Motion

Whether an object is as small as an electron or large as a planet, motion plays an important role in the behavior, interactions, and dynamics between an object and its environment. Particle motion refers to the physical movement of an object or particle along a predefined path within geometric boundaries. This chapter introduces the student to the concept of motion and associated phenomena.

4.1 DEFINITIONS OF PLANAR MOTION

Position defines the location of an object within a specified region and corresponds to the two-dimensional coordinates of an object P:

$$P = P(x, y)$$

Position is measured in units of length and is a scalar quantity.

Distance represents the length or line between two positions covered by the same object

$$r = \text{ change in position} = P(x_f, y_f) - P(x_o, y_o)$$

where $P(x_o, y_o)$ is the original position and $P(x_f, y_f)$ is the final position. Distance is measured in units of length and is a scalar quantity.

Displacement is a directed line segment between the two positions covered by the same object and is defined by a position vector $\mathbf{r}$ of the object:

$$\mathbf{r} = r_x\mathbf{i} + r_y\mathbf{j} = [r(x_f) - r(x_o)]\mathbf{i} + [r(y_f) - r(y_o)]\mathbf{j}$$

Displacement is a vector quantity whose magnitude is the distance determined by

$$|\mathbf{r}| = \sqrt{(r_x)^2 + (r_y)^2}$$

and direction is given by the angle θ that the vector makes with the positive x axis, defined mathematically by

$$\tan\theta = \frac{\Delta y}{\Delta x} = \frac{r(y_f) - r(y_o)}{r(x_f) - r(x_o)}$$

$$\theta = \tan^{-1}\frac{\Delta y}{\Delta x} = \tan^{-1}\frac{r(y_f) - r(y_o)}{r(x_f) - r(x_o)}$$

Displacement is measured in units of length and is a vector quantity.

Speed is the rate of change of object position over time, or how quickly the object moves from $r(x_o, y_o)$ to $r(x_f, y_f)$:

$$v = \frac{\Delta r}{\Delta t} = \frac{\text{change in distance}}{\text{change in time}} = \frac{r(x_f, y_f) - r(x_o, y_o)}{t(x_f, y_f) - t(x_o, x_o)}$$

where $t(x_o, y_o)$ is the time measured at the starting point or origin and $t(x_f, y_f)$ is the time measured at the final point. The variable Δt is the total time taken for the object to travel from $r(x_o, y_o)$ to $r(x_f, y_f)$. Speed is measured in units of length per time and is a scalar quantity.

Velocity is the rate of change of object displacement over time and is given by

$$\mathbf{v} = v_x\mathbf{i} + v_y\mathbf{j} = [v(x_f) - v(x_o)]\mathbf{i} + [v(y_f) - v(y_o)]\mathbf{j}$$

56

Velocity is a vector quantity whose magnitude is the distance determined by

$$|\mathbf{v}| = \sqrt{(v_x)^2 + (v_y)^2}$$

and whose direction is given by the angle θ that the vector makes with the positive x axis, defined mathematically by

$$\tan \theta = \frac{\Delta y}{\Delta x} = \frac{v(y_f) - v(y_o)}{t(x_f) - t(x_o)}$$

$$\theta = \tan^{-1} \frac{\Delta y}{\Delta x} = \tan^{-1} \frac{v(y_f) - v(y_o)}{t(x_f) - t(x_o)}$$

Velocity is measured in units of length per time and is a vector quantity.

Instantaneous velocity is a measurement of velocity at a particular time, in the limit as Δt approaches zero, as opposed to velocity (above) which yields the average velocity of an object over the entire time interval of the trip and is given by

$$\mathbf{v} = \lim_{t \to 0} \frac{\Delta r}{\Delta t}$$

Acceleration is the rate of change of velocity with respect to time and is given by

$$\mathbf{a} = a_x \mathbf{i} + a_y \mathbf{j} = [a(x_f) - a(x_o)]\mathbf{i} + [a(y_f) - a(y_o)]\mathbf{j}$$

Acceleration is a vector quantity whose magnitude is the distance determined by

$$|\mathbf{a}| = \sqrt{(a_x)^2 + (a_y)^2}$$

and whose direction is given by the angle θ that the vector makes with the positive x axis, defined mathematically by

$$\tan \theta = \frac{\Delta y}{\Delta x} = \frac{v(y_f) - v(y_o)}{t(x_f) - t(x_o)}$$

$$\theta = \tan^{-1} \frac{\Delta y}{\Delta x} = \tan^{-1} \frac{v(y_f) - v(y_o)}{t(x_f) - t(x_o)}$$

Acceleration is measured in units of length per time squared and is a vector quantity.

Instantaneous acceleration is a measurement of acceleration at a particular time, in the limit as Δt approaches zero, as opposed to acceleration (above) which yields the average acceleration of an object over the entire time interval of the trip and is given by

$$\mathbf{a} = \lim_{t \to 0} \frac{\Delta v}{\Delta t}$$

ILLUSTRATIVE EXAMPLE 4.1.   X-ray Angiography and Blood Flow Measurement

An important step in the diagnosis and treatment of vessel disease is the qualitative assessment and quantitative measurement of blood flow through the diseased vessels and surrounding territory. In X-ray angiography, standard X-ray images are acquired of the region of interest following injection of an opaque contrast agent, which enters the bloodstream and permits an accurate visual depiction of blood flow through vessels. Using X-ray angiography, blood flow can be approximately measured by employing equations of motion to describe the movement of contrast dye along a linear portion of the blood vessel, as illustrated in Figure 4-1. The blood flow velocity measurement in this application is a crude approximation due, in part, to (1) the pulsatile nature of blood flow, (2) elasticity of the wall of blood vessels, and (3) geometric taper (gradual narrowing) and curvature of blood vessels; but it is nevertheless useful in clinical decisions.

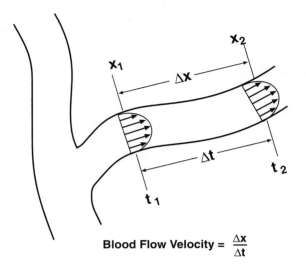

Blood Flow Velocity = $\dfrac{\Delta x}{\Delta t}$

Fig. 4-1

4.2 TYPES OF PLANAR MOTION

Planar motion describes the motion of an object in a two-dimensional coordinate system and can be represented by two individual types of motion: linear motion and projectile motion.

4.2.1 Linear Motion

Linear motion describes the motion of an object directed along either the x axis (horizontal motion) or the y axis (vertical motion) of the coordinate system assuming the acceleration a is constant and directed along the axis of motion. Linear motion is represented by the following equations:

Horizontal Motion (x axis)	Vertical Motion (y axis)
$x = \bar{v}t$	$y = \bar{v}t$
$x = v_{ox}t + \dfrac{1}{2}at^2$	$y = v_{oy}t + \dfrac{1}{2}at^2$
$v_{fx}^2 = v_{ox}^2 + 2ax$	$v_{fy}^2 = v_{oy}^2 + 2ay$
$a = \dfrac{v_{fx} - v_o}{t}$	$a = \dfrac{v_{fy} - v_o}{t}$
$\bar{v} = \dfrac{v_{fx} + v_o}{2}$	$\bar{v} = \dfrac{v_{fy} + v_o}{2}$

Free-fall motion is a specific type of vertical motion and refers to an object dropped from a height and falling under the influence of gravity with a constant acceleration of $g = 9.8 \text{ m s}^{-2}$ (or 980 cm s^{-2} or 32 ft s^{-2}). Because the object is dropped and not thrown, the initial velocity is assumed to be 0.

Solved Problem 4.1. In two Gold Medal performances and world-record feats at the 1996 Olympics, Michael Johnson ran the 200-m dash in 19.32 s and the 400-m dash in 43.49 s. What was Johnson's average velocity in each event?

Solution

Using the equation of motion $x = vt$, we have

$$\bar{v} = \frac{x}{t}$$

The average velocities of Michael Johnson in the two races were

$$\bar{v} = \frac{200\ m}{19.32\ s} = 10.35\ m\ s^{-1} \qquad 200\ m$$

$$\bar{v} = \frac{400\ m}{43.49\ s} = 9.19\ m\ s^{-1} \qquad 400\ m$$

Solved Problem 4.2. Upon being stopped by a patrol officer for exceeding the posted speed limit of 35 mi h^{-1} by 10 mi h^{-1} Marge explained to Officer Einstein that she was 25 mi away and had 30 min to make an urgent appointment. She stated that traveling at the current speed would get her there on schedule. Officer Einstein, having had a physics course or two several years ago, decided to use his knowledge of motion to determine whether she was telling the truth.

(a) Would she have made her appointment on time, continuing at 45 mi h^{-1}?

(b) What would her speed have to be in order for her to make the appointment?

Solution

(a) Using the equation of motion $x = \bar{v}t$, we want to know the time taken to travel 25 mi at a velocity of 45 mi h^{-1}. Therefore,

$$t = \frac{x}{\bar{v}} = \frac{25\ mi}{45\ mi\ h^{-1}} = 0.55\ h = 33\ min$$

Therefore, since she had 30 min to make her appointment, she would have been 3 min late.

(b) To determine the speed she would have had to attain in order to make her appointment, we use the same equation to determine the average velocity required to traverse a distance of 25 mi in 30 min.

$$\bar{v} = \frac{x}{t} = \frac{25\ mi}{0.5\ h} = 50\ mi\ h^{-1}$$

Solved Problem 4.3. A bus traveling from Los Angeles to New York City covered the first fourth of the total distance at a velocity of 55 km h^{-1}, the second fourth of the total distance at a velocity of 80 km h^{-1}, the third fourth of the total distance at a velocity of 60 km h^{-1}, and the final fourth of the total distance at a velocity of 40 km h^{-1}. Determine the average velocity of the bus.

Solution

From the equation of motion $x = \bar{v}t$,

$$t = \frac{x}{\bar{v}}$$

$$t_1 + t_2 + t_3 + t_4 = \frac{x_1}{\bar{v}_1} + \frac{x_2}{\bar{v}_2} + \frac{x_3}{\bar{v}_3} + \frac{x_4}{\bar{v}_4}$$

From the problem, the distance for each leg of the trip is equal, ie.,

$$x_1 = x_2 = x_3 = x_4 = \frac{x}{4}$$

Therefore,

$$\bar{v} = \frac{x}{t} = \frac{x}{\dfrac{x/4}{\bar{v}_1} + \dfrac{x/4}{\bar{v}_2} + \dfrac{x/4}{\bar{v}_3} + \dfrac{x/4}{\bar{v}_4}}$$

$$= \frac{x}{(x/4)\left[\dfrac{1}{\bar{v}_1} + \dfrac{1}{\bar{v}_2} + \dfrac{1}{\bar{v}_3} + \dfrac{1}{\bar{v}_4}\right]}$$

$$= \frac{1}{\dfrac{1}{4\bar{v}_1} + \dfrac{1}{4\bar{v}_2} + \dfrac{1}{4\bar{v}_3} + \dfrac{1}{4\bar{v}_4}}$$

$$= \frac{1}{\dfrac{\bar{v}_2\bar{v}_3\bar{v}_4 + \bar{v}_1\bar{v}_3\bar{v}_4 + \bar{v}_1\bar{v}_2\bar{v}_4 + \bar{v}_1\bar{v}_2\bar{v}_3}{4\bar{v}_1\bar{v}_2\bar{v}_3\bar{v}_4}}$$

$$= \frac{4\bar{v}_1\bar{v}_2\bar{v}_3\bar{v}_4}{\bar{v}_2\bar{v}_3\bar{v}_4 + \bar{v}_1\bar{v}_3\bar{v}_4 + \bar{v}_1\bar{v}_2\bar{v}_4 + \bar{v}_1\bar{v}_2\bar{v}_3}$$

Substituting the values for the velocity into the above expression gives

$$\bar{v} = 55.3 \text{ km h}^{-1}$$

Solved Problem 4.4. The position of a 20-kg object in planar motion is described by the equation

$$x = 5 + 4t + 3t^2 + 2t^3$$

where x is the position, in meters, and t is the time, in seconds.

(1)   What are the units of 5, 4, 3, and 2?

(2)   Determine the velocity of the object as a function of time.

(3)   Determine the average velocity of the object over the time interval $t = 0$ to $t = 5$.

(4)   Determine the instantaneous velocity of the object at $t = 3$ s.

(5)   Determine the acceleration of the object as a function of time.

(6)   Determine the instantaneous acceleration of the object at $t = 2$ s.

Solution

(1)   The units for each of the components in the expression for the position must result in meters. Thus the units are as follows:

      5:   m

      4:   m s^{-1}

      3:   m s^{-2}

      2:   m s^{-3}

(2)   The velocity of the object is determined by the time derivative of the position or

$$\mathbf{v} = \frac{dx}{dt} = 4 + 6t + 6t^2$$

(3)   The average velocity of the object is

$$\bar{v} = \frac{\Delta x}{\Delta t} = \frac{x(t = 5\text{ s}) - x(t = 0\text{ s})}{5\text{ s} - 0\text{ s}} = \frac{350\text{ m} - 5\text{ m}}{5\text{ s} - 0\text{ s}} = \frac{345\text{ m}}{5\text{ s}} = 69 \text{ m s}^{-1}$$

(4)   As determined in (2), the velocity of the object is

$$\mathbf{v} = \frac{dx}{dt} = 4 + 6t + 6t^2$$

The instantaneous velocity is the velocity at a particular time, in this case $t = 3$ s, is

$$\mathbf{v}(t = 3 \text{ s}) = \frac{dx}{dt} = 4 + 6(3) + 6(3)^2 = 76 \text{ m s}^{-1}$$

(5)   The acceleration of the object is determined by the second time derivative of the position or the time derivative of the velocity:

$$\mathbf{a} = \frac{d^2\mathbf{x}}{dt^2} = \frac{d\mathbf{v}}{dt} = 6 + 12t$$

(6)   The instantaneous acceleration of the object at $t = 2$ s is

$$\mathbf{a}(t = 2 \text{ s}) = 6 + 12(2) = 30 \text{ m s}^{-2}$$

Solved Problem 4.5.   In a particular relay race, four runners run equidistant sprints or one-fourth the circumferential distance of a 5-km race track. The total time for the relay race was 9.5 min. Runner 1 ran at an average velocity of 10.6 m s^{-1}, runner 2 ran at an average velocity of 9.8 m s^{-1}, runner 3 ran at an average velocity of 11.2 m s^{-1}, and runner 4 ran at an average velocity of 10.8 m s^{-1}.

(1)   What is the displacement of each runner?

(2)   What is the total displacement from all four runners combined?

(3)   At what time did each runner complete her or his leg of the relay race?

Solution

(1)   The magnitude of the displacement is simply the distance covered by each runner, or

$$x_1 = x_2 = x_3 = x_4 = \frac{5}{4} \text{ km} = 1.25 \text{ km}$$

However, since displacement is a vector quantity, direction must also be determined. Since the runners ran around an elliptical race track, two of the runners actually had negative displacements in that they were running in opposite direction to the two other runners. Therefore, the displacement for each runner is

$$x_1 = 1.25 \text{ km} \qquad x_2 = -1.25 \text{ km} \qquad x_3 = -1.25 \text{ km} \qquad x_4 = 1.25 \text{ km}$$

(2)   The total or net displacement from all four runners combined is the verctor sum of the individual displacement from each runner, or

$$x_T = x_1 + x_2 + x_3 + x_4 = 1.25 \text{ km} - 1.25 \text{ km} - 1.25 \text{ km} + 1.25 \text{ km} = 0$$

(3)   The time for each runner to complete his or her leg of the relay race can be determined from the equation of motion

$$x = \bar{v}t$$

Runner 1 $$t = \frac{x}{\bar{v}} = \frac{1.25 \text{ km}}{10.6 \text{ m s}^{-1}} = 117.9 \text{ s}$$

Runner 2 $$t = \frac{x}{\bar{v}} = \frac{-1.25 \text{ km}}{9.8 \text{ m s}^{-1}} = -127.6 \text{ s}$$

Runner 3 $$t = \frac{x}{\bar{v}} = \frac{-1.25 \text{ km}}{11.2 \text{ m s}^{-1}} = -111.6 \text{ s}$$

Runner 4 $$t = \frac{x}{\bar{v}} = \frac{1.25 \text{ km}}{10.8 \text{ m s}^{-1}} = 115.7 \text{ s}$$

The minus signs correspond to the negative displacement with respect to the choice of the positive direction. We convert to minutes:

Runner 1:   1.96 min

Runner 2:   2.13 min

Runner 3:   1.86 min

Runner 4:   1.93 min

Solved Problem 4.6.   A baseball player hits a ball into centerfield 380 ft from home plate and begins to immediately run the bases at a constant velocity of 25 ft s^{-1}. At the moment the batter reaches first base, the center fielder throws the ball at an average velocity of 60 ft s^{-1} toward the second baseman. Does the runner make it safely to second base? (The bases are arranged in a diamond formation separated 90 ft apart.)

Solution

This problem requires analysis of the center fielder and the base runner on an individual basis. In considering the center fielder, we need to know the distance to second base. Since the center fielder is 380 ft from home plate and each plate is separated by 90 ft, the distance from home plate to second base, denoted by $x_{H\text{-}S}$, is

$$x_{H\text{-}S} = \sqrt{(x_{H\text{-}F})^2 + (x_{F\text{-}S})^2} = \sqrt{(90\ \text{ft})^2 + (90\ \text{ft})^2} = \sqrt{16,200\ \text{ft}^2} = 127.3\ \text{ft}$$

where $x_{H\text{-}F}$ is the distance from home plate to first base and $x_{F\text{-}S}$ is the distance from first base to second base. So the center fielder must throw the ball a distance $x_{\text{CF}} = 380\ \text{ft} - 127.3\ \text{ft} = 252.7\ \text{ft}$ at an average velocity $\bar{v}_{\text{CF}} = 60\ \text{ft s}^{-1}$. The time required for the ball to reach second base is

$$t_{\text{CF}} = \frac{x_{\text{CF}}}{\bar{v}_{\text{CF}}} = \frac{252.7\ \text{ft}}{60\ \text{ft s}^{-1}} = 4.2\ \text{s}$$

The question now becomes, Can the base runner go from first base to second base at an average velocity of 25 ft s^{-1} in less than 4.2 s? The time required for the base runner t_{BR} is

$$t_{\text{BR}} = \frac{x_{\text{BR}}}{\bar{v}_{\text{BR}}} = \frac{90\ \text{ft}}{25\ \text{ft s}^{-1}} = 3.6\ \text{s}$$

Therefore, the base runner makes it safely to second base by 0.6 s.

Solved Problem 4.7.   An object is dropped from a height of 75 m. (1) Determine the distance traveled (a) during the first second of motion and (b) during the last second of motion; (2) the final velocity at which the object strikes the ground; (3) the total time of the free-fall drop; and (4) the time the object will travel in (a) the first meter and (b) the final meter.

Solution

(1)   The distance traveled by the object can be determined from the equation of motion:

$$y = v_o t + \frac{1}{2} a t^2$$

Since the particle is dropped from rest, $v_o = 0$ and $a = g$. Therefore, the total distance traveled by the object in the first second of motion is

$$y_1 = \frac{1}{2} g t^2 = \frac{1}{2} (9.8\ \text{m s}^{-2})(1\ \text{s})^2 = 4.9\ \text{m}$$

To determine the final second of motion, the total time taken for the object to fall the entire distance is gotten by rearranging the equation of motion used in the previous problem:

$$t = \sqrt{\frac{2y}{g}} = \sqrt{\frac{2 \cdot 75\ \text{m}}{9.8\ \text{m s}^{-2}}} = \sqrt{\frac{150\ \text{m}}{9.8\ \text{m s}^{-2}}} = 3.9\ \text{s}$$

As the object reaches its last second of motion, it has traveled a distance during $t_f = t - t_1 = 3.9\,\text{s} - 1.0\,\text{s} = 2.9\,\text{s}$ equal to

$$y = \frac{1}{2}gt^2 = \frac{1}{2}(9.8\,\text{m s}^{-2})(2.9\,\text{s})^2 = 41.2\,\text{m}$$

Thus, the total distance traveled during the final second of motion is

$$y_2 = y - y_1 = 75\,\text{m} - 41.2\,\text{m} = 33.8\,\text{m}$$

(2)   The final velocity of the object can be determined from the equation of motion

$$v_f^2 = v_o^2 + 2ay$$

where $v_o = 0$ since it was dropped from rest and $a = g$. Also, from the text of the problem, $y = 75$ m. Assuming that down is positive and substituting into the above equation of motion, we have

$$v_f^2 = 2gy = 2(9.8\,\text{m s}^{-2})(75\,\text{m}) = 1470\,\text{m}^2\,\text{s}^{-2}$$
$$v_f = 38.3\,\text{m s}^{-1}$$

(3)   The total time for the free-fall drop requires the following equation of motion:

$$a = \frac{v_f - v_o}{t}$$

The final velocity, as calculated above, is $38.3\,\text{m s}^{-1}$, $v_o = 0$, and $a = g$. Making the appropriate substitutions yield

$$t = \frac{v_f - v_o}{g} = \frac{38.3\,\text{m s}^{-1} - 0}{9.8\,\text{m s}^{-2}} = 3.9\,\text{s}$$

(4)   The time taken for the object to drop the first meter is given by the equation of motion

$$y = \frac{1}{2}gt^2 \Rightarrow t = \sqrt{\frac{2y}{g}} = \sqrt{\frac{2 \cdot 1\,\text{m}}{9.8\,\text{m s}^{-2}}} = 0.45\,\text{s}$$

Solved Problem 4.8.   An object is dropped from a tall building. (*a*) How far has the object fallen in (1) 1 s, (2) 2 s, and (3) 5 s? Assume $g = 9.8\,\text{m s}^{-2}$. (*b*) Repeat the calculations, assuming the same building is located on the moon ($g = 1.6\,\text{m s}^{-2}$), Mars ($g = 3.8\,\text{m s}^{-2}$), and Jupiter ($g = 25\,\text{m s}^{-2}$).

Solution

The distance that the object falls upon being dropped from the tall building is

$$y = v_o t + \frac{1}{2}at^2$$

Since the object is dropped, $v_o = 0$ and $a = g$. Therefore, the distance equation can be simplified:

$$y = \frac{1}{2}gt^2$$

(a)   The distances that the object has fallen are (1) 4.9 m, (2) 19.6 m, and (3) 122.5 m.

(*b*)   On the moon: (1) 0.8 m, (2) 3.2 m, and (3) 20 m.

On Mars: (1) 1.9 m, (2) 7.6 m, and (3) 47.5 m.

On Jupiter: (1) 12.5 m, (2) 50 m, and (3) 312.5 m.

Solved Problem 4.9.   In learning the art of juggling, a prospective juggler begins with the elementary exercise of tossing two balls in the air, one at a time, from one hand. At a point during this exercise, one ball (ball *A*) is thrown vertically upward while a previously thrown ball (ball *B*) begins its descent from rest toward the hand. At what time will ball *A* meet ball *B*, i.e., will they pass each other in flight? (Assume g is acting downward.)

Solution

The vertical distance traveled by ball A is

$$y_A = v_{oA}t - \frac{1}{2}gt^2$$

The distance traveled by ball B is

$$y_B = \frac{1}{2}gt^2$$

The vertical distance between balls A and B is

$$y_{AB} = y_T - (y_A + y_B)$$

where y_T is the maximum vertical distance attained by either ball and

$$y_A + y_B = v_{oA}t - \frac{1}{2}gt^2 + \frac{1}{2}gt^2 = v_{oA}t$$

Therefore,

$$y_{AB} = y_T - v_A t$$

Ball A and ball B meet when $y_{AB} = 0$, or

$$0 = y_T - v_A t$$

$$y_T = v_A t$$

Solving for t, the time that ball A will meet ball B, gives

$$t = \frac{y_T}{v_A}$$

4.2.2 Projectile Motion

Projectile motion is the motion of a projectile or an object within the plane of a two-dimensional coordinate system, as shown in Figure 4-2. Projectile motion follows a parabolic path where (1) the horizontal component of motion proceeds at constant velocity, that is, $v_f = v_o = v$ and, as a consequence, $a = 0$; and (2) the vertical component of motion proceeds at variable velocity and with acceleration $a = g$ directed downward.

The trajectory or changing y coordinate of the motion of the projectile is defined as

$$y = (\tan \theta_o)x - \frac{g}{2v_o^2 \cos^2 \theta_o}x^2$$

The horizontal range or maximum distance along the x axis traveled by the projectile is

$$R = \frac{v_o^2}{g}\sin 2\theta_o$$

Since both coordinates are involved in projectile motion, the equations of motion describing the position and motion along both the x and y components are as follows:

	x component	y component
Position	$x = (v_o \cos \theta_o)t$	$y = (v_o \sin \theta_o)t - \frac{1}{2}gt^2$
Velocity	$v_x = v_o \cos \theta_o$	$v_y = v_o \sin \theta_o - gt$

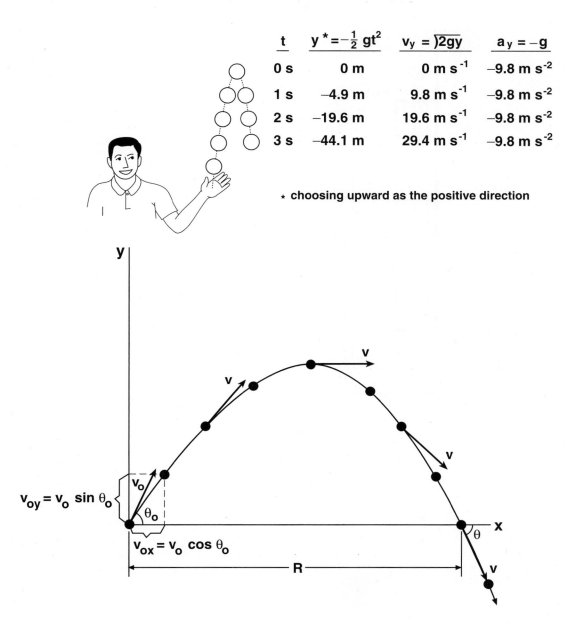

t	$y^* = -\frac{1}{2}gt^2$	$v_y = \sqrt{2gy}$	$a_y = -g$
0 s	0 m	0 m s^{-1}	−9.8 m s^{-2}
1 s	−4.9 m	9.8 m s^{-1}	−9.8 m s^{-2}
2 s	−19.6 m	19.6 m s^{-1}	−9.8 m s^{-2}
3 s	−44.1 m	29.4 m s^{-1}	−9.8 m s^{-2}

* choosing upward as the positive direction

$v_{oy} = v_o \sin \theta_o$

$v_{ox} = v_o \cos \theta_o$

Fig. 4-2

In summary, in projectile motion:

- Position vector **r** varies in time for both x and y coordinates.
- Velocity vector **v** is always tangent to the parabolic path and points in the direction of the trajectory.
- Acceleration vector **a** is always downward with no horizontal component.

ILLUSTRATIVE EXAMPLE 4.2. Criminal Investigation of a Fall

Often in mystery and detective stories as well as in real criminal investigations, a detective finds a body positioned under an open window a certain distance from a tall building. Although a lot of physical evidence and eyewitness reports are

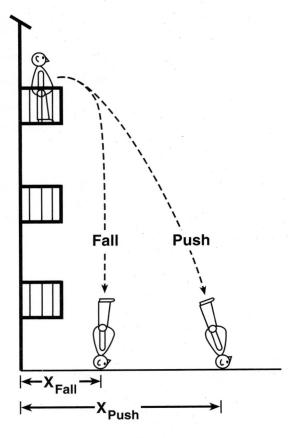

Fig. 4-3

required to positively identify the sequence of events leading to the fall, an important part of the investigation is to try to determine whether the victim fell or was pushed from the known or suspected height.

This question can be answered by employing equations relevant to projectile motion. If the person fell, the trajectory would be similar to that represented by (Fall) in Figure 4-3 and the person would land very close to the building. If the person were pushed, the person would be subject to an initial horizontal velocity, broadening the trajectory as represented by (Push) in Figure 4-3. Let's consider, for the moment, the scenario in which the victim fell from a known height. The initial velocity of the body in the y direction is $v_{oy} = 0$. Also, the acceleration of a body in free fall is $a_y = -g$. The time taken for the body to reach the ground can be determined from the equation of motion

$$y = v_{oy}t - \frac{1}{2}gt^2$$

or, since $v_{oy} = 0$,

$$y = -\frac{1}{2}gt^2$$

where y is the height from which the fall originated and becomes negative. Therefore, the above equation can be rewritten as

$$-h = -\frac{1}{2}gt^2$$

Solving for the time of the fall gives

$$t = \sqrt{\frac{2h}{g}}$$

The distance along the x axis resulting from a fall is given by

$$x_{\text{Fall}} = v_{ox}t = v_{ox}\sqrt{\frac{2h}{g}}$$

The velocity v_{ox}, although unknown, is small in cases of a fall and can be approximated through experimental trials with a weighted dummy similar to the build of the person. If the object falls a marked distance, i.e., greater than x_{Fall}, from a building, given the height, the fall becomes suspicious and should be investigated further.

Solved Problem 4.10.   What time is needed for a projectile to complete its trajectory and strike the ground?

Solution

The equation of motion required to solve the problem is

$$x = (v_o \cos\theta_o)t$$

where x now becomes the horizontal range or maximum distance attained by the projectile.

$$\frac{v_o^2}{g}\sin 2\theta_o = (v_o \cos\theta_o)t$$

Using the trigonometric relation $\sin 2\theta_o = 2\sin\theta_o \cos\theta_o$ gives

$$\frac{2v_o^2}{g}\sin\theta_o \cos\theta_o = (v_o \cos\theta_o)t$$

Solving for t yields

$$t = \frac{2v_o}{g}\sin\theta_o$$

Solved Problem 4.11.   A football is kicked from a tee with an initial velocity of 20 m s^{-1} and an initial angle of elevation of $40°$. Assuming $g = 9.8 \text{ m s}^{-2}$, determine (1) the vertical and horizontal components of the velocity, (2) the football's total time in flight, (3) the maximum height attained by the football, (4) the horizontal range or total distance traveled by the football, and (5) the final horizontal and vertical velocities of the football prior to striking the ground.

Solution

In this problem, let y represent the vertical height and choose up as the positive direction. Therefore, it is known that $g = -9.8 \text{ m s}^{-2}$.

(1)   The horizontal (v_{ox}) and vertical (v_{oy}) components of the velocity are

$$v_{ox} = v_o \cos 40° = (20 \text{ m s}^{-1})(0.766) = 15.3 \text{ m s}^{-1}$$
$$v_{oy} = v_o \sin 40° = (20 \text{ m s}^{-1})(0.643) = 12.9 \text{ m s}^{-1}$$

(2)   The equation of motion which will solve this problem is

$$y = v_{oy}t + \frac{1}{2}gt^2$$

At the instance of kickoff and at the time the football strikes the ground, $y = 0$. The total time of flight is given by the difference between the values of time at the two points when $y = 0$, and the equation becomes

$$0 = (12.9 \text{ m s}^{-1})t + (-4.9 \text{ m s}^{-2})t^2$$

Since the above equation is a quadratic, the solution of t yields two values:

$$t = \frac{-b \pm \sqrt{b^2 - 4ac}}{2a}$$

where $a = -4.9$ m s^{-2}, $b = 12.9$ m s^{-1}, and $c = 0$. Thus, $t = 0$ at kickoff, and $t = 2.63$ s at the end of the flight. The total time in flight is the difference between these two times, or $t = 2.63$ s.

(3)   The maximum height attained by the football can be determined from the equation of motion

$$v_{fy}^2 = v_{oy}^2 + 2gy$$

where $v_{fy} = 0$, $v_{oy} = 12.9$ m s^{-1}, and $g = -9.8$ m s^{-2}. Substituting the appropriate values yields

$$y = \frac{-v_{oy}^2}{-2g} = 8.5 \text{ m}$$

(4)   The horizontal distance traveled by the football is given by the equation of motion $x = \bar{v}t$, where $\bar{v}$ is the average velocity along the horizontal axis ($v_{ox} = 15.3$ m s^{-1}) and t is the total time in flight given by the solution in part (1) of this problem ($t = 2.63$ s). Therefore, the horizontal distance is

$$x = (15.3 \text{ m s}^{-1})(2.63 \text{ s}) = 40.2 \text{ m}$$

(5)   The horizontal velocity of the football remains constant, so $v_x = v_{ox} = 15.3$ m s^{-1}. The vertical velocity is given by the equation of motion

$$v_{fy} = v_{oy} + gt = 12.9 \text{ m s}^{-1} + (-9.8 \text{ m s}^{-2})(2.63 \text{ s})$$
$$= 12.9 \text{ m s}^{-1} + (-25.7 \text{ m s}^{-1}) = -12.8 \text{ m s}^{-1}$$

The negative velocity implies that the velocity is directed downward.

Solved Problem 4.12.   During batting practice, a baseball player struck a baseball at an angle of 20°, imparting to the baseball an initial velocity of 30 m s^{-1}. (1) Determine the maximum height of the baseball during its flight. (2) Calculate the time needed to reach the maximum height. (3) Determine the maximum distance or range of the baseball.

Solution

(1)   The maximum height of the baseball can be determined by the equation of motion

$$v_{fy}^2 = v_{oy}^2 + 2ay$$

where $v_{fy}^2 = 0$;

$$v_{oy} = v_o \sin\theta = 30 \text{ m s}^{-1} \sin 20° = 10.3 \text{ m s}^{-1} \qquad v_{oy}^2 = 105.3 \text{ m}^2 \text{ s}^{-2}$$

Gravity is acting downward, implying $a = -g = -9.8$ m s^{-2}. Substituting into the equation of motion, we can determine the maximum height y:

$$0 = 105.3 \text{ m}^2 \text{ s}^{-2} + 2(-9.8 \text{ m s}^{-2})y$$

Solving for y yields

$$y = 5.4 \text{ m}$$

(2)   The time needed for the baseball to reach its maximum height is given by the equation of motion

$$v_{fy} = v_{oy} + at$$

From the previous part,

$$v_{fy} = 0$$
$$a = -g = -9.8 \text{ m s}^{-2}$$
$$v_{oy} = 10.3 \text{ m s}^{-1}$$

Solving for t from the equation of motion and making the appropriate substitutions give

$$t = \frac{v_{fy} - v_{oy}}{-g} = \frac{0 - 10.3 \text{ m s}^{-1}}{-9.8 \text{ m s}^{-2}} = 1.05 \text{ s}$$

(3)   The maximum distance or range of the baseball is given by

$$x = \bar{v} t_T$$

where $\bar{v} = v_{ox} = v_o \cos \theta = 30 \text{ m s}^{-1} \cos 20° = 28.2 \text{ m s}^{-1}$ and t_T is the total time of flight of the baseball $(2 \cdot t = 2 \cdot 1.05 \text{ s} = 2.1 \text{ s})$. Making substitutions gives

$$x = (28.2 \text{ m s}^{-1})(2.1 \text{ s}) = 59.2 \text{ m}$$

Supplementary Problems

4.1.   If the time derivative of velocity is acceleration, what is the time derivative of speed?

Solution

0

4.2.   The velocity of a car increases uniformly from 30 to 70 m s^{-1} in 8 s. Determine the (1) average velocity of the car, (2) distance traveled by the car, and (3) acceleration of the car.

Solution

(1) $v = 50 \text{ m s}^{-1}$; (2) $x = 400 \text{ m}$; (3) $a = 5 \text{ m s}^{-2}$

4.3.   A bus starting from rest eventually attains an acceleration of 4 m s^{-2}. Calculate the (1) time taken to travel 200 m and (2) velocity at this time.

Solution

(1) $t = 10 \text{ s}$; (2) $v_f = 40 \text{ m s}^{-1}$

4.4.   A car travels at a velocity of 60 km h^{-1} during the first third of its total driving time, at 55 km h^{-1} during the second third of the total driving time, and at 35 km h^{-1} during the final third of the total driving time. Determine the average velocity of the car.

Solution

$v = 50 \text{ km h}^{-1}$

4.5.   In a 1500-m race, a runner maintains an average velocity of 3.9 m s^{-1}. How long will it take the runner to complete the race?

Solution

$t = 6.4 \text{ min}$

4.6.   A car completed the 1447-mi trip from Los Angeles to Dallas in 25 h. What was its average velocity? Express the velocity in km h^{-1}, m s^{-1}, and ft s^{-1}.

Solution

$v = 57.9 \text{ mi h}^{-1} = 93.2 \text{ km h}^{-1} = 25.8 \text{ m s}^{-1} = 84.8 \text{ ft s}^{-1}$

4.7.   The velocity of a car increases from 35 to 70 km h^{-1} in 5 s. Calculate the acceleration of the car.

Solution

$a = 2.5 \times 10^4 \text{ km h}^{-2}$

4.8.   The position, in meters, of an object is described by the vector

$$R(t) = 4t\mathbf{i} + 3t^2\mathbf{j} + 2t^3\mathbf{k}$$

(1)   Determine the velocity at $t = 3$ s.

(2)   Determine the acceleration at $t = 3$ s.

(3)   Determine the time at which $v(t)$ is perpendicular to $a(t)$.

Solution

(1) $v(3) = 57 \text{ m s}^{-1}$; (2) $a(3) = 36.5 \text{ m s}^{-2}$; (3) velocity v is not perpendicular to a

4.9.   An object in motion can be described by the vector

$$R(t) = A \sin kt\mathbf{i} + B \cos kt\mathbf{j}$$

Determine the (1) velocity and (2) acceleration.

Solution

(1) $v(t) = kA \cos kt\mathbf{i} - kB \sin kt\mathbf{j}$

(2) $a(t) = -k^2 A \sin kt\mathbf{i} - k^2 B \cos kt\mathbf{j}$

4.10.   If both masses are dropped at the same instant, determine the difference in time required to strike the ground for two projectiles, one of mass M and the other of mass $2M$.

Solution

They strike at the same time since the equations of motion for a projectile are independent of mass.

4.11.   A golf ball is struck from a tee with a velocity of 120 ft s^{-1} at an angle of 42° above the horizontal (x axis). Determine the magnitude and direction of the velocity vector after (1) 1 s, (2) 1.5 s, (3) 2.0 s, and (4) 3.0 s in flight.

Solution

At $t = 0.5$ s:	$v = 112.4$ ft s^{-1}	$\theta = 25.1°$
At $t = 1$ s:	$v = 106.6$ ft s^{-1}	$\theta = 17.2°$
At $t = 1.5$ s:	$v = 102.9$ ft s^{-1}	$\theta = 8.7°$

4.12.   For a projectile in flight, what are the magnitude and direction of velocity?

Solution

$$|v| = \sqrt{v_o^2 + g^2 t^2 - (2v_o \sin \theta_o) gt} \qquad \theta = \tan^{-1} \frac{v_o \sin \theta_o - gt}{v_o \cos \theta_o}$$

4.13.   A child propels a 25-g toy car with a velocity of 20 m s^{-1} up a ramp inclined at 30° from the $+x$ axis. Determine (1) the time taken to return to its starting point and (2) the distance attained by the toy car.

Solution

(1) $t = 5$ s; (2) $x = 50$ ft

4.14.   Standing on the fourth-floor balcony of his apartment building at a height of 15 m, Huey throws, in a horizontal direction, a water balloon at his cousin Louie with a velocity of 20 m s^{-1}. Determine (1) the time during which the balloon is in motion, (2) the distance from the apartment building where the balloon will strike the ground, (3) the final velocity with which it strikes the ground, and (4) the angle formed by the trajectory of the balloon with the horizontal at the point where it strikes the ground. (Neglect the resistance of air.)

Solution

(1) $t = 1.75$ s; (2) $x = 35$ m; (3) $v = 26.4$ m s^{-1}; (4) $\phi = 40.6°$

4.15. A baseball is thrown horizontally with a velocity of $v = 30$ m s^{-1}. Determine the normal and tangential acceleration of the stone after 1 s of motion.

Solution

$$a_t = 3.04 \text{ m s}^{-2} \qquad a_n = 9.32 \text{ m s}^{-2}$$

4.16. A stone is thrown with a velocity of 15 m s^{-1} at an angle $\theta = 35°$ to the horizontal (along the horizontal axis). Determine (1) the height y the stone will attain, (2) the range or horizontal distance from the point where the stone is thrown to the point where the stone strikes the ground, and (3) the time that the stone is in motion.

Solution

(1) $y = 3.77$ m; (2) $x = 13.17$ m; (3) 1.76 s

4.17. A stone is thrown with a velocity $v = 20$ m s^{-1} at an angle $\theta = 30°$ to the horizon. At what time will the stone reach its maximum height?

Solution

$t = 1.02$ s

4.18. During a practice session, a tennis player hits a ball which leaves the racquet at a velocity of 40 m s^{-1} at an angle $\theta = 30°$ to the horizon. The tennis ball strikes a wall at a distance $x = 10$ m. What is the velocity of the tennis ball as it strikes the wall?

Solution

$v = 38.6$ m s^{-1} $\qquad \theta = 31.9°$

4.19. A penny is dropped from a tall building 10 m above ground level. Determine (1) the time required for the penny to reach the ground and (2) the velocity of the penny as it strikes the ground.

Solution

(1) $t = 1.43$ s; (2) $v_f = 14$ m s^{-1}

4.20. A stone is thrown straight upward with an initial velocity of 40 m s^{-1}. Determine (1) the maximum height the stone will attain and (2) the time required for the stone to return to ground level.

Solution

(1) $y = 81.6$ m; (2) $t = 8.2$ s

4.21 To get to the other side of a 0.2-mi river, Chicken Little paddled his canoe at an average speed $v_{CL} = 4$ mi h^{-1} but was confronted with an undercurrent directed perpendicular to him at a speed $v_{uc} = 2$ mi h^{-1}. Determine (1) the resultant velocity of Chicken Little's canoe with respect to the shore and (2) the distance downstream traveled by Chicken Little.

Solution

(1) $v = 4.5$ mi h^{-1}, $\theta = 26.6°$; (2) $x = 0.1$ mi

4.22.   Referring to Problem 4.21 and Chicken Little's attempt to paddle his canoe across a 0.2-mi river, we now assume that he is aware of the undercurrent traveling perpendicular to him at 2 mi h^{-1} and wants to compensate for the undercurrent in his travel so that he lands on the other side at the point directly opposite him. At what velocity (magnitude and direction) must Chicken Little travel to accomplish such a feat?

Solution

$\mathbf{v} = 3.46$ mi h^{-1} $\theta = 30°$

Chapter 5

Statics

In the previous chapter, particle motion was discussed and described in terms of a set of equations. In this chapter, we are concerned more with why the objects are placed into motion. Statics is the study of the resultant effects of external forces exerted on physical objects.

5.1 DEFINITIONS OF STATICS

Force—a mechanical "push" or "pull" exerted on an object. Force is a vector quantity whose units are the newton (N), which is equivalent to kg m s^{-2} or MLT^{-2}.

Inertia—inherent property of all matter that resists motion. Inertia is directly proportional to the quantity of mass.

Mass—amount of inertia that an object possesses. Mass is a scalar quantity whose units are kilograms or M.

Weight—force exerted or pulled on an object by the earth. Weight is related to an object's mass by a gravitational acceleration constant g:

$$W = mg$$

The gravitational acceleration constant g is 9.8 m s^{-2}. Weight is a vector quantity whose SI units are newtons.

Solved Problem 5.1. Basketball player Shaquille O'Neal weighs 1467.8 N (330 lb). Determine his mass.

Solution

The mass of an object can be determined from the relationship

$$W = mg$$

Therefore,

$$m = \frac{W}{g} = \frac{1467.8 \text{ N}}{9.8 \text{ m s}^{-2}} = 149.8 \text{ kg}$$

Solved Problem 5.2. Determine the downward force exerted by a 75-kg person (1) on earth and (2) on Mars. (Assume $g_{\text{Mars}} = 3.3$ m s^{-2}.)

Solution

(1) The downward force is the weight of the person given as

$$W = mg = (75 \text{ kg})(9.8 \text{ m s}^{-2}) = 735 \text{ N}$$

(2) Again, the downward force is the weight of the person, but the gravitational acceleration is that of Mars:

$$W = mg = (75 \text{ kg})(3.3 \text{ m s}^{-2}) = 247.5 \text{ N}$$

5.2 NEWTON'S LAWS OF MOTION

5.2.1 Newton's First Law: All Mass Contains Inertia

An object at rest or in motion will remain at rest or in motion unless acted on by an external force. Two or more external forces can act on an object without directly affecting its physical state as long as they negate or cancel each other.

ILLUSTRATIVE EXAMPLE 5.1. Blood Flow

For blood to flow through a particular artery or vein, an external force must be applied to initiate and maintain motion. This external force is the pressure gradient applied between two ends of the vessel. Blood flows in the direction from higher pressure to lower pressure and will continue in motion unless acted on by another force, such as a vessel obstruction (stenosis or atherosclerotic blockage) to prevent or substantially hinder further progress of blood flow.

5.2.2 Newton's Second Law: $\Sigma F = ma$

An external force applied to an object of mass m sufficient to overcome its inertia will result in acceleration of the object along the direction of force. The relationship between force, mass, and acceleration is given by

$$Force = mass \times acceleration$$
$$\mathbf{F} = m\mathbf{a}$$

Note the linear relationship between force and acceleration. Since it is entirely possible for more than one force to act on an object, Newton's second law can be restated as

$$\Sigma \mathbf{F} = m\mathbf{a}$$

where $\Sigma \mathbf{F}$ refers to the sum of all forces acting on the object.

ILLUSTRATIVE EXAMPLE 5.2. Blood Flow Revisited

It was stated in Illustrative Example 5.1 that an external force was required to initiate blood flow. Newton's second law allows us to quantitatively approximate the specific magnitude of force or pressure needed to accelerate a given mass or volume of blood. The mass of blood is determined by the volume of blood multiplied by its density.

5.2.3 Newton's Third Law: Law of Action and Reaction

For every force acting on an object, the object responds with a force equal in magnitude yet opposite in direction. Given a force $\mathbf{F}_A$ acting on an object, the object responds with a force $\mathbf{F}_B$ such that $\mathbf{F}_A = -\mathbf{F}_B$.

ILLUSTRATIVE EXAMPLE 5.3. Skeletal Mechanics of Standing

The elementary process of standing requires direct interaction with the environment, that is, the ground. For a person standing and exerting a force equal to her or his weight on the ground, the ground is acting upward with a force (normal force) equal in magnitude to that of the person's weight but in an opposite direction, as shown in Figure 5-1. It is like looking into a mirror. This can be visualized in a sense by a person standing on a mirror who sees himself or herself "exerting" a force (weight) of equal magnitude but in the opposite direction (into the mirror) (Figure 5-2).

ILLUSTRATIVE EXAMPLE 5.4. Locomotion of an Octopus

The octopus is a highly developed mollusk (cephalopod) that spends the majority of the day in seclusion behind rock crevices along the ocean flood. When the octopus emerges from its hideout to hunt for prey, it moves along in a gliding motion, propelled by a jet of water thrust from its gill chamber. The jet of water thrust exerts a force equal in magnitude and opposite in direction to the desired motion, allowing the octopus to move as a result of this force. If alarmed, the octopus can escape at high speeds by increasing the magnitude and direction of the water thrust.

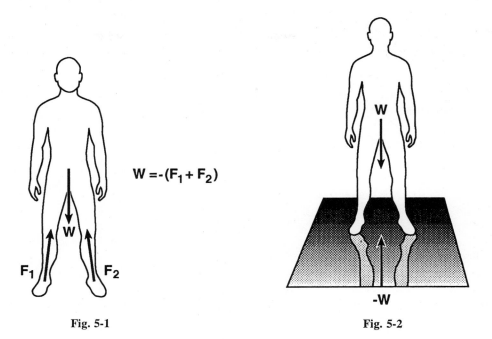

Fig. 5-1   Fig. 5-2

5.3 STATE OF TRANSLATIONAL EQUILIBRIUM

As stated by Newton's second law, an unbalanced force exerted on an object will result in an acceleration in the direction of the applied force, as given by the relation $\Sigma\mathbf{F} = m\mathbf{a}$. However, in translational equilibrium, all external forces acting on an object must cancel each other so that $\Sigma\mathbf{F} = 0$, that is, no change in object position is observed.

Solved Problem 5.3. A 2.5-kg steel ball is suspended on one end of a massless thread. Determine the tension in the thread if (1) it is raised with a uniform acceleration of 7 m s^{-2} and (2) it is lowered with a uniform acceleration of 7 m s^{-2}.

Solution

(1) The upward force generated by the tension T in the thread is counteracted by the weight of the steel ball W acting downward. Since the applied force to the thread results in an acceleration,

$$\Sigma F_x = 0$$
$$\Sigma F_y = ma = T - W$$

Solving for the tension T gives

$$T = W + ma = mg + ma = m(g + a) = 2.5 \text{ kg}(7 \text{ m s}^{-2} + 9.8 \text{ m s}^{-2}) = 42 \text{ N}$$

(2) The downward force generated by the weight is counteracted by the upward force of the tension. The equation of motion now becomes

$$\Sigma F_y = ma = W - T$$

or, solving for T, we get

$$T = W - ma = mg - ma = m(g - a) = 2.5 \text{ kg}(9.8 \text{ m s}^{-2} - 7 \text{ m s}^{-2}) = 7 \text{ N}$$

Solved Problem 5.4. An automobile weighing $5 \times 10^5 \text{ N}$ suddenly brakes and comes to a complete stop after traveling 30 m in 6 s. Determine (1) the initial speed of the automobile and (2) the braking force.

Solution

(1)   From Newton's second law

$$F = ma$$

where F is the braking force, m is the mass of the automobile, and a is the deceleration (negative acceleration) of the automobile. Since the automobile is uniformly decelerated, we can use the equation of motion

$$x = v_o t + \tfrac{1}{2}at^2 \qquad (1)$$
$$v_f = v_o + at \qquad (2)$$

From Eq. (2), $v_f = 0$ and

$$0 = v_o + at$$
$$a = -\frac{v_o}{t} \qquad (3)$$

Inserting Eq. (3) into Eq. (1) yields

$$x = v_o t + \frac{1}{2}\left(-\frac{v_o}{t}\right)t^2 = v_o t - \frac{1}{2}v_o t = \frac{1}{2}v_o t$$

Solving for v_o gives

$$v_o = \frac{2x}{t} = \frac{2 \cdot 30\,\text{m}}{6\,\text{s}} = 10\,\text{m s}^{-1}$$

(2)   Substituting into Eq. (1) gives

$$x = \left(\frac{2x}{t}\right)t + \frac{1}{2}(-a)t^2 = \frac{1}{2}at^2$$

Solving for a in terms of x and t gives

$$a = \frac{2x}{t^2}$$

Substituting into $F = ma$ yields

$$F = m\left(\frac{2x}{t^2}\right) = \left(\frac{5 \times 10^5\,\text{N}}{9.8\,\text{m s}^{-2}}\right)\left(10\,\frac{\text{m}}{\text{s}}\right) = 5.1 \times 10^5\,\text{N}$$

Solved Problem 5.5.   Determine the thrust force required to vertically launch a 650-kg rocket at an acceleration of 8.5 m s^{-2}.

Solution

The net force required to produce the acceleration is

$$F_{net} = ma = (650\,\text{kg})(8.5\,\text{m s}^{-2}) = 5525\,\text{N}$$

There is a simultaneous gravitational force or weight W acting downward given by

$$W = mg = (650\,\text{kg})(9.8\,\text{m s}^{-2}) = 6370\,\text{N}$$

The net force F_{net} acting on the rocket is

$$F_{net} = F_{thrust} - W$$

Therefore, the rocket engine must generate a thrust force equal to

$$F_{thrust} = F_{net} + W = 5525\,\text{N} + 6370\,\text{N} = 11,895\,\text{N}$$

5.4   FRICTION

Friction is a force generated by the properties of the interface between the object and the surface and acts to resist motion. A frictional force is defined as

$$\mathbf{F}_f = \mu N$$

where μ is a coefficient of friction and N is the normal force exerted by the surface on the object. Friction is independent of the sliding speed of the object and the area of contact between the object and surface and is directly proportional to the normal force exerted by the object on the surface.

There are, in effect, two types of frictional forces corresponding to the physical state of motion of an object. If the object is stationary, then a frictional force (static friction) is acting on the object to prevent motion, described by

$$\mathbf{F}_{f,s} \leq \mu_s N$$

where μ_s is the coefficient of static friction. Force f_s can increase to a maximum value of $\mu_s N$ to ensure static equilibrium of the object. Once this force is overcome and the object is set in motion, the object is subject to kinetic (sliding) friction which is a force that acts to retard or slow down motion of the object and is defined as

$$\mathbf{F}_{f,k} = \mu_k N$$

Static friction is generally greater in magnitude than kinetic friction.

ILLUSTRATIVE EXAMPLE 5.5.   Osteoarthritis and Friction at Skeletal Joints

Skeletal joints of the human body permit movement and motion necessary for performing common daily tasks, such as walking, running, turning around, and lifting the arm. When two bones are connected at a joint, the ends of the bones do not touch each other directly but are covered by cartilage which permits low-friction movement and are surrounded by a space filled with synovial fluid, a lubricating fluid having the consistency of water. The coefficient of friction in many human joints ranges from 0.005 to 0.02. Let's consider, for the moment, the frictional force exerted on the hip since it is involved in the most common of daily activities—walking. Depending on the motion and corresponding speed, the frictional force exerted on the hip is approximately

$$\mathbf{F}_f = 2.5\mu W$$

where W is the weight of the person. Therefore, for a normal hip of a 75-kg person, the exerted frictional force is

$$\mathbf{F}_f = (2.5)(0.005)(75 \text{ kg})(9.8 \text{ m s}^{-2}) = 9.18 \text{ N}$$

In osteoarthritis, immunological alterations occur, compounded by the aging process, which adversely affect the composition of the synovial fluid, thereby diminishing the lubricating capacity of the synovial fluid and ultimately permitting direct contact between the two connecting bones. Without the lubricant, the force generated by normal movements is translated to heat energy, which acts to further destroy the joint, in addition to creating the symptoms of inflammation, swelling, and pain. Assuming a coefficient of friction of 0.5, the frictional force exerted on an arthritic hip is

$$\mathbf{F}_f = (2.5)(0.5)(75 \text{ kg})(9.8 \text{ m s}^{-2}) = 918 \text{ N}$$

Solved Problem 5.6.   In explaining the concept of friction, a physics professor holds a 0.05-kg eraser against a chalkboard. Assuming the coefficient of friction between the eraser and the chalkboard is 0.25, what force must be exerted by the professor to allow downward movement of the eraser at a constant speed?

Solution

There are, in essence, two forces acting on the system. The weight or gravitational force is acting downward, and the frictional force is opposing the direction of motion. The weight of the eraser is

$$W = mg = (0.05 \text{ kg})(9.8 \text{ m s}^{-2}) = 0.49 \text{ N}$$

Therefore, the force that must be exerted by the professor to permit motion is the normal force

$$F_n = \frac{F_f}{\mu} = \frac{0.49 \text{ N}}{0.25} = 1.96 \text{ N}$$

Solved Problem 5.7.   Jay is pulling his sister Kay on a sled with a force of $T = 220$ N at an angle of 40° with the ground. If the combined mass of Kay and the sled is 55 kg and a frictional force of $F_f = 50$ N is exerted on the sled, determine Kay's acceleration.

Solution

First we need to analyze motion along the x axis:

$$T_x = T \cos \theta = (220 \text{ N})(\cos 40°) = 168.5 \text{ N}$$

The net force exerted on the sled is

$$F_{\text{net}} = T_x - F_f = 168.5 \text{ N} - 50 \text{ N} = 118.5 \text{ N}$$

Thus, Kay's acceleration is

$$\mathbf{a} = \frac{\mathbf{F}}{m} = \frac{118.5 \text{ N}}{55 \text{ kg}} = 2.2 \text{ m s}^{-2}$$

5.5   FREE-BODY DIAGRAMS

Free-body diagrams are often used in problems in statics, allowing one to isolate the object, to identify visually and analyze the magnitude and direction of all externally applied forces, and to quantitatively determine the resultant force acting on an object in equilibrium. In performing free-body diagrams,

- Isolate the body in an imaginary coordinate system where the body represents the origin.
- Represent the magnitude and angles for all forces acting on the body.
- Resolve all forces into their x and y components.
- Add all force components to determine the resultant x and y components.
- Use the resultant x and y components to determine the magnitude and direction of the resultant vector representing all forces acting on the object.

In three common types of statics problem, an object is (1) subjected to a force along a horizontal or inclined surface, (2) suspended from a fixed surface by one or more strings or cords, or (3) suspended by a pulley assembly of strings or cords.

(1)   Object subjected to a force along a horizontal or inclined surface

In the first case, there are four forces that should be readily identifiable from a particular problem and accounted for in the schematic diagram:

(1) $\mathbf{F}_{\text{ext}}$—the external force (parallel to the surface and acting along the direction of motion)

(2) $\mathbf{F}_f$—the frictional force (parallel to the surface and acting against or opposite the direction of motion)

(3) $W = mg$—the weight of the object (perpendicular to the surface and acting downward)

(4) N—the normal force exerted by the surface (perpendicular to the surface and acting upward)

The objective in solving problems of this nature is to determine the acceleration of the object. Let's consider a block of mass m sliding down a ramp inclined at an angle θ, as shown in Figure 5-3. Since the block is assumed to be moving, the coefficient of kinetic friction between the block and the surface of the ramp is μ_k. We begin

Problem

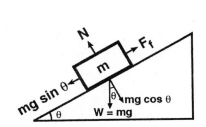

Free-Body Diagram

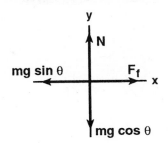

Equations

If block is at rest:

x component
$$\Sigma F_x = 0$$
$$F_f - mg \sin \theta = 0$$

y component
$$\Sigma F_y = 0$$
$$N - mg \cos \theta = 0$$

where $F_f = \mu_s N$

If block is in motion:

x component
$$\Sigma F_x = ma_x$$
$$F_f - mg \sin \theta = ma_x$$

y component
$$\Sigma F_y = ma_y$$
$$N - mg \cos \theta = 0$$

where $F_f = \mu_k N$

Fig. 5-3

by introducing a free-body diagram of the object on the ramp, as shown in Figure 5-3. The equations of motion based on Newton's second law can now be derived in terms of both x and y components. From

$$\Sigma \mathbf{F}_x = ma_x$$

we can write the equation of motion

$$\mathbf{F}_f - \mu_k mg \cos \theta = ma_x$$

However, since motion occurs only in the x direction,

$$\Sigma \mathbf{F}_y = ma_y = 0$$

Thus, the acceleration of the object is determined from $\Sigma \mathbf{F}_x = ma_x$:

$$a = a_x = \frac{\mathbf{F}_f - \mu_k mg \cos \theta}{m}$$

If the block were either pulled down or pushed up the ramp, then F_{ext} could easily be implemented in the free-body diagrams and factored into the calculations.

ILLUSTRATIVE EXAMPLE 5.6.　Skeletal Mechanics: Raising the Arm

An arm can be raised primarily due to the forces exerted by the deltoid muscle. The deltoid muscle connects at the upper end of the shoulder, extends over the upper arm bone (humerus), and attaches near the elbow. Three forces are involved in

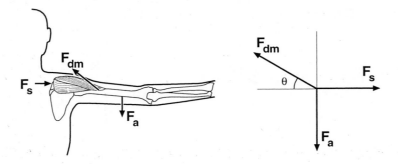

$\mathbf{F_a}$ = weight of the arm acting downward

$\mathbf{F_{dm}}$ = force of the deltoid muscle acting at an angle θ

$\mathbf{F_s}$ = force exerted by the shoulder on the humerus

Fig. 5-4

raising the arm: (1) the force exerted by the shoulder $\mathbf{F}_s$ on the humerus acting along the positive x axis, (2) the weight of the arm $\mathbf{F}_a$ acting downward, and (3) the force of the deltoid muscle $\mathbf{F}_{dm}$ acting at an angle of approximately 15° along the negative x axis, as shown in Figure 5-4. Therefore, assuming that the weight of the arm or $\mathbf{F}_a$ is 13 N, we can determine expressions for the forces $\mathbf{F}_{dm}$ and $\mathbf{F}_s$ from the following equations of static equilibrium:

$$\Sigma F_x: \qquad \mathbf{F}_{dm} \cos 15° = \mathbf{F}_s$$
$$\Sigma F_y: \qquad \mathbf{F}_{dm} \sin 15° = \mathbf{F}_a = 13.3 \text{ N}$$

Therefore, from the above equations:

$$\mathbf{F}_{dm} = \frac{13.3 \text{ N}}{0.259} = 51.4 \text{ N}$$
$$\mathbf{F}_s = \mathbf{F}_{dm} \cos 15° = (51.4 \text{ N})(0.965) = 49.6 \text{N}$$

ILLUSTRATIVE EXAMPLE 5.7.   Skeletal Mechanics: The Hip

Motion of the leg is controlled by a group of three independent muscles collectively known as the *hip abductor muscles* attached to the pelvis. Their principal action is to swing the leg sideways relative to the hip. To determine the resultant force acting on the pelvis, we must first identify the magnitude and direction of the forces exerted by the three hip abductor muscles, as shown in Figure 5-5. These forces can be approximated by the following vectors:

F1:   75 N at an angle of 86° from the positive x axis

F2:   220 N at an angle of 78° from the positive x axis

F3:   100 N at an angle of 48° from the positive x axis

Using component analysis, the resultant force can be determined from

$$\Sigma F_x = 75 \text{ N} \cdot \cos 86° + 220 \text{ N} \cdot \cos 78° + 100 \text{ N} \cdot \cos 48°$$
$$= 5.2 \text{ N} + 45.7 \text{ N} + 66.9 \text{ N} = 117.8 \text{ N}$$
$$\Sigma F_y = 75 \text{ N} \cdot \sin 86° + 220 \text{ N} \cdot \sin 78° + 100 \text{ N} \cdot \sin 48°$$
$$= 74.8 \text{ N} + 215.2 \text{ N} + 74.3 \text{ N} = 364.3 \text{ N}$$
$$R = \sqrt{(\Sigma F_x)^2 + (\Sigma F_y)^2} = \sqrt{(117.8 \text{ N})^2 + (364.3 \text{ N})^2} = 382.9 \text{ N}$$
$$\theta = \tan^{-1} \frac{\Sigma F_y}{\Sigma F_x} = \tan^{-1} \frac{364.3 \text{ N}}{117.8 \text{ N}} = 72.1°$$

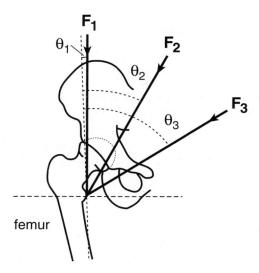

Fig. 5-5

Solved Problem 5.8.   An orderly pushes a patient in a wheelchair up an inclined ramp. Identify the forces acting on the wheelchair.

Solution

There are four forces that influence the motion of the wheelchair up the ramp, and they can be represented by vectors, as shown in Figure 5-6:

(1)   The combined weight of the patient and wheelchair W, acting downward

(2)   The frictional force F_f of the surface of the ramp acting against the direction of motion

(3)   The external force F_{ext} exerted by the orderly pushing the wheelchair up the ramp

(4)   The reaction or normal force N of the inclined ramp acting upward to support the combined weight of the wheelchair and patient

Solved Problem 5.9.   An orderly is transporting a patient in a wheelchair; the mass of the orderly is 90 kg, the mass of the patient is 60 kg, and the mass of the wheelchair is 15 kg. If the orderly exerts a force of 125 N, what is the acceleration produced? (Assume friction is negligible).

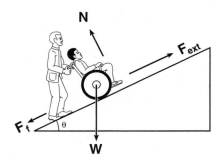

Fig. 5-6

Solution

We begin with the equation $\mathbf{F} = m\mathbf{a}$. Solving for the acceleration $\mathbf{a}$ yields

$$\mathbf{a} = \frac{\mathbf{F}}{m}$$

Since the orderly is pushing the wheelchair and accelerating with the system (that is, the patient and the wheelchair), the mass under acceleration by the exerted force of the orderly is the sum of the masses of the orderly, patient, and wheelchair, or

$$m = 90\,\text{kg} + 60\,\text{kg} + 15\,\text{kg} = 165\,\text{kg}$$

Making the appropriate substituents yields

$$\mathbf{a} = \frac{125\,\text{N}}{165\,\text{kg}} = 0.76\,\text{m}\,\text{s}^{-2}$$

Solved Problem 5.10. Determine the force required by a tow truck to move a 2000-kg car up a 7° inclined ramp. Also determine the force exerted by the road against the car.

Solution

We must first identify all forces involved in towing the car: (1) weight of the car $\mathbf{F}_{car}$ which acts vertically downward, (2) the force exerted by the ramp agains the car $\mathbf{F}_{ramp}$, and (3) the force of the tow cord which acts parallel to the ramp $\mathbf{F}_{tow}$. Thus, the free-body diagram can be drawn, as illustrated in Figure 5-7.

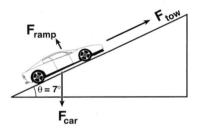

Fig. 5-7

Applying Newton's second law, we get

$$\Sigma F_x: \qquad F_{ramp} \cos 83° = F_{tow} \cos 7°$$
$$0.122 F_{ramp} = 0.993 F_{tow}$$
$$\Sigma F_y: \qquad F_{ramp} \sin 83° + F_{tow} \sin 7° = F_{car}$$
$$0.993 F_{ramp} + 0.122 F_{tow} = 1.96 \times 10^4\,\text{N}$$

The problem is asking us to solve for F_{tow} and F_{ramp} which can easily be done since we have two equations. From ΣF_x:

$$F_{ramp} = 8.14 F_{tow}$$

Substituting into ΣF_y gives

$$\Sigma F_y: \qquad F_{ramp} \sin 83° + F_{tow} \sin 7° = F_{car}$$
$$8.14 F_{tow} + 0.122 F_{tow} = 1.96 \times 10^4\,\text{N}$$
$$8.26 F_{tow} = 1.96 \times 10^4\,\text{N}$$
$$F_{tow} = 2373\,\text{N}$$

Substituting the expression for F_{tow} into that for ΣF_x yields $F_{ramp} = 1.93 \times 10^4\,\text{N}$.

Solved Problem 5.11.   Brian is taking his extended teddy bear family in two wagons to the park for a picnic, as shown in Figure 5-8. Brian is pulling the first wagon of mass 8 kg with a force F_B, at an angle $\theta = 35°$ to the horizon, which is connected with a massless string to a second wagon of mass 5 kg. Both wagons are subjected to a frictional force. Determine (1) the force exerted by Brian in pulling the two connected wagons, (2) the acceleration of the wagons, and (3) the tension in the cord connecting the two wagons.

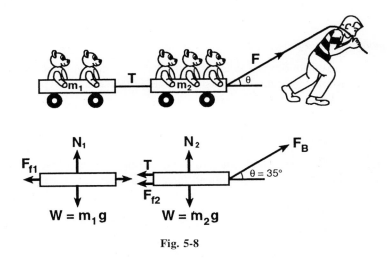

Fig. 5-8

Solution

The first step in solving a problem of this nature is to draw free-body diagrams for each wagon and derive equations of motion.

Wagon 1:	ΣF_x:	$F_B \cos \theta - F_{f1} - T = 0$
	ΣF_y:	$F_B \sin \theta + N_1 - m_1 g = 0$

where $F_{f1} = \mu N_1$

Wagon 2:	ΣF_x:	$T - F_{f2} = 0$
	ΣF_y:	$N_2 - m_2 g = 0$

where $F_{f2} = \mu N_2$

(1)   From the equations for wagon 1, F_B can be determined according to the following:

$$N_1 = m_1 g - F_B \sin \theta$$

Therefore, f_1 can be determined:

$$F_{f1} = \mu N_1 = \mu(m_1 g - F_B \sin \theta)$$
$$F_B \cos \theta = F_{f1} + T = \mu(m_1 g - F_B \sin \theta) + \mu m_2 g$$
$$= \mu m_1 g + \mu m_2 g - \mu F_B \sin \theta$$
$$F_B \cos \theta + \mu F_B \sin \theta = \mu g(m_1 + m_2)$$
$$F_B(\cos \theta + \mu \sin \theta) = \mu g(m_1 + m_2)$$

Solving for F_B gives

$$F_B = \frac{\mu g(m_1 + m_2)}{\cos \theta + \mu \sin \theta}$$

(2)   The acceleration of the system (wagon 1 + wagon 2) can be determined from the results in (1):

$$\mathbf{a} = \frac{\mathbf{F}_B}{m_T} = \frac{\dfrac{\mu g(m_1 + m_2)}{\cos\theta + \mu\sin\theta}}{m_1 + m_2} = \frac{\mu g}{\cos\theta + \mu\sin\theta}$$

(3)   From the equations for wagon 2, T can be determined:

$$T = F_{f2} = \mu N_2 = \mu m_2 g$$

(2)   Object suspended from a fixed surface by one or more strings or cords

In the second case, an object of mass m is suspended by two cords which are attached at angles θ_1 and θ_2 to a rigid surface, as illustrated in Figure 5-9. Two forces are involved in this type of problem and are depicted schematically in the free-body diagram:

(1)   Tension of both cords T_1 and T_2, acting upward at their respective angles

(2)   Weight of the suspended object W acting downward

The primary objective of this type of problem is to determine the tension in each of the cords. The equations of static equilibrium can be derived by resolving all forces into their x and y components.

$$\Sigma F_x: \qquad T_2\cos\theta_2 - T_1\cos\theta_1 = 0$$
$$\Sigma F_y: \qquad T_1\sin\theta_1 + T_2\sin\theta_2 - W = 0$$

Since typically the weight of the object is given as well as the angles of attachment for the two cords in problems of this type, we have two equations (ΣF_x and $\Sigma \mathbf{F}_y$) to solve for two unknowns T_1 and T_2. From ΣF_x:

$$T_2\cos\theta_2 = T_1\cos\theta_1$$
$$T_1 = \frac{T_2\cos\theta_2}{\cos\theta_1}$$

Substituting into ΣF_y gives

$$\frac{T_2\cos\theta_2}{\cos\theta_1}\cdot\sin\theta_1 + T_2\sin\theta_2 = W$$
$$T_2\cos\theta_2\tan\theta_1 + T_2\sin\theta_2 = W$$
$$T_2 = \frac{W}{\cos\theta_2\tan\theta_1 + \sin\theta_2}$$

Problem   **Free-Body Diagram**   **Equations**

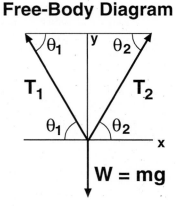

x component
$$\Sigma F_x = 0$$
$$T_2\cos\theta_2 - T_1\cos\theta_1 = 0$$

y component
$$\Sigma F_y = 0$$
$$T_1\sin\theta_1 + T_2\sin\theta_2 - W = 0$$

Fig. 5-9

Solved Problem 5.12.   A framed photograph is hung on a nail by a wire fastened to the back of the photograph as in Figure 5-10. Assuming that the tension T in each segment of the wire is 4.6 N and directed at an angle of 45° from the horizon, determine (1) the reaction force exerted by the nail and (2) the weight of the framed photograph.

Solution

(1)   To solve this problem, we must first determine the x and y components of forces:

$$\Sigma F_x: \qquad -T\cos 45° = T\cos 45°$$
$$\Sigma F_y: \qquad T\sin 45° + T\sin 45° = W$$

The reaction force F_R exerted by the nail is, in effect, the y component of T counteracting the weight of the photograph. Therefore, from ΣF_y

$$F_R = T\sin 45° + T\sin 45° = 2T\sin 45° = 2\cdot 4.6\,\text{N}\cdot 0.707 = 6.5\,\text{N} \qquad \text{in positive-}y\text{ direction}$$

(2)   The weight of the framed photograph W is equal in magnitude to that of the reaction force F_R but opposite in direction, or

$$W = 6.5\,\text{N} \qquad \text{in negative }y\text{ direction}$$

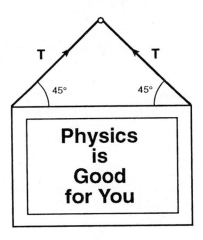

Fig. 5-10

Solved Problem 5.13.   Using the information in Illustrative Example 5.4, determine the magnitude and direction of F_{dm}, given that $F_a = 40\,\text{N}$ and $F_s = 150\,\text{N}$.

Solution

In essence, F_{dm} is the resultant vector following the summation of F_a and F_s. Thus,

$$F_{\text{dm}} = \overline{\big)(F_a)^2 + (F_s)^2} = \overline{\big)(40\,\text{N})^2 + (150\,\text{N})^2} = 155\,\text{N}$$

$$\theta = \tan^{-1}\frac{F_s}{F_a} = \tan^{-1}\frac{150\,\text{N}}{40\,\text{N}} = 75.1°$$

(3)   Object suspended by a pulley assembly of strings or cords

In the free-body diagram for the third case, two forces that should be represented are (Figure 5-11)

(1)   The weight of the object (acting downward)

(2)   The tension in the string or cord (acting upward)

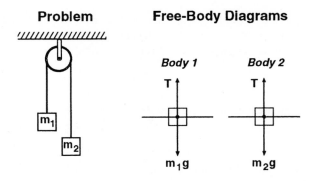

Fig. 5-11

The strings are typically assumed to be massless and can sometimes translate their force through pulleys which are assumed to be frictionless.

The objective in solving problems of this nature is to determine (1) the tension in the cord and (2) the acceleration of the system. First we must isolate each object and draw a free-body diagram so that an equation of motion can be derived for each object. The basis for the equations of motion is Newton's second law, or **F** = m**a**:

Object 1: $\Sigma F_x = 0$ $\Sigma F_y = T - m_1 g = m_1 a$

Object 2: $\Sigma F_x = 0$ $\Sigma F_y = T - m_2 g = -m_2 a$

We now have two equations to use to solve for two unknowns, that is, the tension in the string T and the acceleration of the system a:

$$T = \frac{2m_1 m_2}{m_1 + m_2} g$$

$$a = \frac{m_2 - m_1}{m_2 + m_1} g$$

Solved Problem 5.14.   Use free-body diagrams to determine the tension in the string and the acceleration of the system in Figure 5-12.

Solution

The free-body diagrams for each of the two blocks are displayed in Figure 5-13. First we must determine the equations of motion based on Newton's second law:

Object 1: $\Sigma F_x = T = m_1 a$ $\Sigma F_y = N - m_1 g = 0$

Object 2: $\Sigma F_x = 0$ $\Sigma F_y = T - m_2 g = m_2 a$

Again, we have two equations and two unknowns which yield expressions for the parameters T and a:

$$T = \frac{m_1 m_2}{m_1 + m_2} g$$

$$a = \frac{m_2}{m_1 + m_2} g$$

Free-Body Diagrams

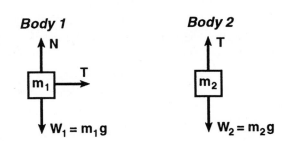

Equations of Motion

Body 1

$$\Sigma F_x = T - m_1 a = 0$$
$$\Sigma F_y = N - m_1 g = 0$$

Body 2

$$\Sigma F_x = 0$$
$$\Sigma F_y = T - m_2 g = m_2 a$$

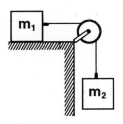

Fig. 5-12   Fig. 5-13

Solved Problem 5.15. As an extension to Solved Problem 5.14, suppose that mass m_1 is now subject to a frictional force. The coefficient of kinetic or sliding friction between m_1 and the surface is μ_k. Determine the tension in the string and the acceleration of the system.

Solution

First, let us assume that the upward direction is positive. The free-body diagrams for each of the two blocks in this problem are displayed in Figure 5-14. The equations of motion for each block can be derived from Newton's second law:

$$m_1: \qquad \Sigma F_x = T - F_f = m_1 a \qquad\qquad (1)$$
$$\Sigma F_y = N - m_1 g = 0 \qquad\qquad (2)$$

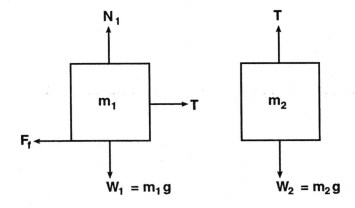

Fig. 5-14

Since $F_f = \mu_k N = \mu m_1 g$, from Eq. (1)

$$T = F_f + m_1 a = \mu_k m_1 g + m_1 a = m_1 (\mu_k g + a) \tag{3}$$

In the case of m_2, the motion is directed downward and thus $m_2 g > T$. The force generated by the tension in the cord acts on m_1 to the right and prevents m_2 from falling freely. Therefore, the equations of motion for m_2 are

$$m_2: \qquad \Sigma F_x = 0 \tag{4}$$
$$\Sigma F_y = T - m_2 g = -m_2 a \tag{5}$$

From Eq. (5),

$$T = m_2 g - m_2 a = m_2 (g - a) \tag{6}$$

Subtracting Eq. (6) from Eq. (3) yields

$$m_1 (\mu_k g + a) - m_2 (g - a) = 0$$

Solving for a yields

$$a = \frac{m_2 - \mu_k m_1}{m_1 + m_2} g \tag{7}$$

Substituting Eq. (7) into Eq. (6) gives

$$T = m_2 (g - a) = m_2 \left(g - \frac{m_2 - \mu_k m_1}{m_1 + m_2} g \right)$$

$$= \frac{m_1 m_2 g (1 + \mu_k)}{m_1 + m_2}$$

ILLUSTRATIVE EXAMPLE 5.8.   Traction Systems

A potential problem in the healing process of a fractured thigh bone (femur) is the tendency for the major leg muscles to compress or pull together the two segments of the broken bone at the fracture point, ultimately resulting in a limp. This problem can be remedied by placing the leg in traction. Traction systems stabilize the broken limb and prevent unnecessary movements which could hinder the healing process.

In one system known as the *Russell system* of traction, shown in Figure 5-15, the objective is to generate a force along the axis of the femur. As a consequence, the lower leg (tibia) is stabilized by a traction cord attached to weights through a pulley system. The force exerted on the leg by the traction cord acting along the x axis is due solely to the weight acting downward ($-y$ axis) (assuming negligible friction in the pulley).

Solved Problem 5.16.   A leg is stabilized by a traction device as illustrated in Figure 5-16. Assuming that the tension in the cord is 15 N, determine the (1) horizontal force, (2) vertical force, and (3) magnitude and direction of the resultant force.

Solution

(1)   The horizontal force is the x component of the resultant force, given by

$$\Sigma F_x = 15 \text{ N} \cos 60° + 15 \text{ N} \cos 30° = 7.5 \text{ N} + 12.9 \text{ N} = 20.4 \text{ N}$$

(2)   Similarly, the vertical force is the y component of the resultant force and is given by

$$\Sigma F_y = 15 \text{ N} \sin 60° - 15 \text{ N} \sin 30° = 12.9 \text{ N} - 7.5 \text{ N} = 5.4 \text{ N}$$

(3)   The magnitude and direction of the vertical force can be determined from the solutions in (1) and (2):

$$F = \overline{(\Sigma F_x)^2 + (\Sigma F_y)^2} = \overline{(20.4 \text{ N})^2 + (5.4 \text{ N})^2} = 21.1 \text{ N}$$

$$\theta = \tan^{-1} \frac{5.4 \text{ N}}{20.4 \text{ N}} = 14.8°$$

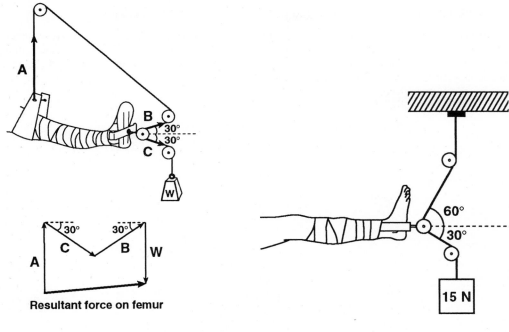

Resultant force on femur

Fig. 5-15   **Fig. 5-16**

5.6   CENTER OF GRAVITY

Although each and every atom and molecule composing a given object is subject to the gravitational force of the earth, it becomes convenient to consider a single point or location within the object where the gravitational force is concentrated. The *center of gravity* of an object is that point where the force of gravity is considered to act.

5.7   CENTER OF MASS

Up to now, our discussion of physical principles has been appplied to a particle or point source where the mass is concentrated at a single point. However, the mass of many objects ranging from a molecule to the human body is distributed over the object. For example, the human body is composed of a number of extended objects (i.e., arms, legs, head, chest, and abdomen), all of which have weight and therefore possess mass. To simplify problems involving the statics of rigid bodies, it becomes convenient to identify a single point where the total mass can be considered to be concentrated. The *center of mass* of an object is the point where the total mass is concentrated, and it can be determined from knowledge of the mass of the extended object and its distance from a central point defined as an origin. In two-dimensional coordinates, the center of mass is given by

$$x_{\text{CM}} = \frac{\Sigma xm}{\Sigma m} = \frac{x_1 m_1 + x_2 m_2 + x_3 m_3 + \cdots + x_n m_n}{m_1 + m_2 + m_3 + \cdots + m_n}$$

$$y_{\text{CM}} = \frac{\Sigma ym}{\Sigma m} = \frac{y_1 m_1 + y_2 m_2 + y_3 m_3 + \cdots + y_n m_n}{m_1 + m_2 + m_3 + \cdots + m_n}$$

Usually, the center of mass and center of gravity are located at the same point and vary only for extremely large objects where the acceleration due to gravity may vary from point to point within the object.

ILLUSTRATIVE EXAMPLE 5.9.   Center of Mass of a Water Molecule

A molecule of water, defined chemically as H_2O, consists of two hydrogen atoms bonded to an oxygen atom, according to the structural arrangement depicted in Figure 5-17. The origin is defined as the oxygen atom, and symmetry of the hydrogen atom reveals a distribution only along the x axis. The masses of the atoms in the water molecule are given in terms of the atomic mass unit, which is the conventional atomic weight. The mass of the hydrogen atom is 1.0 u and positioned at 0.96 angstroms (Å) from the oxygen atom. The mass of the oxygen atom is 16.0 u. Thus, the center of mass of the water molecule is

$$x_{CM} = \frac{(1.0\,u)(0.96\,\text{Å}\cos 52°) + (1.0\,u)(0.96\,\text{Å}\cos 52°) + (1.0\,u)(0)}{1.0\,u + 1.0\,u + 16.0\,u} = \frac{1.18\,u\,\text{Å}}{18.0\,u} = 0.065\,\text{Å}$$

Similarly, $y_{CM} = 0.084$ Å. The center of mass of the water molecule is much closer to the oxygen atom than either of the hydrogen atoms because of its comparatively larger size.

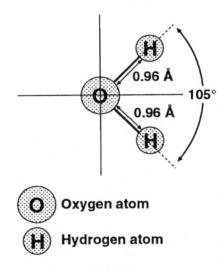

Fig. 5-17

5.8   NEWTON'S UNIVERSAL LAW OF GRAVITATION

A force of attraction exists between two bodies of mass m_1 and m_2 that is directly proportional to the product of the masses and inversely proportional to the distance between their centers of mass r, or

$$F = \frac{Gm_1 m_2}{r^2}$$

where G is the gravitational constant $= 6.67 \times 10^{-11}\,\text{N m}^2\,\text{kg}^{-2}$.

Solved Problem 5.17.   Given that the mass and diameter of the earth are 6×10^{24} kg and 1.28×10^7 m, respectively, determine the gravitational acceleration of an object at the earth's surface. (Assume $G = 6.67 \times 10^{-11}\,\text{N m}^2\,\text{kg}^{-2}$.)

Solution

Since force is always the product of mass and acceleration, the expression for the gravitational force can be rewritten as

$$F = ma = m \cdot \frac{Gm}{r^2}$$

where r is the earth's radius, or one-half the diameter. Therefore, the gravitational acceleration of an object near the earth's surface is

$$a = \frac{Gm}{r^2} = \frac{6.67 \times 10^{-11} \text{ N m}^2 \text{ kg}^{-2} \cdot 6 \times 10^{24} \text{ kg}}{(6.4 \times 10^6 \text{ m})^2}$$
$$= 9.77 \text{ N kg}^{-1} = 9.77 \text{ [kg m s}^{-2}\text{][kg}^{-1}\text{]}$$
$$= 9.77 \text{ m s}^{-2}$$

Supplementary Problems

5.1. Calculate the weight of (1) an electron of mass $m_e = 9.11 \times 10^{-31}$ kg and (2) the earth of mass $m_E = 5.98 \times 10^{24}$ kg.

Solution

(1) 8.93×10^{-30} N; (2) 5.86×10^{25} N

5.2. For a person weighing 85 N, calculate the mass of the person (1) on earth where $g = 9.8$ m s^{-2} and (2) on the moon where $g = 1.67$ m s^{-2}.

Solution

(1) 8.67 kg; (2) 50.89 kg

5.3. A 2-kg mass is acted on by a force with components $\mathbf{F}_x$ and $\mathbf{F}_y$. If $\mathbf{F}_x = 5$ N and $\mathbf{F}_y = 0$ N, find the acceleration of the mass.

Solution

2.5 m s^{-2}

5.4. An elevator with four passengers has a total mass of 1600 kg. The elevator passes the fourth floor at a velocity of 15 m s^{-1} and continues with constant acceleration until it stops on the ground floor, after it has traveled a distance of 28 m. Determine the tension T in the supporting cable.

Solution

2.2×10^4 N

5.5. In an extension of Solved Problem 5.11, Brian decides to attach a third wagon of teddy bear family friends to the two wagons he initially is transporting to the park for a picnic. In this case, however, the sidewalk surface is frictionless. Assuming that he is pulling horizontally to the ground and accelerating at a rate of 0.1 m s^{-2} and that each wagon weighs 19.8 N, determine the tension in each cord.

Solution

$T_1 = 0.6$ N; $T_2 = 0.4$ N; $T_3 = 0.2$ N

5.6. In replacing a washing machine, the Mayflag appliance repairer was pulling a new 850-N washing machine over a surface with a coefficient of friction equal to 0.4 into a new house. If the repair person exerts a force F_{RP} of 600 N in a horizontal direction with a massless rope, determine (1) the reaction force F_R to the force exerted by the repair person, (2) the resultant force F_{WM} on the washing machine, and (3) the acceleration of the washing machine.

Answer

(1) $-850\,\text{n}$; (2) $510\,\text{N}$; (3) $5.9\,\text{m}\,\text{s}^{-2}$

5.7.   What is the gravitational force of attraction between 80-kg Doug and 65-kg Patti who are separated by $5\,\text{m}$?

Answer

$9.63 \times 10^{-9}\,\text{N}$

5.8.   Determine the speed that a satellite must maintain to travel in a circular orbit at an altitude of $15{,}000\,\text{km}$ above the earth. (Assume that the radius of the earth is $6400\,\text{km}$ and acceleration due to gravity is inversely proportional to the radius squared.)

Answer

$4329\,\text{m}\,\text{s}^{-1}$

5.9.   Two forces that act on a 100-kg crate are given in terms of components:

Force 1: $F_{x,1} = 55\,\text{N}$ $F_{y,1} = 60\,\text{N}$
Force 2: $F_{x,2} = 30\,\text{N}$ $F_{y,2} = 75\,\text{N}$

Determine the acceleration and direction of the object.

Answer

$1.59\,\text{m}\,\text{s}^{-2}$, $57.8°$

5.10.   Given that the mass and mean radius of the moon are $7.4 \times 10^{22}\,\text{kg}$ and $1.74 \times 10^{6}\,\text{m}$, respectively, determine the gravitational acceleration on the moon. (Assume that the gravitational constant is $G = 6.67 \times 10^{-11}\,\text{N}\,\text{m}^2\,\text{kg}^{-2}$.)

Answer

$1.63\,\text{m}\,\text{s}^{-2}$

5.11.   An external force is applied to a block at rest on a horizontal surface according to the diagram in Figure 5-18. (1) What are the horizontal and vertical components of F? (2) Derive the equations of motion based on Newton's second law. (3) What is the frictional force acting against the block? (4) Determine the acceleration of the block.

Answer

(1) $F_x = F\cos\theta$ (toward the right); $F_y = -F\sin\theta$ (in a downward direction)

(2)   The equations of motion are

$$\Sigma F_x: \qquad F\cos\theta - F_f = ma_x$$
$$\Sigma F_y: \qquad N - mg - F\sin\theta = 0$$

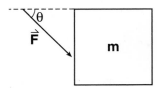

Fig. 5-18

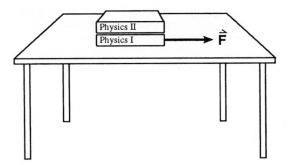

Fig. 5-19

(3)   The frictional force is

$$F_f = \mu N = \mu(mg + F \sin \theta)$$

(4)   The acceleration of the block is

$$a_x = \frac{1}{m}[F \cos \theta - \mu(mg + F \sin \theta)] = \frac{1}{m}[F(\cos \theta - \mu \sin \theta) - \mu mg]$$

5.12.   A book entitled *Physics I* of mass 20 kg rests on the smooth surface of the tabletop, as shown in Figure 5-19. On top of this book is placed another book, *Physics II*, of mass 25 kg. If the coefficient of friction between the books is $\mu_s = 0.25$, determine the maximum horizontal force that can be applied to *Physics I* without causing *Physics II* to slip on the upper surface of *Physics I*.

Solution

110.3 N

5.13.   An external force of 10 N is exerted on a 40-N crate. Determine the acceleration of the crate.

Solution

2.4 m s^{-2}

5.14.   What is the resultant force acting on a car traveling at a constant velocity of 55 mi h^{-1}?

Solution

Zero

5.15.   A 420-N neon sign is suspended by three cords as shown in Figure 5-20. Determine T_2 and θ.

Solution

$T_2 = 209.9$ N; $\theta = -30.9°$

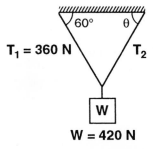

Fig. 5-20

5.16. Annie is drying her clothes of weight 4 N by hanging them in the center of a clothesline in her backyard, as shown in Figure 5-21. If the weight of the clothes causes the clothesline to sag by 20° from the horizontal, determine the tension in the clothesline on either side of the clothes.

Solution

5.85 N

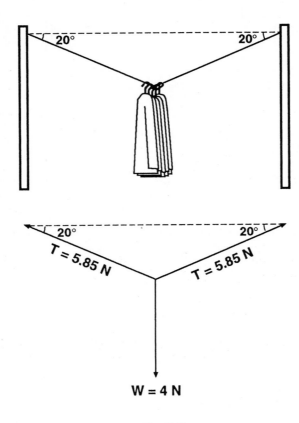

W = 4 N

Fig. 5-21

5.17. Consider the arrangement of blocks in Figure 5-22. The two blocks of mass $M_1 = 2.0$ kg and $M_2 = 3.5$ kg are placed on a table surface with a coefficient of friction $\mu = 0.3$. The two blocks are connected to a third block of mass $M_3 = 8.0$ kg which accelerates the entire system of blocks. Determine (1) the tension in the cord and (2) the acceleration of the system.

Solution

(1) 41.5 N; (2) 4.61 m s^{-2}

5.18. A block of mass $M_1 = 5.0$ kg is being pulled with a force $\mathbf{F} = 25$ N at an angle $\theta = 25°$ by a rope on a surface with a coefficient of friction of 0.15, as shown in Figure 5-23. This block is connected to a second block of mass $M_2 = 4.0$ kg with a cord positioned over a pulley. (1) Determine the horizontal and vertical components of the force $\mathbf{F}$. (2) What is the acceleration of the system? (3) What is the tension in the rope?

Solution

(1)
$$F_x = -F \cos \theta$$
$$F_y = F \sin \theta$$

(2) −2.48 m s^{-2}; (3) 29.3 N

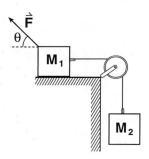

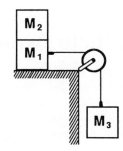

Fig. 5-22   **Fig. 5-23**

5.19.   Two forces $\mathbf{F}_1$ and $\mathbf{F}_2$ are acting simultaneously on a crate of mass $M = 85$ kg. Force $\mathbf{F}_1$ has a magnitude of 20 N
and is directed 30° below the horizontal, and $\mathbf{F}_2$ has a magnitude of 7 N and is directed 45° above the horizontal.
(1) Determine the magnitude and direction of the resultant force. (2) Determine the magnitude and direction of the
acceleration of the crate.

Solution

(1) 22.8 N; $-12.9°$ below the horizontal; (2) 0.27 m s^{-2}

5.20.   In moving a large box, an upward force of 500 N is applied to a 45-kg mass placed on a 20° inclined ramp whose
coefficient of kinetic friction is 0.35. Determine the acceleration of the box.

Solution

3.86 m s^{-2} (upward)

5.21.   Two blocks of weight $W_1 = 9.8$ N and $W_2 = 19.6$ N are connected by a cord with tension T_1. The system is being
pulled by a force T_2, as shown in Figure 5-24. Assuming the system accelerates at 0.25 m s^{-2}, determine the forces
T_1 and T_2.

Solution

$T_1 = 0.25$ N; $T_2 = 0.75$ N

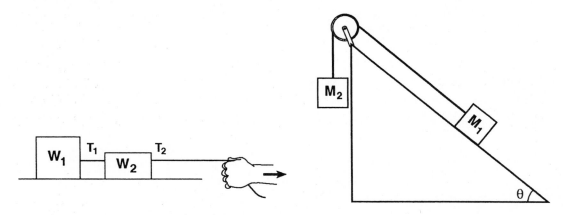

Fig. 5-24   **Fig. 5-25**

5.22.   Two blocks of mass $M_1 = 30$ kg and $M_2 = 5$ kg are connected by a cord on an inclined ramp of $\theta = 30°$ with M_2
suspended over a pulley, as shown in Figure 5-25. The block M_1 is subjected to a kinetic frictional force with
$\mu_k = 0.10$. Determine the acceleration of the blocks and the tension in the cord.

Solution

2.1 m s^{-2}; 59.5 N

Particle Dynamics: Work, Energy, Power

A force applied to an object results in an acceleration of the object in the direction of the force. This is guaranteed by Newton's second law and, in effect, explains why the object is set into motion. How the object responds to this force, ultimately resulting in motion, is the basis of this chapter. Chapter 6 describes the dynamics of an object in terms of work done, energy expended, and power output as a result of an external force.

6.1 DEFINITIONS OF PARTICLE DYNAMICS

Work represents the physical effects of an external force applied to a particle or object that results in a net displacement in the direction of the force. Work is a scalar quantity and has units of joules (J) or newton-meters (N m).

6.1.1 Work Done by a Constant Force

Work done by a constant force is equal to the scalar produce of force **F** and displacement **d** in the direction of the applied force and is given by

$$W = \mathbf{F} \cdot \mathbf{d} = |\mathbf{F}|\,|\mathbf{d}|\cos\theta$$

where θ is the angle between force and displacement.

Solved Problem 6.1. For 25, 50, and 100-kg objects, calculate the work done in (1) lifting the object a vertical distance of 50 cm from the ground, (2) pushing the object up a 2.5-m ramp inclined at 70° above the horizontal, and (3) pushing the object down a 2.5-m ramp inclined at 70° above the horizontal.

Solution

(1) By definition, work is equal to force multiplied by displacement. Two forces are acting on the object: (1) a downward gravitational force, F_d, exerted by the earth and equivalent to the object's weight, $W = mg$ and (2) an upward external force, F_u, exerted by the person in lifting the object. The work done by gravity on the object is:

$$W_{\text{down}} = F_d d \cos 180° = \text{mgd}(-1) = -\text{mgd}$$

The work done by the person on the object is:

$$W_{\text{up}} = F_u d \cos 0° = \text{mgd}(+1) = \text{mgd}$$

Thus, the net work, regardless of the object weight, is

$$W_{\text{net}} = W_{\text{down}} + W_{\text{up}} = 0$$

(2) The work required to push an object of mass m up an inclined plane is

$$\text{Work} = Fd\cos\theta = (25\,\text{kg})(9.8\,\text{m s}^{-2})(2.5\,\text{m})(\cos 160°) = -575\,\text{N m}$$
$$\text{Work} = Fd\cos\theta = (50\,\text{kg})(9.8\,\text{m s}^{-2})(2.5\,\text{m})(\cos 160°) = -1151\,\text{N m}$$
$$\text{Work} = Fd\cos\theta = (100\,\text{kg})(9.8\,\text{m s}^{-2})(2.5\,\text{m})(\cos 160°) = -2302\,\text{N m}$$

where θ is equal to $90° + 70° = 160°$.

(3) Similar to that in (2):

$$\text{Work} = Fd\cos\theta = (25\,\text{kg})(9.8\,\text{m s}^{-2})(2.5\,\text{m})(\cos 20°) = 575\,\text{N m}$$
$$\text{Work} = Fd\cos\theta = (50\,\text{kg})(9.8\,\text{m s}^{-2})(2.5\,\text{m})(\cos 20°) = 1151\,\text{N m}$$
$$\text{Work} = Fd\cos\theta = (100\,\text{kg})(9.8\,\text{m s}^{-2})(2.5\,\text{m})(\cos 20°) = 2302\,\text{N m}$$

where θ is equal to $90° - 70° = 20°$.

Solved Problem 6.2. Calculate the work done by a mother who exerts a force of 8 N at an angle of 40° below the horizontal x axis in pushing her baby in a carriage a distance of 20 m.

Solution

By definition, work can be calculated according to

$$W = Fd\cos\theta = (8\,\text{N}\cos(-40°))(20\,\text{m}) = (8\,\text{N})(0.766)(20\,\text{m}) = 122.6\,\text{J}$$

6.1.2 Work Done by Pressure

Work (W) done by pressure P in moving a volume of fluid V through a finite displacement within a vessel is given by

$$W = P\Delta V$$

where the pressure P = force per unit area and ΔV is the change of volume over the displacement.

Solved Problem 6.3. Derive the expression for the work done by pressure, using the expression for work done by a constant force.

Solution

Work done by a constant force is given as

$$\text{Work} = \text{force} \cdot \text{displacement}$$

or

$$W = F \cdot d$$

The force exerted by a pressure on a fluid confined within a vessel of cross-sectional area A is defined as

$$\text{Force} = \text{pressure} \cdot \text{area}$$

or

$$F = P \cdot A$$

Substituting into the expression for W yields

$$\text{Work} = \text{pressure} \cdot \text{area} \cdot \text{displacement}$$

Since area changes over a given displacement, combining dimensions of measurement gives

$$\text{Work} = \text{pressure} \cdot (\Delta\ \text{area} \cdot \text{displacement}) = \text{pressure} \cdot \Delta\ \text{volume}$$

or

$$\text{Work} = P\Delta V$$

Solved Problem 6.4. If the pressure P in a balloon is given by $P = 200\,\text{N/m}^2$, find the work done in blowing up the balloon from $V_i = 0$ to $V_f = 1.5\,\text{m}^3$.

Solution

The work performed in blowing up the balloon can be determined by

$$W = P\Delta V = (200\,\text{N/m}^2)(\Delta V) = 200 \cdot 1.5 = 300\,\text{J}$$

6.1.3 Work Done by a Variable Force

Work done by a variable force represents the area defined by the force curve or function within the prescribed limits of position x_o and x_f and is equal to the integral of the curve or function representing the force over the differential interval of the displacement dx:

$$W = \int_{x_o}^{x_f} F(x) \, dx$$

where x_o represents the original point of displacement and x_f represents the final point of displacement.

Solved Problem 6.5.   An object of mass m, constrained to motion along the x axis, is subjected to a variable force $F = 5x^3$ N, causing a displacement from rest to $x = 2$ m, where x is in meters and the force is exerted along the $+x$ axis. Determine the work done on the object by the force.

   Solution

   Since we have a variable force, the work is calculated according to

$$W = \int_0^{2m} 5x^3 \, dx = \frac{5x^4}{4}\bigg|_0^{2m} = \frac{5 \cdot 16}{4} - 0 = 20\,\text{N m}$$

Solved Problem 6.6.   Suppose that the force generated by an oscillating mass of a pendulum of length, L, during a complete cycle is defined by

$$F = F_o \tan \theta$$

Determine the work done for displacement over 1 cycle.

   Solution

   The work done by a variable force can be obtained by

$$W = \int_{x_1}^{x_2} F \cos \theta \, dx = \int_0^{2\pi} (F_o \tan \theta) \cos \theta (L \, d\theta) = \int_0^{2\pi} F_o L \sin \theta \, d\theta = -F_o L \cos \theta \bigg|_0^{2\pi} = -F_o L + F_o L = 0$$

ILLUSTRATIVE EXAMPLE 6.1.   Cardiac Stress and Treadmill Exercise

   In the diagnosis of patients suspected of heart disease, the physician must assess cardiac function while the patient is at rest (heart beating at a normal pace) and under stress (heart beating rapidly after exercise). This permits the physician to delineate and identify abnormal regions of the heart by comparing the two distinct physiological conditions. To simulate conditions of stress, the patient exercises by walking a treadmill, to increase heartbeat and sustain an adequately high level of cardiac stress. Average times required for the patient to reach this level of stress are typically between 20 to 30 min, although increasing the speed and angle of inclination of the treadmill can substantially reduce this time.

   The work done by the patient in achieving this level of cardiac stress can be determined by analyzing the forces involved, as shown in Figure 6-1:

$$\Sigma \mathbf{F}_x = ma_x = \mathbf{F}_{ext} - mg \sin \alpha - \mathbf{F}_f = \mathbf{F}_{ext} - mg \sin \alpha - \mu_k N$$
$$\Sigma \mathbf{F}_y = ma_y = mg \cos \alpha - N = 0$$

where $\mathbf{F}_{ext}$ is the external force exerted by the patient, mg is the weight of the patient, N is the normal force exerted upward by the treadmill, α is the angle of inclination of the treadmill, and μ_k is the coefficient of static friction. Force $\mathbf{F}_{ext}$ is the muscular force exerted by the patient in running to maintain motion along the positive x axis. Since the patient's motion is only in the x direction parallel to the treadmill, we are concerned only with $\Sigma \mathbf{F}_x$. The total upward force is $\mathbf{F}_{ext}$, and the total downward force is $mg \sin \alpha + \mathbf{F}_f$. Thus, the net force acting along the positive x axis is

$$\mathbf{F}_{x,\text{net}} = \mathbf{F}_{ext} - (mg \sin \alpha + \mathbf{F}_f) = \mathbf{F}_{ext} - (mg \sin \alpha + \mu_k mg \cos \alpha)$$

The work expended by the patient on the treadmill can be defined as

$$W = \mathbf{F} \cdot \mathbf{d} = \mathbf{F}_{ext} - (mg \sin \alpha + \mu_k mg \cos \alpha)] \cdot d$$
$$= |[\mathbf{F}_{ext} - (mg \sin \alpha + \mu_k mg \cos \alpha)]| \cdot |\mathbf{d}| \cos \theta$$

where θ is the angle between the net force and the resultant displacement in the direction of the force.

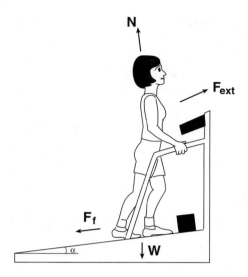

Fig. 6-1

Solved Problem 6.7.   A 50-kg patient exercises on the treadmill, exerting a constant force of 500 N while running at a constant velocity of $4\,\mathrm{m\,s^{-1}}$ at an inclination angle of the treadmill $\alpha = 30°$ for 5 min. The coefficient of kinetic friction of the treadmill is 0.45. Determine the work done by the patient.

Solution

The work done by the patient can be determined from the equation

$$W = \mathbf{F}d\cos\theta = (T - mg\sin\theta - \mu_k N)d\cos\theta$$
$$= |(500\,\mathrm{N} - 50\,\mathrm{kg}\cdot 9.8\,\mathrm{m\,s^{-2}}\cdot\sin 30° - 0.45\cdot 50\,\mathrm{kg}\cdot 9.8\,\mathrm{m\,s^{-2}}\cdot\cos 30°)|\cdot|\mathbf{d}|\cos\theta$$
$$= 64.1 Nd\cos\theta$$

The displacement of the patient can be determined from the equation of motion:

$$d = vt = (4\,\mathrm{m\,s^{-1}})(5\ \mathrm{min})(60\,\mathrm{s\,min^{-1}}) = 1200\,\mathrm{m}$$

Thus, the work can now be determined:

$$W = (64.1\,\mathrm{N}\cdot 1200\,\mathrm{m})\cos 120° = -3.8\times 10^4\,\mathrm{J}$$

ILLUSTRATIVE EXAMPLE 6.2.   Work Done by Normal and Diseased Hearts

The heart is an elastic, muscular double pump whose rhythmic contractions provide the force needed to circulate the blood throughout the network of blood vessels comprising the circulatory system. Anatomically, the heart, as shown in Figure 6-2, is composed of four chambers. Two smaller chambers (right atrium and left atrium) contract systematically to eject blood into the two larger chambers (right ventricle and left ventricle).

The work done by the heart is the sum of work done by the left and right ventricles, whose primary purpose is to pump blood into the systemic (body) and pulmonary (lung) circulation, respectively. Since the problem involves a pressure exerted against a known volume of blood, work is determined by

$$W = P\Delta V$$

During each heartbeat in the resting condition of a normal human, the left ventricle pumps approximately $5\,\mathrm{L\,min^{-1}}$ $(= 83\,\mathrm{cm^3\,s^{-1}})$ of blood under an average pressure of $100\,\mathrm{mmHg}$ $(= 1.333\times 10^4\,\mathrm{N\cdot m^{-2}})$. The right ventricle pumps the

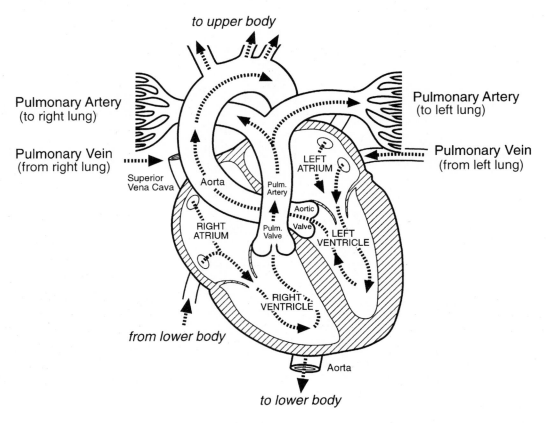

Fig. 6-2

same volume of blood but at an average pressure one-fifth that of the left ventricle, that is, 20 mmHg ($= 0.266 \times 10^4 \, \text{N} \cdot \text{m}^{-2}$). Therefore, the work of the left and right ventricles can be determined by

Left ventricle: $W_{\text{LV}} = P_{\text{LV}} \Delta V_{\text{LV}} = (1.333 \times 10^4 \, \text{N} \cdot \text{m}^{-2})(83 \times 10^{-6} \, \text{m}^3) = 1.11 \, \text{J}$

Right ventricle: $W_{\text{RV}} = P_{\text{RV}} \Delta V_{\text{RV}} = (0.266 \times 10^4 \, \text{N} \cdot \text{m}^{-2})(83 \times 10^{-6} \, \text{m}^3) = 0.22 \, \text{J}$

The total average work done by the heart, W_T during a single contraction is

$$W_T = W_{\text{LV}} + W_{\text{RV}} = 1.11 \, \text{J} + 0.22 \, \text{J} = 1.33 \, \text{J}$$

In clinical applications, the work done by the left ventricle can be illustrated by a pressure-volume curve, as shown in Figure 6-3, where the area of the curve is the product of the pressure of the left ventricle and the volume of ejected blood and is calculated from

$$W = \int_{V_i}^{V_f} P \, dV$$

where W is the work (in ergs), P is the pressure (in dyn cm^{-2}), V_i and V_f are the initial and final volumes of the left ventricle during the cardiac cycle, respectively.

In a diseased heart, atherosclerosis, typically referred to as hardening of the arteries, involves the formation of calcified fatty or lipid deposits along the walls of the aorta and major arteries, acting to increase the size and stiffness of the vessel wall. Increases in the size and stiffness of the vessel wall (1) reduce the lumen or inner diameter of the blood vessels through which blood may flow and (2) diminish the compliance or elasticity of the blood vessels, adversely affecting their ability to expand and accommodate large volumes of blood upon ejection from the left ventricle. As a consequence, internal regulatory mechanisms signal the heart to exert more pressure and hence perform more work upon each heartbeat, depending on the severity and extent of atherosclerotic wall thickening.

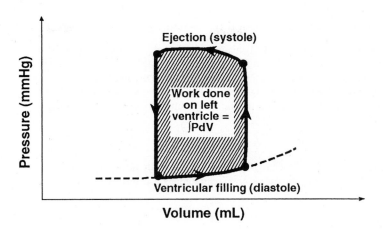

Fig. 6-3

Solved Problem 6.8. During extreme exercise, the blood output from each ventricle increases fourfold, and the pressure exerted by each ventricle increases by 50 percent. Calculate the work done by the heart after exercise.

Solution

The work done by the heart during extreme exercise can be determined from the work done by the left and right ventricles at the increased values of pressure and volume:

Left ventricle: $P_{LV} = 2.0 \times 10^4 \, \text{N} \cdot \text{m}^{-2}$ $V_{LV} = 3.32 \times 10^{-4} \, \text{m}^3$

Right ventricle: $P_{RV} = 0.4 \times 10^4 \, \text{N} \cdot \text{m}^{-2}$ $V_{RV} = 3.32 \times 10^{-4} \, \text{m}^3$

Therefore, the work of the left and right ventricles can be determined by

Left ventricle: $W_{LV} = P_{LV} \Delta V_{LV} = (2.0 \times 10^4 \, \text{N} \cdot \text{m}^{-2})(3.32 \times 10^{-4} \, \text{m}^3) = 6.64 \, \text{J}$

Right ventricle: $W_{RV} = P_{RV} \Delta V_{RV} = (0.4 \times 10^4 \, \text{N} \cdot \text{m}^{-2})(3.32 \times 10^{-4} \, \text{m}^3) = 1.33 \, \text{J}$

The total average work done by the heart W_T during a single contraction under exercise conditions is

$$W_T = W_{LV} + W_{RV} = 6.64 \, \text{J} + 1.33 \, \text{J} = 7.97 \, \text{J}$$

Solved Problem 6.9. How much work is done by the resting heart in an average lifetime of 80 years?

Solution

The volume of blood ΔV or cardiac output ejected from the heart at each heartbeat is 83 mL. Assuming each heartbeat occurs every 1.0 s, the total volume of blood ejected from the heart over a lifetime of 80 yr is

$$\Delta V = 83 \, \frac{\text{mL}}{\text{s}} \cdot \frac{3600 \, \text{s}}{1 \, \text{h}} \cdot \frac{24 \, \text{h}}{1 \, \text{d}} \cdot \frac{365 \, \text{d}}{1 \, \text{yr}} \cdot 80 \, \text{yr} = 2.1 \times 10^{11} \, \text{mL}$$

The work of the left and right ventricles can be determined by

Left ventricle: $W_{LV} = P_{LV} \Delta V_{LV} = (1.333 \times 10^4 \, \text{N} \cdot \text{m}^{-2})(2.1 \times 10^5 \, \text{m}^3) = 2.8 \times 10^9 \, \text{J}$

Right ventricle: $W_{RV} = P_{RV} \Delta V_{RV} = (0.266 \times 10^4 \, \text{N} \cdot \text{m}^{-2})(2.1 \times 10^5 \, \text{m}^3) = 0.56 \times 10^9 \, \text{J}$

The total average work done by the heart W_T during an average lifetime is

$$W_T = W_{LV} + W_{RV} = 2.8 \times 10^9 \, \text{J} + 0.56 \times 10^9 \, \text{J} = 3.36 \times 10^9 \, \text{J}$$

ILLUSTRATIVE EXAMPLE 6.3.   Work Done by Breathing

Work done by breathing is a complicated problem and involves the contribution of a host of respiratory muscles. During respiration, as air is taken in and the lungs expand to accommodate the increase in air volume, the work done by these muscles results in (1) the storage of potential energy by the surrounding elastic tissues of the lungs and chest, (2) dissipation of thermal energy by the incoming airflow, and, (3) forces of resistance generated by tissue movement. The pressure P driving respiratory flow is related to the lung volume V by the differential equation

$$P = k_1 V + k_2 \frac{dV}{dt} + k_3 \left(\frac{dV}{dt}\right)^2 \qquad (1)$$

where dV/dt is the rate of airflow or changing lung volume with respect to time. Since respiration is a cyclical or repetitive process, the flow rate dV/dt can be approximated by a sinusoidal function:

$$\frac{dV}{dt} = A \sin \omega t \qquad (2)$$

where A is amplitude and ω is the angular frequency of breathing. The total volume of air inspired during a given breath is known as the *tidal volume*, and it can be determined by integrating the flow rate over one-half cycle (corresponding to the intake of air or inspiration):

$$V_{\text{tidal}} = \int_0^{\pi/\omega} \frac{dV}{dt} \cdot dt = \int_0^{\pi/\omega} A \sin \omega t = -\frac{1}{\omega} A \cos \omega t$$

Work done by pressure of inspiration exerted on a lung volume V_{tidal}, given by Eqs. (1) and (2), can be determined by substituting into the expression

$$W = \int P dv$$

and integrating over the limits of one half-cycle, i.e., from 0 to π/ω. In clinical practice, the work done by breathing is determined by generating P-V (pressure-volume) curves depicting inspiration and expiration and quantitating values of work from these data, as shown in Figure 6-4.

ILLUSTRATIVE EXAMPLE 6.4.   Hooke's Law and the Work of a Spring

Hooke's law describes the force F required to stretch a spring a distance x and is given by

$$F = -kx$$

where k is the force constant of the spring. The minus sign corresponds to the opposite direction where k is the force constant of the spring. The minus sign corresponds to the opposite direction with which the restoring force of the spring acts against

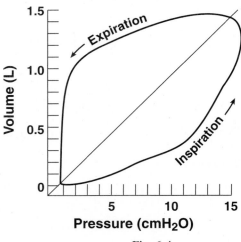

Fig. 6-4

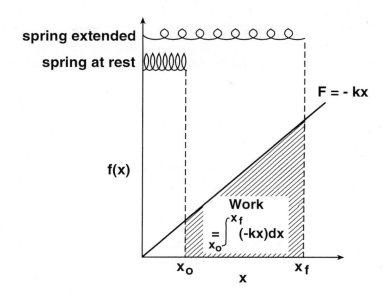

Fig. 6-5

the external stretching force. The area under the curve in Figure 6-5 represents the work done by the external stretching force and can be determined quantitatively by integrating Hooke's law over a given region from $x = x_o$ to $x = x_f$:

$$W = \frac{1}{2}kx_f^2 - \frac{1}{2}kx_o^2$$

The minus sign was omitted from the integral by reversal of the direction along which the work was performed. Assuming that x_o represents the origin $(x = 0)$, then the work done by the spring is

$$W = \frac{1}{2}kx^2$$

To extend this concept, suppose we are only interested in obtaining the work (or area under the curve of Hooke's law) over a defined region, say, from $x = 0$ to $x = 4$. In this case, the work is determined in a similar fashion with the substitution of the numerical limits in the evaluation of the integral:

$$W = \frac{1}{2}k(4)^2 - \frac{1}{2}k(0)^2 = 8k$$

6.1.4 Power

Power is the rate at which work is performed or energy is expended. It represents the time required to do work and is defined by the relation

$$P = \frac{W}{t}$$

where P is power, W is work, and t is time. Power is a scalar quantity and given in units of watts (W), where $1\,\text{W} = 1\,\text{J}\,\text{s}^{-1}$.

ILLUSTRATIVE EXAMPLE 6.5.   Power Output of the Heart

The power of the heart represents the time rate at which the work is done and can be discussed as an extension to Illustrative Example 6.2. The work done by each ventricle of the resting heart was calculated as follows:

Left ventricle: $W_{LV} = 1.11$ J
Right ventricle: $W_{RV} = 0.22$ J

with the total average work done by the heart W_T during a single contraction being

$$W_T = W_{LV} + W_{RV} = 1.11 \text{ J} + 0.22 \text{ J} = 1.33 \text{ J}$$

The power output of the heart can be calculated over the time elapsed during each heartbeat $t_c = 1.0$ s by

$$\text{Power output} = \frac{W_{LV} + W_{RV}}{t_c} = \frac{1.11 \text{ J} + 0.22 \text{ J}}{1 \text{ s}} = 1.33 \text{ W}$$

Solved Problem 6.10.   In transporting a patient in a wheelchair up an inclined ramp at an angle of $20°$ with the horizontal ground, an orderly exerts a force of 1500 N, resulting in a velocity of 3 m s^{-1}. Calculate the power produced by the orderly.

Solution

The power produced by the orderly is

$$\text{Power} = \frac{\text{force} \cdot \text{displacement}}{\text{time}} = \text{force} \cdot \frac{\text{displacement}}{\text{time}} = \text{force} \cdot \text{velocity}$$

Making the appropriate substitutions yields

$$P = 1500 \text{ N} \cdot 3 \text{ m s}^{-1} = 4500 \text{ W}$$

Solved Problem 6.11.   Determine the power generated by a person of mass 60 kg climbing the staircase in a corporate building to an office 50 m from the ground level while (*a*) leisurely walking in a time $t = 20$ s and (*b*) running at a time $t = 5$ s.

Solution

The power output can be determined from the relation

(*a*) $P = \dfrac{W}{t} = \dfrac{mgh}{t} = \dfrac{(60 \text{ kg})(9.8 \text{ m s}^{-2})(50 \text{ m})}{20 \text{ s}} = 1470 \text{ J}$

(*b*) $P = \dfrac{W}{t} = \dfrac{mgh}{t} = \dfrac{(60 \text{ kg})(9.8 \text{ m s}^{-2})(50 \text{ m})}{5 \text{ s}} = 5880 \text{ J}$

6.2   MECHANICAL EFFICIENCY

Mechanical efficiency is defined as the ratio of amount of useful work expended to the amount of energy required, or

$$\text{Mechanical efficiency} = \frac{\text{amount of useful work expended}}{\text{amount of energy required}}$$

ILLUSTRATIVE EXAMPLE 6.6.   Mechanical Efficiency of the Heart

The mechanical efficiency of the heart is less than 0.1. In its resting state, the rate of energy turnover of the heart is high; i.e., the heart uses energy at a rate of about 10 times its power output of approximately 12 W. Since the total metabolic rate for a resting person is on the order of 100 W, the power output of the heart constitutes a small fraction of the total power output.

6.3 ENERGY OF A PARTICLE: KINETIC AND POTENTIAL

Energy represents the ability to do work. Since energy represents the ability to do work, units of energy are the same as those of work, i.e., joules. An object possesses two forms of energy, depending on its physical state:

Potential energy is the stored energy of an object at rest and represents the ability of an object to do work against gravity. Potential energy (PE) is dependent upon the position of the object and is given by

$$PE = mgh$$

where m is the mass of the object, g is the gravitational acceleration, and h is the height of the object with respect to the ground surface.

Kinetic energy is the energy of an object in motion. For an object of mass m moving at a velocity v, the kinetic energy (KE) is

$$KE = \frac{1}{2}mv^2$$

The *total mechanical energy E* is the sum of the kinetic energy and potential energy of an object, or

$$E = PE + KE = mgh + \frac{1}{2}mv^2$$

ILLUSTRATIVE EXAMPLE 6.7. Dynamics of Bumblebee Flight

As a bumblebee approaches a suitable flower for nectar and begins feeding, it remains airborne or hovers about the flower during the entire feeding process, as shown in Figure 6-6. The mass of the bumblebee is approximately 0.25 g, and its corresponding weight is W ($= mg$). The average force exerted by the two wings during a single downward movement or stroke is

$$\mathbf{F} = \mathbf{F}_{\text{wing 1}} + \mathbf{F}_{\text{wing 2}} = W + W = 2W$$

Each downward stroke involves movement of the wing through a vertical distance d. Assuming that each wing of length 1.0 cm sweeps a total angle of 80° during each stroke, the linear distance d traversed by each wing is

$$d = r\theta$$

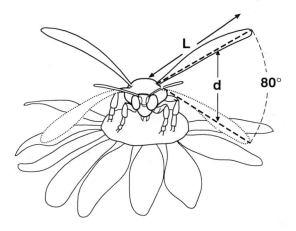

Fig. 6-6

where r is the length of the wing and θ is the angle swept by the wing in radians. Converting the angle to radians gives

$$\theta = (80°)\left(\frac{2\pi\,\text{rad}}{360°}\right) = 1.4\,\text{rad}$$

$$d = (1.0\,\text{cm})(1.4\,\text{rad}) = 1.4\,\text{cm}$$

Thus, the work done by the bumblebee in hovering about the flower is

$$\text{Work} = 2Wd = 2mgd = 2(0.25\,\text{g})(980\,\text{cm s}^{-2})(1.4\,\text{cm}) = 686\ \text{erg per downward stroke}$$

Assuming that the bumblebee averages 100 downward strokes per second, the power output of the bumblebee is

$$\text{Power} = \frac{\text{work}}{\text{time}} = \left(686\,\frac{\text{erg}}{\text{stroke}}\right)\left(100\,\frac{\text{strokes}}{\text{s}}\right) = 6.86 \times 10^4\,\text{erg s}^{-1}$$

Solved Problem 6.12.   A 60-kg mother squats down and picks up her 15-kg child from the crib and places the child on a table 0.8 m from the crib. Calculate the work done by the mother.

Solution

The work done by the mother is

$$W = \text{PE}$$

In addition to raising the child 0.8 m, the mother raises her center of gravity 0.8 m in transition from the squatting position to the upright position. Therefore,

$$W = \text{PE} = mgh = (60\,\text{kg} + 15\,\text{kg})(9.8\,\text{m s}^{-2})(0.8\,\text{m}) = 588\,\text{J}$$

Solved Problem 6.13.   Calculate the energy required for a runner of mass 70 kg to accelerate from rest to (1) $2\,\text{m s}^{-1}$, (2) $5\,\text{m s}^{-1}$, and (3) $10\,\text{m s}^{-1}$.

Solution

The kinetic energy of the runner can be calculated from

(1) $\text{KE} = \dfrac{1}{2}mv^2 = \dfrac{1}{2}(70\,\text{kg})(2\,\text{m s}^{-1})^2 = 140\,\text{J}$

(2) $\text{KE} = \dfrac{1}{2}mv^2 = \dfrac{1}{2}(70\,\text{kg})(5\,\text{m s}^{-1})^2 = 875\,\text{J}$

(3) $\text{KE} = \dfrac{1}{2}mv^2 = \dfrac{1}{2}(70\,\text{kg})(10\,\text{m s}^{-1})^2 = 3500\,\text{J}$

Conservation of energy states that energy can be neither created nor destroyed, but can be transformed to other forms of energy. In other words, the following must hold:

$$E_{\text{before}} = E_{\text{after}}$$

$$\left(mgh + \frac{1}{2}mv^2\right)_{\text{before}} = \left(mgh + \frac{1}{2}mv^2\right)_{\text{after}}$$

A force is conservative if it satisfies the following properties:

(1)   The work done by an external force on an object is zero whose displacement as a result of the force is zero.

(2)   The work done by an external force on an object is independent of the path followed.

Examples of conservative forces are the gravitational force near the earth's surface, the elastic force of a spring, and the electrostatic force of an electric charge. An example of a nonconservative force is the force of friction.

ILLUSTRATIVE EXAMPLE 6.8.   Conversion of Potential Energy to Kinetic Energy during a Fall

At rest, a person of mass m possesses potential energy, which is available to do work on the body, given by

$$PE = mgh$$

where mg is the body weight and h is the distance from the center of gravity (hip) to the ground. For an average person of mass 70 kg, the weight is 686 N; and assuming that h is 1.5 m, the potential energy strored by the person is 1029 J. During a fall, the person is in motion, and the potential energy is converted to kinetic energy. For a person at greater heights, the potential energy stored by a person becomes correspondingly greater, hence the potential for greater bodily harm upon a fall. As one can easily extrapolate, the greater the distance involved in the fall, the greater the kinetic energy and hence velocity with which the person will strike the ground.

The *work-energy theorem* states that the work done on an object by all external forces is directly related to the energy expended by the object. As an object in motion expends energy, the expended energy will be either transformed to another form of energy or used to perform work. The work-energy theorem can be stated in equation form as

$$\text{Work} = \text{KE}_f - \text{KE}_o = \frac{1}{2}mv_f^2 - \frac{1}{2}mv_o^2 = \Delta\text{KE}$$

Solved Problem 6.14.   Nick is a skateboarder of mass 50 kg who skates down the curved surface of a "bowl" and acquires a velocity of 15 m s^{-1} before he travels on a horizontal platform a distance of 20 m, as shown in Figure 6-7. The coefficient of friction on the horizontal surface is 0.35, while the curved surface is frictionless. Following this horizontal surface, Nick ascends an identical curved surface on the opposite side of the bowl. How high or what value of h does Nick ascend the opposing curved surface of the bowl before coming to a complete stop?

Solution

The initial kinetic energy of Nick is

$$\text{KE} = \frac{1}{2}mv^2 = \frac{1}{2}(50 \text{ kg})(15 \text{ m s}^{-1})^2 = 5625 \text{ J}$$

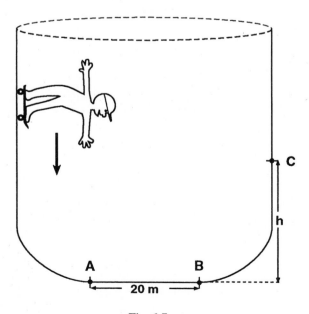

Fig. 6-7

This is the kinetic energy at point A in Figure 6-7. The work done by friction along the horizontal surface of the bowl is

$$W_f = -fd = -\mu mgd = -(0.35)(50 \text{ kg})(9.8 \text{ m s}^{-2})(20 \text{ m}) = -3430 \text{ J}$$

Using the work-energy theorem, we can determine the kinetic energy at point B by

$$W_f = \text{KE}_B - \text{KE}_A = \text{KE}_B - 5625 \text{ J} = -3430 \text{ J}$$

Solving for KE_B gives

$$\text{KE}_B = -3430 \text{ J} + 5625 \text{ J} = 2195 \text{ J}$$

Since the curved surface is frictionless, it is possible to equate the kinetic energy of Nick at point B with his gravitational potential energy at point C:

$$mgh = \text{KE}_B = 2195 \text{ J}$$

Therefore, the height to which Nick ascends prior to coming to a complete stop is

$$h = \frac{2195 \text{ J}}{50 \text{ kg} \cdot 9.8 \text{ m s}^{-2}} = 4.47 \text{ m}$$

Solved Problem 6.15.   Jennifer of mass $m = 20$ kg slides down a slide of height $h = 3.5$ m and length $L = 7.0$ m, as shown in Figure 6-8. Determine (1) the kinetic energy of Jennifer at the bottom of the slide and (2) the final velocity of Jennifer at the bottom of the slide. Assume the coefficient of friction μ exerted on Jennifer during the entire trip down the slide is 0.05.

Fig. 6-8

Solution

The potential energy of Jennifer at the top of the slide is transformed to kinetic energy and work against frictional forces W_f and can be stated in equation form as

$$\text{PE} = \text{KE} + W_f$$

$$mgh = \frac{1}{2}mv^2 + F_f L$$

where F_f is the frictional force. From elementary trigonometric relations,

$$h = L \sin \phi$$

where ϕ can be determined by

$$\sin\phi = \frac{h}{L}$$

and

$$F_f = \mu N \cos\phi = \mu mg \cos\phi$$

where μ is the coefficient of friction and ϕ is the angle of inclination of the slide.

(1)   The kinetic energy of Jennifer is given by

$$KE = \frac{1}{2}mv^2 = mgh - F_f L$$
$$= mgL(\sin\phi - \mu\cos\phi)$$

We now need a numerical value for ϕ:

$$\phi = \sin^{-1}\frac{h}{L}\sin^{-1}\frac{3.5\,\text{m}}{7.0\,\text{m}} = 30°$$

We can also determine $\cos\phi = \cos 30° = 0.866$, and the kinetic energy is

$$KE = (20\,\text{kg})(9.8\,\text{m s}^{-2})(7.0\,\text{m})[0.5 - (0.05)(0.866)] = 626.6\,\text{J}$$

(2)   The final velocity of Jennifer can be determined from the expression for kinetic energy:

$$KE = \frac{1}{2}mv^2 \Rightarrow v = \sqrt{\frac{2\cdot KE}{m}} = \sqrt{\frac{2\cdot 626.7\,\text{J}}{20\,\text{kg}}} = 7.9\,\text{m s}^{-1}$$

Solved Problem 6.16.   A baseball is thrown directly upward with an initial speed of $15\,\text{m s}^{-1}$. Determine its height above ground when its speed is $5\,\text{m s}^{-1}$. (Ignore air resistance, and assume down is the positive direction.)

Solution

During its trip upward, the ball loses kinetic energy while it is gaining gravitational potential energy, or

$$KE_{loss} = GPE_{gain}$$
$$\frac{1}{2}mv_f^2 - \frac{1}{2}mv_a^2 = mgh_f - mgh_o = mg(h_f - h_o) = mgh$$

where h is the height we were asked to determine. This can be simplified by dividing through by m:

$$\frac{1}{2}(v_f^2 - v_o^2) = gh$$

where $v_f = 5\,\text{m s}^{-1}$, $v_o = 15\,\text{m s}^{-1}$, and $g = 9.8\,\text{m s}^{-2}$. Making the appropriate substitutions yields

$$h = \frac{v_f^2 - v_o^2}{2g} = \frac{(5\,\text{m s}^{-1})^2 - (15\,\text{m s}^{-1})^2}{2\cdot 9.8\,\text{m s}^{-2}} = \frac{-200\,\text{m}^2\,\text{s}^{-2}}{19.6\,\text{m s}^{-2}} = -10.2\,\text{m}$$

Supplementary Problems

6.1.   A 200-N child is pushed up a 5.5-m slide by her father who is exerting a constant force of 330 N. The slide makes an angle of 30° with the horizontal ground. (1) What is the work done by the father? (2) What is the potential energy of the child at the top of the slide? (3) What is the kinetic energy of the child after sliding down the slide in 1.2 s?

Answer

(1)   1571 J; (2) 550 J; (3) 215.8 J

6.2.   As an extension to Problem 6.1, suppose the father pulls the 200-N child up the slide at an angle of 15° above the surface of the slide by an elastic cord. What is the work done by the father?

Answer

1283 J

6.3.   A person who has been diagnosed with chronic arterial hypertension has a heart rate slightly elevated above normal (80 beats min^{-1}) and a systolic blood pressure of 160 mmHg. Assuming that pressure in the right ventricle is one-fifth that of the left ventricle, determine the average work and power output of (1) the left ventricle, (2) the right ventricle, and (3) the heart required to maintain a cardiac output of 80 cm^3.

Answer

(1)   7.07 J, 9.43 W; (2) 1.43 J, 1.91 W; (3) 8.50 J, 11.34 W

6.4.   What is the net work done in lifting a 20-kg barbell 1.5 m from the floor to a table?

Answer

0

6.5.   Determine the amount of energy needed to accelerate a 2000-kg car from rest to 50 m s^{-1}. Assume that the car is riding on a frictionless surface.

Answer

2.5×10^6 J

6.6.   Consider a block of mass m which is subjected to a 15-N force applied along the horizontal. Calculate the work done by the force in pulling the block 10 m on (1) a frictionless surface and (2) a surface with a coefficient of friction of $\mu = 0.5$.

Answer

(1)   150 J; (2) 75 J

6.7.   Kelly is moving and wants to move a box of 25-kg mass a distance of 15 m. Determine the work done if (1) she picks up the box a distance of 1.5 m and walks the distance of 15 m and (2) she pushes the box with a force of 250 N the distance of 15 m.

Answer

(1)   0; (2) 3750 J

6.8.   Suppose that three constant forces

$$F_1 = 3\,\text{N} \qquad F_2 = 4\,\text{N} \qquad F_3 = 6\,\text{N}$$

act on an object to move it a distance of 30 m. What is the total work done?

Answer

390 J

6.9.   A 16-kg marble attached to a 3.0-m-long string is spun in circular motion. (1) Determine the work done by the marble. (2) Explain your results.

Answer

(1)   0

6.10.   A 4-kg particle is subjected to a force, causing the particle to assume a velocity of

$$v_x = 1.5\,\mathrm{m\,s}^{-1} \qquad v_y = 4.0\,\mathrm{m\,s}^{-1} \qquad v_z = 2.0\,\mathrm{m\,s}^{-1}$$

What is the kinetic energy of the particle?

Answer

44.5 J

6.11.   Calculate the kinetic energy of a 3.5-kg object moving with a velocity

$$\mathbf{v} = 10\mathbf{i} + 10\mathbf{j}\,\mathrm{m\,s}^{-1}$$

Answer

350 J

6.12.   Calculate the work needed to be performed on a moving block of mass 10 kg to (1) increase its velocity from 5 ms^{-1} and (2) stop from an initial velocity of 10 m s^{-2}.

Answer

(1)   1000 J; (2) −500 J

6.13.   On his way to work, Fred Flintstone is driving Barney Rubble to work as part of his carpool arrangement in his 800-kg Rockmobile. One morning, as Fred is talking to Barney, Fred suddenly brakes from an initial velocity of 40 m s^{-1} to a stop in 15 m, to avoid striking a stray dinosaur. Determine the stopping force exerted on the Rockmobile by Fred's feet.

Answer

42,667 N

6.14.   A 2.0-g coin was dropped into a 25-m wishing well. Determine (1) the work done on the coin by the gravitational force and (3) the amount of gravitational potential energy lost during the trip.

Answer

(1)   0.49 J; (2) −0.49 J

6.15.   Requesting the latest romance novel from her friend Gladys, seated across the table, Sue pushes the 2.0-kg book toward her friend at an initial speed of 40 cm s^{-1}, and it comes to rest in 1.25 m. Determine the frictional force exerted on the book.

Answer

−0.13 N

6.16.   On takeoff, an airplane of 5000-kg mass ascends to 10,000 m in 6.2 min. What are the work and power performed by the airplane engines?

Solution

4.9×10^8 J; 1.32×10^6 W

6.17. A 20-g bullet projected from a rifle with a muzzle velocity of 900 m s^{-1} strikes a target at the same level with a velocity of 500 m s^{-1}. Determine the work done in overcoming air resistance.

Solution

-5600 J

6.18. An impressive quality of Superman is his ability to stop a bullet with his chest. Determine the average force required to stop a 0.20-kg bullet at a speed of 300 m s^{-1} within a distance of 5 cm.

Solution

1.8×10^5 N

6.19. Karen is at the top of a staircase when she decides to launch her toy rocket upward at an angle $\theta = 25°$ to send it to her brother Kel at the bottom of the staircase 15 m below Karen. If the rocket is launched at a velocity of 5 m s^{-1}, what is the velocity of the rocket once it reaches Kel?

Solution

17.9 m s^{-1}

Chapter 7

Momentum and Impulse

In previous chapters, the resultant effects of external forces exerted on an object, i.e., work, energy, and power, were introduced and discussed. Momentum and impulse represent an extension of the dynamic behavior of a particle and its interactions with its surrounding environment or with other particles.

7.1 DEFINITIONS OF MOMENTUM AND IMPULSE

Linear Momentum **p** is the product of object mass m and its velocity **v** and is defined by

$$\mathbf{p} = m\mathbf{v}$$

Momentum can also be used to restate Newton's second law as

$$\mathbf{F} = m\mathbf{a} = \frac{d\mathbf{p}}{dt} = \frac{d(m\mathbf{v})}{dt}$$

where **F** is the net force exerted on a particle. Momentum is a vector quantity whose direction is that of the velocity and is given in units of kg m s^{-1}.

Impulse **I** is a physical entity that represents the change of momentum of an object in both magnitude and direction and is defined as the force acting on the object multiplied by the time interval during which this force acts:

$$\mathbf{I} = \mathbf{F}t = \Delta\mathbf{p} = m\mathbf{v}_f - m\mathbf{v}_o = m(\mathbf{v}_f - \mathbf{v}_o)$$

Impulse is a vector quantity whose direction is that of the force and is given in units of newton-seconds.

ILLUSTRATIVE EXAMPLE 7.1   Impulsive Force and Injury due to a Fall

A person who falls, regardless of the height, undergoes a conversion from potential energy to kinetic energy prior to striking the ground (see Illustrative Example 6.8). The resultant injury upon impact from a fall depends on the impulsive force exerted on the person. For a person in free fall from a height h, the velocity on impact with the ground, neglecting air resistance, can be determined from the equation of motion

$$v_f^2 = v_o^2 + 2ay$$

where $v_o = 0$, $a = g$, and $y = h$. Making the appropriate substitutions yields

$$v_f = \sqrt{2gh}$$

Thus, the momentum of the person upon impact is

$$\mathbf{p} = m\mathbf{v} = m\sqrt{2gh}$$

Multiplying the above expression for momentum by g/g ($= 1$), we get

$$\mathbf{p} = m\frac{g}{g}\sqrt{2gh} = mg\sqrt{\frac{2gh}{g^2}} = W\sqrt{\frac{2h}{g}}$$

After the impact, the person is at rest and the momentum is zero, i.e.,

$$\mathbf{p}_{\text{before impact}} = \mathbf{p}_{\text{after impact}}$$

leading to the following expression:

$$mv_f - mv_o = W\sqrt{\frac{2h}{g}}$$

By definition, the impulse is defined as

$$\mathbf{I} = \mathbf{F}(\Delta t) = mv_f - mv_o$$

Thus, the average impulsive force exerted on the person upon impact is

$$\mathbf{F} = \frac{mv_f - mv_o}{\Delta t} = \frac{W\sqrt{2h/g}}{\Delta t} = \frac{m}{\Delta t}\sqrt{2gh}$$

The parameter of particular importance is the estimate of duration of impact or collision with the ground. The time of impact depends primarily on the surface texture and the surface area of the body. Impact with a hard surface such as concrete or steel while landing on the feet results in a collision time of approximately 0.02 s while an impact with a padded or grassy surface for a person who brakes the fall by bending the knees and falling on the wrists increases the collision time to about 0.2 s. In fact, skydivers whose parachutes failed to open have survived falls from heights of 3000 m (10,000 ft) by landing on grassy or marshy ground and increasing collision time by breaking the fall.

The damage or injury to the human body can be estimated by comparing the impulsive force generated at impact with the force per unit area required to cause fracture of a bone, which is approximately 1×10^9 dyn cm^2.

Solved Problem 7.1.   What is the momentum of a 1500-kg car traveling at a velocity of 60 km h^{-1}?

Solution

The momentum is the product of the car's mass and velocity. However, as given, the velocity in km h^{-1} must be converted to m s^{-1}:

$$\mathbf{v} = (60 \text{ km h}^{-1})(1000 \text{ m km}^{-1})(0.00028 \text{ h s}^{-1}) = 16.8 \text{ m s}^{-1}$$

Substituting into the expression for momentum yields

$$\mathbf{p} = m\mathbf{v} = (1500 \text{ kg})(16.8 \text{ m s}^{-1}) = 25{,}200 \text{ kg m s}^{-1}$$

Solved Problem 7.2.   A racquetball of 0.02 kg mass strikes a wall at an angle of $\theta_1 = 60°$ to the normal with a velocity $v_1 = 90$ m s^{-1}, and it rebounds elastically within 0.1 s at the same angle, that is, $\theta_2 = \theta_1$, without losing its velocity, so $v_2 = v_1$. Determine the impulse of the force exerted on the wall during the impact.

Solution

From Newton's second law, the impulse is related to the change of momentum by

$$\mathbf{I} = \mathbf{F}\,\Delta t = \Delta\mathbf{p} = m\,\Delta\mathbf{v}$$

where $\Delta\mathbf{v}$ is the resultant difference in the velocity vectors. The magnitudes of the velocity vectors are identical but differ in direction. Thus,

$$\Delta v = v_2 \cos\theta_2 - (-v_1 \cos\theta_1) = v_2 \cos\theta_2 + v_1 \cos\theta_1$$

According to the initial conditions,

$$v_1 = v_2 = v \qquad \text{and} \qquad \theta_1 = \theta_2 = \theta$$

Therefore,

$$\Delta v = v \cos\theta + v \cos\theta = 2v \cos\theta$$

and

$$F\,\Delta t = m\,\Delta v$$
$$F(0.1 \text{ s}) = (0.02 \text{ kg})(2 \cdot 90 \text{ m s}^{-1} \cdot \cos 60°)$$
$$F = 18 \text{ N}$$

Solved Problem 7.3.   Walking at a constant speed of 1.5 m s^{-1} through a construction site, Humphrey runs into a barrier and strikes his head on a steel beam. (1) Assuming that his 3.5-kg head stops within 1.0 cm in 0.01 s, determine the force exerted on his head. (2) Determine the force exerted on his head if the steel beam is padded and his impact time is lengthened to 0.05 s.

Solution

(1)   We first must calculate the impulse, which is also the change in momentum:

$$\mathbf{I} = \mathbf{F}\,\Delta t = \Delta\mathbf{p} = m\,\Delta\mathbf{v} = m(v_f - v_o) = (3.5\text{ kg})(1.5\text{ m s}^{-1}) - (3.5\text{ kg})(0\text{ m s}^{-1}) = 5.25\text{ kg m s}^{-1}$$

Therefore, the force exerted on Humphrey's head is

$$\mathbf{F} = \frac{\Delta\mathbf{p}}{\Delta t} = \frac{5.25\text{ kg m s}^{-1}}{0.01\text{ s}} = 525\text{ N}$$

(2)   Repeating the calculations above with substitution of the new impact time gives

$$\mathbf{F} = \frac{\Delta\mathbf{p}}{\Delta t} = \frac{5.25\text{ kg m s}^{-1}}{0.05\text{ s}} = 105\text{ N}$$

Solved Problem 7.4.   A 0.45-kg baseball is thrown to the batter at 30 m s^{-1} and is hit with a velocity of 20 m s^{-1} in the direction of the pitcher. Assuming that the ball is in contact with the bat for 0.02 s, determine (1) the impulse of the batter and (2) the average force exerted on the ball by the bat.

Solution

Let's begin with the assumption that the pitcher pitches the ball along the positive x direction, and since the batter hits in the direction of the batter, that is in the negative x direction.

(1)   The impulse can be determined by

$$\begin{aligned}
\mathbf{I} = \Delta\mathbf{p} &= m(v_f - v_o)\\
&= (0.45\text{ kg})(-20\text{ m s}^{-1} - 30\text{ m s}^{-1})\\
&= (0.45\text{ kg})(-50\text{ m s}^{-1})\\
&= -22.5\text{ kg m s}^{-1}
\end{aligned}$$

(2)   The average force exerted on the ball by the bat is

$$\mathbf{I} = \mathbf{F}t = m(v_f - v_o)$$

Solving for $\mathbf{F}$ yields

$$\mathbf{F} = \frac{m(v_f - v_o)}{t} = \frac{22.5\text{ kg m s}^{-1}}{0.02\text{ s}} = 1125\text{ N}$$

Solved Problem 7.5.   Determine the force required to stop a 2000-kg car, originally traveling at a speed of 30 m s^{-1}, in 7 s.

Solution

By definition, the impulse is related to the change in linear momentum according to

$$\mathbf{I} = \mathbf{F}\,\Delta t = m\,\Delta\mathbf{v} = mv_f - mv_o$$

Solving for $\mathbf{F}$ gives

$$\mathbf{F} = \frac{(2000\text{ kg})(0\text{ m s}^{-1}) - (2000\text{ kg})(30\text{ m s}^{-1})}{7\text{ s}} = \frac{0 - 60{,}000\text{ kg m s}^{-1}}{7\text{ s}} = 8571\text{ N}$$

Solved Problem 7.6.   A 40-kg object possesses momentum of 320 kg m s^{-1}. What is the kinetic energy of the object?

Solution

The kinetic energy of the object is given by

$$KE = \frac{1}{2}mv^2$$

Also the momentum is defined as

$$\mathbf{p} = m\mathbf{v}$$

Substituting into the expression for KE gives

$$KE = \frac{p^2}{2m} = \frac{(320\text{ kg m s}^{-1})^2}{2 \cdot 40\text{ kg}} = 1280\text{ J}$$

Solved Problem 7.7. A basketball of mass 0.55 kg is shot toward the basket at a speed of 5.2 m s^{-1} at an angle $35°$ from the horizontal axis. Determine (1) the x and y components of momentum and (2) the magnitude and direction of the resultant momentum vector.

Solution

(1) The x and y components of momentum are

$$\mathbf{p}_x = mv\cos\theta = (55\text{ kg})(5.2\text{ m s}^{-1})(\cos 35°) = 0.82\text{ kg m s}^{-1}$$

$$\mathbf{p}_y = mv\sin\theta = (55\text{ kg})(5.2\text{ m s}^{-1})(\sin 35°) = 1.64\text{ kg m s}^{-1}$$

(2) The magnitude and direction of the resultant momentum vector can be determined by

$$\mathbf{p} = \sqrt{(\mathbf{p}_x)^2 + (\mathbf{p}_y)^2} = \sqrt{(0.82\text{ kg m s}^{-1})^2 + (1.64\text{ kg m s}^{-1})^2} = 1.83\text{ kg m s}^{-1}$$

$$\theta = \tan^{-1}\frac{\mathbf{p}_y}{\mathbf{p}_x} = \tan^{-1}\frac{1.64\text{ kg m s}^{-1}}{0.82\text{ kg m s}^{-1}} = 63.4°$$

Solved Problem 7.8. Herb of mass $m = 60$ kg is a stuntman who, in one stunt, falls from a height h off a building and lands on his heels. As a result, he suffers a fractured leg. Assuming that the force F required to cause a bone fracture is 1×10^{10} dyn and that the impact time Δt was 0.01 s, determine the height from which Herb originated his stunt.

Solution

From Illustrative Example 7.1,

$$\mathbf{F} = \frac{m}{\Delta t}\sqrt{2gh}$$

Solving for h, we can determine the height of the fall by appropriate substitution of the values provided:

$$h = \frac{1}{2g}\left(\frac{\mathbf{F}\,\Delta t}{m}\right)^2 = \frac{1}{2\cdot 980\text{ cm s}^{-2}}\left(\frac{1\times 10^{10}\text{ dyn} \cdot 10^{-2}\text{ s}}{70\times 10^3\text{ g}}\right)^2 = 1041\text{ cm} = 10.41\text{ m}$$

7.2 CONSERVATION OF LINEAR MOMENTUM

If the sum or resultant of all external forces acting on a system is zero, then the total linear momentum of the system remains unchanged, or

$$\Sigma\mathbf{p} = \mathbf{p}_1 + \mathbf{p}_2 + \mathbf{p}_3 + \cdots + \mathbf{p}_n = \text{constant}$$

A commn example of the conservation of linear momentum is the collision of two bodies. Two types of collisions are possible, depending on whether linear momentum and kinetic energy are conserved: elastic and inelastic collisions.

7.2.1 Elastic Collisions

An *elastic collision* is a collision in which two bodies approach each other, collide, and part upon impact with a different velocity in a different direction. In an elastic collision, both linear momentum and kinetic energy are conserved. For a system of two bodies which undergo an elastic collision, the following can be stated:

Total linear momentum before collision = total linear momentum after collision

$$m_{1B}\mathbf{v}_{1B} + m_{2B}\mathbf{v}_{2B} = m_{1A}\mathbf{v}_{1A} + m_{2A}\mathbf{v}_{2A}$$

This relation assumes that the two bodies are in one-dimensional motion, i.e., along the x and $-x$ axes. In more realistic collisions, the bodies are traveling in two dimensions. Since linear momentum is conserved, the resultant components of the momentum (or velocity) must be determined and applied according to the conservation laws.

7.2.2 Inelastic Collisions

An *inelastic collision* is one in which two bodies approach each other, collide, stick together, and part as a single body with a different velocity in a different direction. In an inelastic collision, only linear momentum is conserved while some kinetic energy is transformed to internal energy. For a system of two bodies which undergo an inelastic collision, the following can be stated:

Total linear momentum before collision = total linear momentum after collision

$$m_{1B}\mathbf{v}_{1B} + m_{2B}\mathbf{v}_{2B} = (m_{1A} + m_{2A})\mathbf{v}_{12A}$$

The *coefficient of restitution* is the ratio between the relative velocities of the two bodies before and after the collision and is defined by

$$e = \frac{\mathbf{v}_{2A} - \mathbf{v}_{1A}}{\mathbf{v}_{1B} - \mathbf{v}_{2B}}$$

The coefficient of restitution is a pure number and reflects the ability of the two bodies to physically respond following impact. It varies between 0 (bodies stick together, or no rebound) and 1 (the relative velocities before and after impact are equal, or maximum rebound).

Solved Problem 7.9.   A 25-kg toddler standing on a 15-kg stroller jumps off and, in the process, kicks the stroller, imparting a velocity of 4.2 m s^{-1} to the wagon. Determine the velocity of the toddler.

Solution

From the conservation of momentum

$$(mv)_{\text{toddler}} = (mv)_{\text{stroller}}$$
$$m_{\text{toddler}}v_{\text{toddler}} = m_{\text{stroller}}v_{\text{stroller}}$$

Solving for v_{toddler} and making appropriate substitutions give

$$v_{\text{toddler}} = \frac{(15\text{ kg})(4.2\text{ m s}^{-1})}{25\text{ kg}} = 2.5\text{ m s}^{-1}$$

Solved Problem 7.10.   Jovita of mass 30 kg is running at a constant speed of 4.7 m s^{-1} before jumping into a 15-kg wagon. Determine the resultant speed of the wagon moving with Jovita in it.

Solution

From the conservation of momentum

$$(mv)_{\text{before}} = (mv)_{\text{after}}$$
$$m_{\text{before}}v_{\text{before}} = m_{\text{after}}v_{\text{after}}$$

Solving for v_{after} and making appropriate substitutions, we find

$$v_{\text{after}} = \frac{(30 \text{ kg})(4.7 \text{ m s}^{-1})}{30 \text{ kg} + 15 \text{ kg}} = 3.1 \text{ m s}^{-1}$$

Solved Problem 7.11. An 1800-kg minivan (car 1) traveling eastward at a constant velocity of 40 m s^{-1} collides with a 1250-kg sports car (car 2) traveling westward at a constant velocity of 35 m s^{-1}. Upon collision, the two automobiles stick together. Determine the resultant velocity after the collision.

   Solution

For an inelastic collision, the following can be stated:

$$(m_1 v_1)_{\text{before}} + (m_2 v_2)_{\text{before}} = (m_1 + m_2)v$$
$$1800 \text{ kg} \cdot 40 \text{ m s}^{-1} + 1250 \text{ kg} \cdot (-35 \text{ m s}^{-1}) = (1800 \text{ kg} + 1250 \text{ kg})v$$

The minus sign of the velocity of car 2 corresponds to its direction westward (along the negative x axis).

$$72{,}000 \text{ kg m s}^{-1} - 43{,}750 \text{ kg m s}^{-1} = 3050 \text{ kg} \cdot v$$

Solving for v yields

$$v = 9.26 \text{ m s}^{-1}$$

Solved Problem 7.12. An example of an inelastic collision is the ballistic pendulum. The ballistic pendulum consists of a wooden block suspended by a cord secured to a level surface, as shown in Figure 7-1. A projectile or bullet is fired horizontally into the wooden block and becomes embedded in the block. Assume that a 25-g bullet is fired horizontally at a 3-kg block of wood suspended by a cord. If the initial velocity of the bullet upon firing is 110 m s^{-1}, determine the height of the swing.

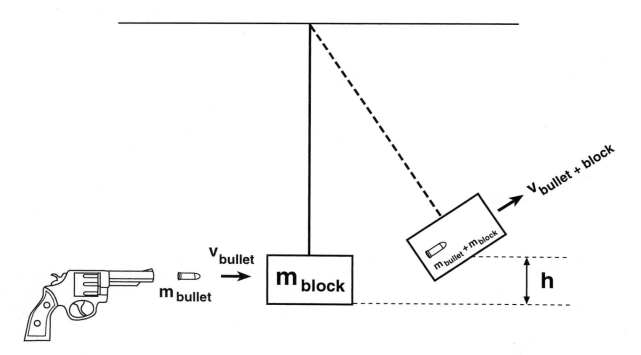

Fig. 7-1

Solution

The ballistic pendulum is in an inelastic collision, and by conservation of momentum, it can be stated that

$$(mv)_{\text{bullet}} + (mv)_{\text{block}} = (m_{\text{block}} + m_{\text{bullet}}) v_{\text{block+bullet}}$$

We need to determine the velocity of the block + bullet system following impact. Making the appropriate substitutions gives

$$(0.025 \text{ kg})(110 \text{ m s}^{-1}) + (3 \text{ kg})(0 \text{ m s}^{-1}) = (0.025 \text{ kg} + 3 \text{ kg}) v_{\text{block+bullet}}$$

Solving for $v_{\text{bullet+block}}$ gives

$$v_{\text{bullet+block}} = 0.91 \text{ m s}^{-1}$$

Upon impact with the block, the bullet possesses kinetic energy which is converted to potential energy during the movement of the pendulum. The height of this swing can be determined through the conservation of energy:

$$\text{PE} = \text{KE}$$
$$(m_{\text{bullet}} + m_{\text{block}}) gh = \frac{1}{2} (m_{\text{bullet}} + m_{\text{block}}) v_{\text{bullet+block}}^2$$

This can be simplified to

$$gh = \frac{1}{2} v_{\text{bullet+block}}^2$$

Solving for h and making substitutions yield

$$h = 4.2 \text{ cm}$$

Solved Problem 7.13.   A ballistic pendulum consists of a 10-g bullet fired into a 10-kg block of wood. If the velocity of the bullet + block is 0.6 m s^{-1}, determine the initial velocity of the bullet.

Solution

Conservation of momentum applied to the system before and after impact can be written as

$$\text{Momentum before impact} = \text{momentum after impact}$$
$$(mv)_{\text{bullet}} + (mv)_{\text{block}} = (mv)_{\text{bullet+block}}$$
$$(0.010 \text{ kg})(v_{\text{bullet}}) + (10 \text{ kg})(0 \text{ m s}^{-1}) = (0.010 \text{ kg} + 10 \text{ kg})(0.6 \text{ m s}^{-1})$$

Solving for v_{bullet} gives

$$v_{\text{bullet}} = 600.6 \text{ m s}^{-1}$$

Solved Problem 7.14.   A 2000-kg car traveling west at 40 m s^{-1} collides with a 4500-kg truck originally at rest. Determine the speed of the truck following impact if the car is stopped at the point of impact.

Solution

Applying the conservation of momentum, we have

$$mv_{\text{before impact}} = mv_{\text{after impact}}$$
$$(mv)_{b,\text{car}} + (mv)_{b,\text{truck}} = (mv)_{a,\text{car}} + (mv)_{a,\text{truck}}$$

where

$$m_{b,\text{car}} = m_{a,\text{car}} = 2000 \text{ kg}$$
$$v_{b,\text{car}} = 40 \text{ m s}^{-1} \qquad v_{a,\text{car}} = 0 \text{ m s}^{-1}$$
$$v_{b,\text{truck}} = 0 \text{ m s}^{-1} \qquad v_{a,\text{truck}} = \text{unknown}$$

Making the appropriate substitutions and solving for $v_{a,\text{truck}}$ yield

$$v_{a,\text{truck}} = 17.8 \text{ m s}^{-1}$$

Solved Problem 7.15. A 0.8-kg ball traveling in the positive x direction with a speed of 70 cm s^{-1} collides with a 0.5-kg ball traveling in the negative x direction with a speed of 50 cm s^{-1}. After collision, they both veer at 30° from the x axis along their respective directions of motion with the 0.8-kg ball attaining a new velocity of 35 cm s^{-1}. (1) Determine the final velocity of the 0.8-kg ball. (2) Is this collision perfectly elastic?

Solution

(1) We want to employ the conservation of momentum in two dimensions and determine the components of the final velocity:

x-axis: $(0.8 \text{ kg})(0.7 \text{ m s}^{-1}) - (0.5 \text{ kg})(0.5 \text{ m s}^{-1}) = 0.8 \text{ kg} (0.35 \text{ m s}^{-1})(\cos 30°) + (0.5 \text{ kg}) v_x$

$$v_x = 0.14 \text{ m s}^{-1}$$

y-axis: $0 = 0.8 \text{ kg} (0.35 \text{ m s}^{-1})(\sin 30°) + (0.5 \text{ kg}) v_y$

$$v_y = 0.28 \text{ m s}^{-1}$$

Thus, the magnitude and direction of the velocity are

$$v = \overline{\sqrt{v_x^2 + v_y^2}} = \overline{\sqrt{(0.14 \text{ m s}^{-1})^2 + (0.28 \text{ m s}^{-1})^2}} = 0.31 \text{ m s}^{-1}$$

$$\theta = \tan^{-1} \frac{v_y}{v_x} = \tan^{-1} \frac{0.28 \text{ m s}^{-1}}{0.14 \text{ m s}^{-1}} = 63.4°$$

(2) To determine whether the collision is perfectly elastic, the kinetic energy is calculated before and after the collision and compared. The collision is perfectly elastic if $KE_{before} = KE_{after}$.
Total KE before collision:

$$\frac{1}{2}(0.8 \text{ kg})(0.7 \text{ m s}^{-1})^2 + \frac{1}{2}(0.5 \text{ kg})(-0.5 \text{ m s}^{-1})^2 = 0.196 \text{ J}$$

Total KE after collision:

$$\frac{1}{2}(0.8 \text{ kg})(0.35 \text{ m s}^{-1})^2 + \frac{1}{2}(0.5 \text{ kg})(-0.31 \text{ m s}^{-1})^2 = 0.073 \text{ J}$$

Therefore, $KE_{before} \neq KE_{after}$, implying that KE is lost as a result of the collision, most likely in the form of heat, and thus the collision is not perfectly elastic.

Solved Problem 7.16. Two balls are involved in a collision where $m_1 = m_2$ but $r_1 = 2r_2$. The initial velocity of ball 1 is 4 m s^{-1} traveling east with that of ball 2 being 8 m s^{-1} traveling west. Calculate the final velocity if the collision is (1) perfectly elastic and (2) inelastic.

Solution

(1) Since this is a one-dimensional collision, ball 1 is in the positive x direction and ball 2 is in the negative x direction. For a perfectly elastic collision, we can use the conservation of momentum and kinetic energy:

Conservation of momentum

$$\text{Momentum before} = \text{momentum after}$$

$$m_{1b} v_{1b} + m_{2b} v_{2b} = m_{1a} v_{1a} + m_{2a} v_{2a}$$

$$m_{1b}(4 \text{ m s}^{-1}) + m_{2b}(-8 \text{ m s}^{-1}) = m_{1a} v_{1a} + m_{2a} v_{2a}$$

Since $m_1 = m_2$, all m quantities cancel, leaving

$$-4 \text{ m s}^{-1} = v_{1a} + v_{2a} \qquad\qquad (1)$$

Conservation of kinetic energy

$$\text{Kinetic energy before} = \text{kinetic energy after}$$

$$\frac{1}{2} m_{1b} v_{1b}^2 + \frac{1}{2} m_{2b} v_{2b}^2 = \frac{1}{2} m_{1a} v_{1a}^2 + \frac{1}{2} m_{2a} v_{2a}^2$$

This equation reduces to

$$40 \text{ m}^2 \text{ s}^{-2} = v_{1a}^2 + v_{2a}^2 \qquad (2)$$

From Eq. (1),

$$v_{2a} = -4 \text{ m s}^{-1} - v_{1a}$$

Substituting into Eq. (2) gives

$$40 \text{ m}^2 \text{ s}^{-2} = v_{1a}^2 + (-4 \text{ m s}^{-1} - v_{1a})^2$$
$$40 \text{ m}^2 \text{ s}^{-2} = v_{1a}^2 + (16 + 8v_{1a} + v_{1a}^2) \text{ m}^2 \text{ s}^{-2}$$

This equation can be rearranged to yield a polynomial (quadratic) equation:

$$2v_{1a}^2 + 8v_{1a} - 24 = 0$$

whose solution is given by

$$v_{1a} = \frac{-b \pm \sqrt{b^2 - 4ac}}{2a}$$

where $a = 2$, $b = 8$, and $c = -24$. Making the substitutions yields

$$v_{1a} = \frac{-8 \pm \sqrt{8^2 - (4)(2)(-24)}}{2(2)} = \frac{-8 \pm 16}{4} \text{ m s}^{-1}$$

The two possible solutions are

$$v_{1a} = \frac{-8 + 16}{4} \text{ m s}^{-1} = 2 \text{ m s}^{-1}$$

$$v_{1a} = \frac{-8 - 16}{4} \text{ m s}^{-1} = -6 \text{ m s}^{-1}$$

Although both solutions could be acceptable, we must analyze conditions of the collision stated in the problem which will assist us in the selection of the proper solution. Since ball 1 is traveling east and is involved in an elastic collision, ball 1 will recoil by traveling in an exactly opposite direction, i.e., the velocity must be negative or different in direction from the initial velocity. Thus the correct solution is

$$v_{1a} = -6 \text{ m s}^{-1}$$

(2) For an inelastic collision, the required equations consist of one:

$$\text{Momentum before} = \text{momentum after}$$
$$m_{1b}v_{1b} + m_{2b}v_{2b} = (m_{1a} + m_{2a})v$$
$$m_{1b}(4 \text{ m s}^{-1}) + m_{2b}(-8 \text{ m s}^{-1} = (m_{1a} + m_{2a})v$$
$$v = -2 \text{ m s}^{-1}$$

Solved Problem 7.17. A 2.5-kg ball (ball A) rolling on a frictionless surface in an eastward direction with a velocity of 7 m s^{-1} collides head on with a 40-kg ball (ball B) moving in the opposite direction at 16 m s^{-1}. Determine the velocity of both ball A and ball B if (1) $e = 0$ and (2) $e = 0.5$.

Solution

For both cases, the conservation of momentum can be applied:

$$\text{Momentum}_{\text{before collision}} = \text{momentum}_{\text{after collision}}$$
$$(2.5 \text{ kg})(7 \text{ m s}^{-1}) + (4.0 \text{ kg})(-16 \text{ m s}^{-1}) = (2.5 \text{ kg})v_{\text{after},A} + (4.0 \text{ kg})v_{\text{after},B}$$
$$-46.5 \text{ kg m s}^{-1} = (2.5 \text{ kg})v_{\text{after},A} + (4.0 \text{ kg})v_{\text{after},B}$$

(1) For $e = 0$ (a perfectly inelastic collision), the coefficient of restitution can be rewritten as

$$0 = \frac{v_{\text{after},B} - v_{\text{after},A}}{7 \text{ m s}^{-1} - (-16 \text{ m s}^{-1})} = \frac{v_{\text{after},B} - v_{\text{after},A}}{23 \text{ m s}^{-1}}$$

Because we have an inelastic collision,

$$v_{\text{after},B} = v_{\text{after},A} = v$$

The momentum equation can then be rewritten as

$$-46.5 \text{ kg m s}^{-1} = (2.5 \text{ kg})v + (4.0 \text{ kg})v = (6.5 \text{ kg})v$$

Solving for v yields

$$v = 7.15 \text{ m s}^{-1}$$

(2)   The coefficient of restitution can be written as

$$0.5 = \frac{v_{\text{after},B} - v_{\text{after},A}}{23 \text{ m s}^{-1}}$$
$$11.5 \text{ m s}^{-1} = v_{\text{after},B} - v_{\text{after},A}$$
$$v_{\text{after},B} = 11.5 \text{ m s}^{-1} + v_{\text{after},A}$$

Substituting into the momentum equation gives

$$-46.5 \text{ kg m s}^{-1} = (2.5 \text{ kg})v_{\text{after},A} + (4.0 \text{ kg})(11.5 \text{ m s}^{-1} + v_{\text{after},A})$$
$$-92.5 \text{ kg m s}^{-1} = (6.5 \text{ kg})v_{\text{after},A}$$

Solving for $v_{\text{after},A}$ gives

$$v_{\text{after},A} = -14.2 \text{ m s}^{-1}$$

Accordingly,

$$v_{\text{after},B} = 11.5 \text{ m s}^{-1} - 14.2 \text{ m s}^{-1} = -2.7 \text{ m s}^{-1}$$

Supplementary Problems

7.1.   Determine the momentum of an 850-N linebacker approaching the quarterback at a speed of 12 ft s^{-1}.

Solution

$$\mathbf{p} = 736.9 \text{ kg m s}^{-1}$$

7.2.   Determine the muzzle velocity of a 20.0-g bullet possessing momentum of 11.0 kg m s^{-1}.

Solution

$$\mathbf{v} = 550 \text{ m s}^{-1}$$

7.3.   If a rocket upon launch exerts a thrust of 500,000 N, determine the time required to generate an impulse of 1.5×10^6 N s.

Solution

$$\Delta t = 3.0 \text{ s}$$

7.4.   A 9.5-kg rifle fires a 15-g bullet with a muzzle velocity of 750 m s^{-1}. Determine the recoil velocity of the rifle.

Solution

$$v = 1.18 \text{ m s}^{-1}$$

7.5. A 2.5-kg object possessing momentum of $\mathbf{p} = 4\mathbf{i} + 3\mathbf{j}$ collides with the ground in 0.01 s before coming to a complete stop. Determine (1) the x and y components of momentum, (2) the magnitude and direction of momentum, (3) the velocity of the object, and (4) the force exerted by the object on the ground.

Answer

(1) $p_x = 4\,\mathrm{kg\,m\,s^{-1}}$ and $p_y = 3\,\mathrm{kg\,m\,s^{-1}}$

(2) $|\mathbf{p}| = 5\,\mathrm{kg\,m\,s^{-1}}$, $\theta = 36.8°$

(3) $\mathbf{v} = 2.0\,\mathrm{m\,s^{-1}}$

(4) $\mathbf{F} = 500\,\mathrm{N}$

7.6. In assembling a wooden toy, a toymaker strikes a nail with a 6.0-kg hammer moving at a speed of $15\,\mathrm{m\,s^{-1}}$, driving the nail through the wood. Assuming the hammer hits the nail in 0.002 s, determine the (1) impulse and (2) average force exerted on the nail.

Answer

(1) $\mathbf{I} = -90\,\mathrm{kg\,m\,s^{-1}}$

(2) $\mathbf{F} = 4.5 \times 10^4\,\mathrm{N}$

7.7. A golfer strikes a 0.05-kg golf ball positioned on a tee with a force of 150 N, and the ball departs with a speed of $40\,\mathrm{m\,s^{-1}}$. Determine the (1) impulse and (2) time of impact.

Answer

(1) $\mathbf{I} = 2\,\mathrm{N\,s}$

(2) $\Delta t = 0.013\,\mathrm{s}$

7.8. A 5-kg ball (ball 1) is rolling on a frictionless surface when it collides elastically with a slightly larger ball (ball 2) of unknown mass initially at rest. If ball 1 rebounds from the collision with a speed one-third its original speed, what is the mass of ball 2?

Answer

$$m = 10\,\mathrm{kg}$$

7.9. A 4-kg ball, initially moving with a speed of $10\,\mathrm{m\,s^{-1}}$ along the positive x axis, inelastically collides with a 7-kg block, initially at rest. Determine the velocity of the ball + block system following the collision.

Answer

$$3.6\,\mathrm{m\,s^{-1}}$$

7.10. A small car of mass $M_1 = 1200\,\mathrm{kg}$ traveling north (along the positive y axis) at a constant speed of $v_1 = 12\,\mathrm{m\,s^{-1}}$ collides with a larger car of mass $M_2 = 2400\,\mathrm{kg}$ traveling east (along the positive x axis) at a constant speed of $v_2 = 8\,\mathrm{m\,s^{-1}}$. Determine the magnitude and direction of the total momentum just prior to the collision.

Answer

$$\mathbf{p} = 2.36 \times 10^4\,\mathrm{kg\,m\,s^{-1}}, \theta = 36.4°$$

7.11. The nucleus of a radioactive atom of mass $M_A = 3.65 \times 10^{-25}\,\mathrm{kg}$, initially at rest, undergoes alpha decay where it spontaneously emits an alpha particle of mass $M_B = 6.65 \times 10^{-27}\,\mathrm{kg}$ at a speed of $v_B = 6.25 \times 10^7\,\mathrm{m\,s^{-1}}$. Determine the recoil velocity of the radioactive nucleus.

Solution

$$v_A = 1.16 \times 10^6 \, \mathrm{m\,s^{-1}}$$

7.12. Playing a game of pool, Elizabeth strikes a cue ball of mass 5.0 kg, imparting it with a velocity of 3.5 m s^{-1}. The cue ball collides with the eight ball, also of mass 5.0 kg and originally at rest. After the collision, the cue ball moves with a velocity of 2.0 m s^{-1} in a direction at an angle $\theta = 40°$ from the positive x axis. Determine the magnitude and direction of the velocity of the eight ball.

Solution

$$v_{8,\mathrm{after}} = 2.34 \, \mathrm{m\,s^{-1}} \qquad \theta = -33.1° \quad \text{from positive } x \text{ axis}$$

7.13. A tennis ball is dropped from a height $h = 1.5$ m and bounces several times off the floor. If the coefficient of restitution is 0.9, how high does the tennis ball rebound from the first impact?

Solution

$$h = 1.2 \, \mathrm{m}$$

7.14. A bowling ball is dropped from a height $h_1 = 1$ m onto a wooden floor and rebounds to a height $h_2 = 0.2$ m. Determine the coefficient of restitution.

Solution

$$e = 0.45$$

Chapter 8

Rotational Motion

Previous chapters considered the motion and resultant dynamics of an object constrained to a two-dimensional or planar coordinate system and traveling along a directed linear path. Rotational motion refers to particle motion in an angular or circular orbit of radius r about a fixed axis, as opposed to linear motion along a particular direction. Although the physics involved in linear and rotational motion is the same, the expressions are different in that motion is characterized in terms of the orbit radius.

8.1 DEFINITIONS OF ROTATIONAL MOTION

Angular Motion

The physical quantities of linear (translational) and angular (rotational) motion are related according to the following expression:

$$\text{Linear quantity} = \text{angular quantity} \times \text{radius}$$

Angular displacement is an angular change in object position in angular motion and represents the arc of an angular segment corresponding to the area swept out by the radius. Angular displacement θ is defined as

$$\theta = \frac{x}{r}$$

where x is the linear displacement of the object and r is the radius of angular motion.

Measurement of angular displacement is given in degrees (deg or °), revolutions (rev), or radians (rad), and they are related according to

$$360° = 1 \text{ rev} = 2\pi \text{ rad}$$

$$1 \text{ rad} = 57.3° = \frac{180°}{\pi \text{ rad}}$$

All units of angular displacement are pure numbers and have no physical dimension.

Angular velocity is the time rate of change in angular displacement or the angle in radians that would be swept by the radius over 1 s, and it is related to linear or tangential velocity v by

$$\omega = \frac{v}{r}$$

Tangential velocity is the velocity of the object if released from rotational motion and directed tangentially from the circular or angular orbit. This is depicted in Figure 8-1. For this relation to hold, the angular entity, or ω, must be expressed in radians. If an object undergoes an angular displacement from θ_o to θ_f during a time t then the average angular speed ω is

$$\omega = \frac{\theta_f - \theta_o}{t}$$

Assuming a constant angular speed, the angular velocity (sometimes referred to as *angular frequency*) ω is defined as the number of orbits or complete trips around the entire circular path in 1 s. Since the distance traveled

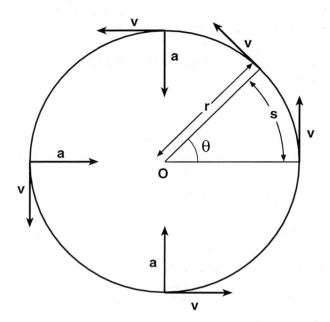

Fig. 8-1

by an object through one orbit or revolution is $2\pi r$, the angular velocity can also be expressed as

$$\omega = \frac{2\pi}{T} = 2\pi\nu \ (\text{rad s}^{-1})$$

where ν is the frequency of rotation and T is the period or time taken to complete 1 rev. Angular velocity is a vector quantity directed tangent to particle motion and has dimensions of T^{-1} given in units of deg s^{-1}, rev s^{-1}, or rad s^{-1}.

Angular acceleration is the time rate of change in angular velocity and is related to linear or tangential acceleration a by

$$\alpha = \frac{a}{r}$$

If the angular motion of an object, originally at an angular velocity ω_o, changes to an angular velocity ω_f during a time t, then the average angular acceleration α is

$$\alpha = \frac{\omega_f - \omega_o}{t}$$

Angular acceleration is a vector quantity directed tangent to particle motion, and it has dimensions of T^{-2} given in units of deg s^{-2}, rev s^{-2}, or rad s^{-2}.

Solved Problem 8.1.   Determine the angular velocity of (1) the daily rotation of the earth, (2) a watch hour hand, (3) a watch minute hand.

   Solution

   (1)   The earth makes a complete rotation or 1 rev each day or 24 h. We want to convert this angular velocity to rad s^{-1}.

$$\omega = \frac{1 \text{ rev}}{24 \text{ h}} \cdot \frac{2\pi \text{ rad}}{1 \text{ rev}} \cdot \frac{1 \text{ h}}{3600 \text{ s}} = 7.27 \times 10^{-5} \text{ rad s}^{-1}$$

(2)   The hour hand of the watch makes a complete revolution in 12 h. Thus, the angular velocity is

$$\omega = \frac{1 \text{ rev}}{12 \text{ h}} \cdot \frac{2\pi \text{ rad}}{1 \text{ rev}} \cdot \frac{1 \text{ h}}{3600 \text{ s}} = 1.45 \times 10^{-4} \text{ rad s}^{-1}$$

(3)   The minute hand of a clock makes a complete rotation in 1 h. Thus, the angular velocity is

$$\omega = \frac{1 \text{ rev}}{1 \text{ h}} \cdot \frac{2\pi \text{ rad}}{1 \text{ rev}} \cdot \frac{1 \text{ h}}{3600 \text{ s}} = 1.74 \times 10^{-3} \text{ rad s}^{-1}$$

Solved Problem 8.2.   In performing a simple experiment on a lazy Susan (rotating circular tray), Claude observed that the linear velocity of a cupcake (cupcake 1) at the edge was 2.5 times greater than the linear velocity of a cupcake (cupcake 2) 5 m closer to the center. Find the radius of the lazy Susan.

Solution

From the information in the problem and the expression relating linear velocity and angular velocity

$$v = \omega r$$

it can be stated that

Cupcake 1: $2.5v = \omega r$

Cupcake 2: $v = \omega(r - 5 \text{ cm})$

where r is the radius of the lazy Susan. Substituting the expression for v, given for cupcake 2, into the expression for cupcake 1 yields

$$2.5\omega(r - 5 \text{ cm}) = \omega r$$

Expanding the relation gives

$$2.5\omega r - 2.5\omega(5 \text{ cm}) = \omega r$$

Eliminating ω and solving for r yield

$$r = \frac{12.5 \text{ cm}}{1.5} = 8.33 \text{ cm}$$

Solved Problem 8.3.   According to Bohr's theory of an atom, the electron moves in an atom of hydrogen along a circular orbit with a constant velocity v. Determine the angular velocity of the electron in orbit about the nucleus and its normal acceleration. The radius of the orbit $r = 0.5 \times 10^{-10}$ m, and the linear velocity of the electron about this orbit is $v = 2.2 \times 10^6 \text{ m s}^{-1}$.

Solution

The angular velocity ω is related to the linear velocity v and radius of angular motion r, by

$$v = \omega r$$

Solving for ω gives

$$\omega = \frac{v}{r} = \frac{2.2 \times 10^6 \text{ m s}^{-1}}{0.5 \times 10^{-10} \text{ m}} = 4.4 \times 10^{16} \text{ rad s}^{-1}$$

The normal angular acceleration α of the electron in this orbit is

$$\alpha = \frac{v^2}{r} = \frac{(2.2 \times 10^6 \text{ m s}^{-1})^2}{0.5 \times 10^{-10} \text{ m}} = \frac{4.84 \times 10^{12} \text{ m}^2 \text{ s}^{-2}}{0.5 \times 10^{-10} \text{ m}} = 9.7 \times 10^{22} \text{ m s}^{-2}$$

Centripetal Acceleration

As an object of mass m positioned on the rim of a rotating wheel of radius r undergoes uniform angular motion, the velocity is constant in magnitude but is continually changing direction along the curvature of the circular orbit. An object whose velocity is changing is also accelerating. This acceleration, called *centripetal* or "center-seeking," is perpendicular, or normal, to the velocity vector and directed toward the center of rotation. The magnitude of centripetal acceleration is defined as

$$a_c = \frac{v^2}{r}$$

and has units typical of acceleration, that is, LT^{-2}.

Centripetal Force

Since the object in rotational motion has mass and is accelerating, there exists a force acting on the object given by Newton's second law:

$$F = ma = m\frac{v^2}{r}$$

Force F is the centripetal force and is directed radially inward to the center of rotation.

ILLUSTRATIVE EXAMPLE 8.1.   Blood Flow in Curved and Tortuous Vessels

The carotid artery, the major blood vessel that supplies blood to the brain, originates from the aorta and passes through the neck, where it branches or bifurcates into two other major arteries which ultimately supply the front (anterior) and back (posterior) circulation of the brain. There exist a number of points along the human carotid artery which exhibit curvature, particularly before and following the major bifurcation apex. The importance of vessel curvature in the carotid artery can be demonstrated by considering the hemodynamic parameter of wall shear stress. The tails of the velocity profile (the points of the profile in contact with the vessel wall) exhibit the maximum wall shear stress. As blood flows through a curved portion of a blood vessel, the velocity profile becomes skewed away from the radius of vessel curvature, as shown in Figure 8-2. This phenomenon occurs in common everyday experiences such as driving a car or bicycle around a curve or sledding in a luge (Figure 8-3). Either example typically involves circular motion of an object or person which is maintained by a centripetal or "center-seeking" force, given by

$$F_{\text{centripetal}} = ma = \frac{mv^2}{r}$$

As r decreases (e.g., the curve becomes sharper), the centripetal force increases accordingly. The centripetal force is counteracted by a centrigugal force of the same magnitude, which is always directed outward from and perpendicular to

Fig. 8-2

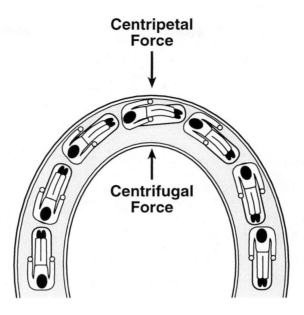

Fig. 8-3

the axis of rotation. As a car or sled encounters a curve, it leans into the curve to counteract the centrifugal force. It is the centrifugal force acting against the blood particles that causes a shift in the center of mass and hence the velocity profile. Once the blood flow passes the curve, the influence of the centrifugal force is removed and the velocity profile returns to its axisymmetric form.

ILLUSTRATIVE EXAMPLE 8.2.   Ultracentrifuge

The centrifuge is an instrument that uses centripetal force to separate macromolecules of different mass suspended in solution. In a centrifuge, a test tube containing the sample solution is placed into a rotor and is subjected to a high rotational speed, typically from 5000 to 50,000 rotations per minute. The macromolecules in the sample are subjected to three forces, as shown in Figure 8-4:

(1)   A centrifugal force $F_c = m\omega^2 r$ acting to propel the macromolecule away from the axis of rotation toward the bottom of the test tube, where m is the mass of the macromolecule, ω is the angular velocity, and r is the radius of motion

(2)   A frictional force $F_f = fv$ due to the motion of the particle through the fluid acting toward the axis of rotation, where f is a frictional coefficient and v is the velocity of the macromolecule in the medium (solution)

(3)   A buoyant force $F_b = (m/\rho)\rho_o \omega^2 r$, produced by the weight of the fluid displaced by the macromolecule, where ρ is the density of the macromolecule and ρ_o is the density of the medium

The centrifugal force is typically greater in magnitude than both the frictional and buoyant forces combined, causing the macromolecules in the solution to sediment toward the bottom of the test tube, in effect, separating the solute from the solvent.

Solved Problem 8.4.   Consider an athlete running around a circular track with a radius of 30 m at a uniform velocity of 10.5 m s^{-1}. Determine the (1) period and (2) centripetal acceleration of the athlete.

Solution

(1)   The period of the athlete is given by

$$T = \frac{2\pi r}{v} = \frac{2\pi \cdot 30 \text{ m}}{10.5 \text{ m s}^{-1}} = 17.9 \text{ s}$$

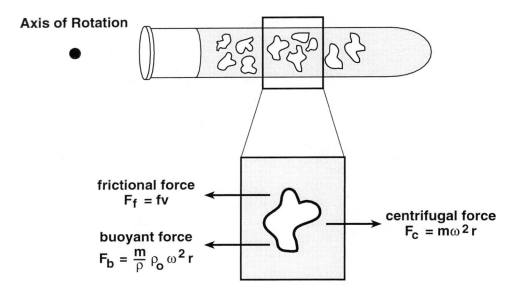

Fig. 8-4

(2)   The centripetal acceleration of the athlete is

$$a = \frac{v^2}{r} = \frac{(10 \text{ m s}^{-1})^2}{30 \text{ m}} = \frac{100 \text{ m}^2 \text{ s}^{-2}}{30 \text{ m}} = 3.33 \text{ m s}^{-2}$$

Solved Problem 8.5.   An artificial satellite is deployed into orbit 2.5×10^6 m above the surface of the earth. Determine the speed that the satellite must maintain to enter a circular orbit. (Assume the radius of the earth is 6.4×10^6 m.)

Solution

Since the orbit is steady, the gravitational attraction (force) by the earth must equal the centrifugal force due to the circular motion of the satellite:

$$m\frac{GM}{r^2} = m\frac{v^2}{r}$$

Solving for v^2 gives

$$v^2 = \frac{GM}{r} = \frac{(6.67 \times 10^{-11} \text{ N m}^2 \text{ kg}^{-2})(6 \times 10^{24} \text{ kg})}{(6400 + 2500) \times 10^3 \text{ m}^2} = 4.49 \times 10^7 \text{ m}^2 \text{ s}^{-2}$$
$$v = 6.7 \times 10^3 \text{ m s}^{-1}$$

Solved Problem 8.6.   What is the period of the satellite described in Solved Problem 8.5?

Solution

The period of the satellite in circular orbit is related to the velocity by

$$v = \frac{2\pi r}{T} \Rightarrow T = \frac{2\pi r}{v} = \frac{2 \cdot 3.14 \cdot (6.4 \times 10^6 \text{ m} + 2.5 \times 10^6 \text{ m})}{6.7 \times 10^3 \text{ m s}^{-1}} = 8.3 \times 10^3 \text{ s or 2.31 h}$$

Solved Problem 8.7.   Darlene is swinging a sling consisting of a 2.5-kg stone attached to a string 60 cm long. If the string can withstand a maximum tension of 380 N, what is the greatest speed attainable by the sling?

Solution

The centripetal force exerted by the string on the stone is given by

$$F = \frac{mv^2}{r}$$

Solving for v yields

$$v = \sqrt{\frac{Fr}{m}}$$

where v is the velocity of the stone, F is the centripetal force, r is the orbit radius of the stone and is equal to the length of the string, and m is the mass of the stone. Substituting known values gives

$$v = \sqrt{\frac{(380 \text{ N})(0.6 \text{ m})}{2.5 \text{ kg}}} = 9.5 \text{ m s}^{-1}$$

Solved Problem 8.8. Murray is driving at a constant speed when, all of a sudden, he notices a concrete barrier a distance D ahead. He is suddenly confronted with two options: rapid application of the brakes or sharp turn to avoid the barrier. Which is the safer option?

Solution

If Murray opts to slam on the brakes, the car will come to a complete stop when the kinetic energy of the car is converted to work done against friction. We can thus state

$$\text{KE} = \text{work}$$

$$\frac{1}{2}mv^2 = F_f x$$

where F_f is the force due to friction and x is the displacement of the car following application of the brakes. Solving for x yields

$$x = \frac{mv^2}{2F_f}$$

Assuming the car stops prior to colliding with the barrier, then the distance x must be less than the distance between the car and the concrete barrier, or

$$x = \frac{mv^2}{2F_f} \leq d$$

or, solving for F_f, we have

$$F_f \geq \frac{mv^2}{2d} \qquad\qquad (1)$$

If Murray decides to suddenly veer around the barrier, the centrifugal force that will be exerted on the car is the frictional force, traveling in an arc of a circle with radius R. Thus, the frictional force is given by

$$F_f = \frac{mv^2}{R}$$

Assuming that the car avoids the barrier, then the arc radius of the circle that the car must attain in order to avoid collision is

$$R = \frac{mv^2}{F_f} \leq d$$

or, solving for F_f yields

$$F_f \geqslant \frac{mv^2}{d} \qquad (2)$$

Comparing Eqs. (1) and (2) reveals that sudden application of the brakes requires half the frictional force of that of veering to avoid the barrier and is thus the safest option for Murray.

Solved Problem 8.9.   Linda is spinning her new toy, consisting of a 2.5-kg ball attached to the end of a 1.75-m cord, at 5.0 rev s^{-1}. If Linda's arm is 0.60 m in length, determine (1) the period of rotation, (2) the linear speed of the ball, and (3) the tension in the cord.

Solution

(1)   To find the period of rotation, we must first convert the angular velocity from rev s^{-1} to rad s^{-1}. Thus,

$$\omega = 5.0 \frac{\text{rev}}{\text{s}} \cdot \frac{2\pi\,\text{rad}}{1\,\text{rev}} = 10\pi \frac{\text{rad}}{\text{s}}$$

The period is related to the angular velocity by

$$T = \frac{2\pi}{\omega} = \frac{2\pi}{10\pi\,\text{rad s}^{-1}} = 0.2\,\text{s}$$

(2)   The linear velocity v is the product of angular velocity ω and the radius of the circular orbit, or

$$v = \omega \cdot r$$

where $\omega = 10\pi\,\text{rad s}^{-1}$ and the radius is the combined length of the cord and Linda's arm, or $r = 1.75\,\text{m} + 0.6\,\text{m} = 2.35\,\text{m}$. Thus, the linear velocity is

$$v = (10\pi\,\text{rad s}^{-1})(2.35\,\text{m}) = 73.8\,\text{m s}^{-1}$$

(3)   The tension in the cord is the centripetal force F_c exerted on the ball to keep it in circular motion.

$$T = F_c = \frac{mv^2}{r} = \frac{(2.5\,\text{kg})(73.8\,\text{m s}^{-1})^2}{2.35\,\text{m}} = 5794\,\text{N}$$

Solved Problem 8.10.   In a particular experiment, an ultracentrifuge with a radial distance of 10.0 cm is operated at 40,000 rev min^{-1}. Determine (1) the angular velocity of the centrifuge in rad s^{-1} and (2) the centrifugal force exerted on the macromolecule of mass M at a point 2.5, 5.0, 7.5, and 10.0 cm from the center of rotation.

Solution

(1)   The angular velocity of the centrifuge, given as 40,000 rev min^{-1}, can be converted to rad s^{-1} by

$$\omega = 40,000\,\text{rev min}^{-1} \cdot \frac{2\pi\,\text{rad}}{1\,\text{rev}} \cdot \frac{1\,\text{min}}{60\,\text{s}} = 4186.6\,\text{rad s}^{-1}$$

(2)   The centripetal force is given by $F_c = M\omega^2 r$. By substituting the appropriate values,

$$r = 2.5\,\text{cm} \qquad F_c = M\omega^2 r = M(4186.6\,\text{rad s}^{-1})^2(2.5\,\text{cm}) = 4.383 \times 10^7\,\text{cm s}^{-2} \times M$$

$$r = 5.0\,\text{cm} \qquad F_c = M\omega^2 r = M(4186.6\,\text{rad s}^{-1})^2(5.0\,\text{cm}) = 8.765 \times 10^7\,\text{cm s}^{-2} \times M$$

$$r = 7.5\,\text{cm} \qquad F_c = M\omega^2 r = M(4186.6\,\text{rad s}^{-1})^2(7.5\,\text{cm}) = 1.315 \times 10^8\,\text{cm s}^{-2} \times M$$

$$r = 10.0\,\text{cm} \qquad F_c = M\omega^2 r = M(4186.6\,\text{rad s}^{-1})^2(10.0\,\text{cm}) = 1.753 \times 10^8\,\text{cm s}^{-2} \times M$$

Solved Problem 8.11.   Suppose that a test tube containing a known solution is inserted into a centrifuge. Determine the magnitude and direction of the resultant force acting on the particles of unit mass suspended in the solution if the centrifugal acceleration is (1) 10 g, (2) 100 g, (3) 1000 g, and (4) 10,000 g. (Neglect the contributions of the frictional and buoyant forces acting on the particles in solution.)

Solution

There are, in effect, two forces acting on the particles subjected to the centrifugal force in the centrifuge. The centrifugal force F_c is acting radially or horizontally from the rotation axis, and the force of gravity F_g is acting downward. Thus, the resultant force F_{res} is

$$F_{res} = \overline{)(F_c)^2 + (F_g)^2}$$

(1) For a centrifugal acceleration of 10 g, the resultant force acting on the particles is

$$F_{res} = \overline{)(10\ g)^2 + (-1\ g)^2} = \overline{)101\ g^2} = 10.0\ g$$

$$\theta_{res} = \tan^{-1}\left(\frac{-1\ g}{10\ g}\right) = -5.7° \text{ or } 357.3° \qquad \text{from positive } x \text{ axis}$$

(2) For a centrifugal acceleration of 100 g, the resultant force acting on the particles is

$$F_{res} = \overline{)(100\ g)^2 + (-1\ g)^2} = \overline{)10,001\ g^2} = 100\ g$$

$$\theta_{res} = \tan^{-1}\left(\frac{-1\ g}{100\ g}\right) = -0.57° \text{ or } 359.4° \qquad \text{from positive } x \text{ axis}$$

(3) For a centrifugal acceleration of 1000 g, the resultant force acting on the particles is

$$F_{res} = \overline{)(1000\ g)^2 + (-1\ g)^2} = \overline{)1,000,001\ g^2} = 1000\ g$$

$$\theta_{res} = \tan^{-1}\left(\frac{-1\ g}{1000\ g}\right) = -0.057° \text{ or } 359.9° \qquad \text{from positive } x \text{ axis}$$

(4) For a centrifugal acceleration of 10,000 g, the resultant force acting on the particles is

$$F_{res} = \overline{)(10,000\ g)^2 + (-1\ g)^2} = \overline{)100,000,001\ g^2} = 10,000\ g$$

$$\theta_{res} = \tan^{-1}\left(\frac{-1\ g}{10,000\ g}\right) = -0.0057° \text{ or } 359.9° \qquad \text{from positive } x \text{ axis}$$

Solved Problem 8.12.   Alexandra of mass 35 kg is swinging within a circular arc of 4.5-m radius on a tire suspended by a rope secured to a large tree. If her horizontal speed is 5 m s^{-1} at the lowest point of the arc, determine the tension in the rope.

Solution

At the lowest point of the arc, the forces acting on Alexandra can be represented in equation form as

$$T - mg = ma$$

where a is the centripetal acceleration a_c:

$$a_c = \frac{v^2}{r}$$

$$T = m\frac{v^2}{r} + mg$$

Substituting appropriate values gives

$$T = (35\ kg)\frac{(5\ m\ s^{-1})^2}{4.5\ m} + (35\ kg)(9.8\ m\ s^{-2}) = 194.4\ N + 343.0\ N = 537.4\ N$$

Equations of Rotational Motion

The equations of motion of an object in rotational motion are analogous to those describing an object in linear motion.

Rotational Motion	**Linear Motion**
$\theta = \omega t$	$x = vt$
$\theta = \omega_o t + \dfrac{1}{2}\alpha t^2$	$x = v_o t + \dfrac{1}{2}at^2$
$\omega_f^2 = \omega_o^2 + 2\alpha\theta$	$v_f^2 = v_o^2 + 2ax$
$\alpha = \dfrac{\omega_f - \omega_o}{t}$	$a = \dfrac{v_f - v_o}{t}$
$\omega = \dfrac{\omega_f + \omega_o}{2}$	$a = \dfrac{v_f + v_o}{2}$

Solved Problem 8.13. A rotating flywheel reduces its velocity from 450 to 150 rev min^{-1} during 2 min. Find the angular acceleration of the braking flywheel and the number of revolutions it completes during this time.

Solution

The angular acceleration can be determined from the equation of motion:

$$\alpha = \frac{\omega_f - \omega_o}{t}$$

Substituting appropriate values yields

$$\alpha = \frac{\omega_f - \omega_o}{t} = \frac{150 \text{ rev min}^{-1} - 450 \text{ rev min}^{-1}}{2 \text{ min}} = -150 \text{ rev min}^{-2}$$

Converting to rad s^{-2} gives

$$-150 \text{ rev min}^{-2} \cdot \frac{2\pi \text{ rad}}{1 \text{ rev}} \cdot \frac{1 \text{ min}^2}{3600 \text{ s}^2} = -0.26 \text{ rad s}^{-2}$$

The negative acceleration corresponds to the braking motion of the rotating flywheel.

The number of revolutions made during this time can be determined from the equation of motion:

$$\theta = \omega t = \frac{\omega_f + \omega_o}{2} t = \frac{450 \text{ rev min}^{-1} + 150 \text{ rev min}^{-1}}{2}(2 \text{ min}) = 600 \text{ rev}$$

Solved Problem 8.14. A rotating fan, originally operating at a velocity of 1100 rev min^{-1}, is switched off and undergoes 90 rev before it comes to a complete stop. How much time has elapsed from the moment the fan was turned off to the moment the fan stopped?

Solution

This problem can be solved by the equation of motion:

$$\theta = \omega t = \frac{\omega_f + \omega_o}{2} t = \frac{1100 \text{ rev min}^{-1} + 0}{2} t = (550 \text{ rev min}^{-1})t = 90 \text{ rev}$$

Solving for t gives

$$t = \frac{90 \text{ rev}}{550 \text{ rev min}^{-1}} = 0.16 \text{ min} = 9.8 \text{ s}$$

Solved Problem 8.15. A fan with a radius $r = 20$ cm rotates with a constant angular acceleration $\alpha = 3.5$ rad s^{-2}. For a point on the rim, find at the end of the first second after motion has begun (1) the angular velocity and (2) the linear velocity.

Solution

For uniform rotational motion, the angular velocity ω is related to time t by the equation of motion:

$$\omega_f = \omega_o + \alpha t$$

where ω_o is 0, and the equation can be simplified to

$$\omega_f = \alpha t$$

At the end of the first second,

$$\omega_f = (3.5 \text{ rad s}^{-2})(1 \text{ s}) = 3.5 \text{ rad s}^{-1}$$

The linear velocity can be determined from $v = \omega r$, or

$$v = (3.5 \text{ rad s}^{-1})(20 \text{ cm}) = 70 \text{ cm s}^{-1}$$

Solved Problem 8.16. A wheel of radius $r = 20$ cm rotates with a constant angular acceleration $\alpha = 4.5$ rad s^{-2}. At the end of the first second of motion, determine (1) the angular velocity, (2) the linear velocity, (3) the tangential acceleration, (4) the normal acceleration, (5) the total acceleration, and (6) the angle formed by the direction of the total acceleration with the wheel radius.

Solution

(1)   The equation of motion that relates the angular velocity ω to time t is

$$\omega = \omega_o + \alpha t$$

Since $\omega_o = 0$,

$$\omega = \alpha t$$

At the end of the first second,

$$\omega = (4.5 \text{ rad s}^{-2})(1 \text{ s}) = 4.5 \text{ rad s}^{-1}$$

(2)   Since $v = \omega r$, the linear velocity is also proportional to time, and

$$v = (4.5 \text{ rad s}^{-1})(0.2 \text{ m}) = 0.9 \text{ m s}^{-1}$$

(3)   The tangential acceleration $a_t = \alpha r$ does not depend on t, that is, is constant during motion, and is

$$a_t = (4.5 \text{ rad s}^{-2})(0.2 \text{ m}) = 0.9 \text{ m s}^{-2}$$

(4)   The normal acceleration a_n is given by

$$a_n = \omega^2 r = \alpha^2 t^2 r$$

So, for $t = 1$ s,

$$a_n = (4.5 \text{ rad s}^{-2})^2 (1 \text{ s})^2 (0.2 \text{ m}) = 4.1 \text{ m s}^{-2}$$

(5)   The total acceleration a_T increases with time according to

$$a_T = \overline{)(a_t)^2 + (a_n)^2} = \overline{)(0.9 \text{ m s}^{-2})^2 + (4.1 \text{ m s}^{-2})^2} = 4.2 \text{ m s}^{-2}$$

(6)   The angle formed by the direction of the total acceleration with wheel radius is given by

$$\sin \phi = \frac{a_t}{a_n}$$

At $t = 0$, $a = a_t$ and is directed tangentially. At $t = \infty$, $a = a_n$ and is directed normally. Therefore,

$$\sin \phi = \frac{0.9 \text{ m s}^{-2}}{4.2 \text{ m s}^{-2}} = 0.21$$

$$\phi = \sin^{-1} 0.21 = 12.4°$$

Supplementary Problems

8.1. On a merry-go-round, Gordie passes his mother every 4.2 s. Determine the angular velocity of Gordie.

Solution

$\omega = 1.5 \text{ rad s}^{-1}$

8.2. A track and field athlete, running on a circular track, completed 15 laps around the track in a total time of 5.3 min. Determine the angular velocity of the athlete.

Solution

$\omega = 0.3 \text{ rad s}^{-1}$

8.3. If the athlete in the previous problem is running on a track 5.0 m in radius, determine the linear velocity of the athlete.

Solution

$v = 1.5 \text{ m s}^{-1}$

8.4. A car driving on an interstate comes across un unbanked curve of radius 40 m. Determine the maximum speed at which the car can make the turn without overturning. Assume the coefficient of friction is 0.75.

Solution

$v = 16.6 \text{ m s}^{-1}$

8.5. A boy swings a 20-kg pail of water attached to a 1.5-m cord in a vertical circle. Determine the minimum velocity of rotation of the pail required to prevent the water from spilling out of the pail.

Solution

$v = 3.8 \text{ m s}^{-1}$

8.6. A satellite is launched into a low-lying orbit near the surface of the earth. Assuming the radius of the earth = 6400 km, determine the speed that the satellite must maintain in order to remain in orbit.

Solution

$v = 7.9 \times 10^3 \text{ m s}^{-1}$

8.7. A uniformly accelerated flywheel reaches an angular velocity of $\omega = 50 \text{ rad s}^{-1}$ in $N = 20$ rev following the beginning of rotation. Determine the angular acceleration of the flywheel.

Solution

$\alpha = 9.96 \, \text{rad s}^{-2}$

8.8. A rotor of a centrifuge is 50 cm in diameter and rotates at 60,000 rev min^{-1}. Determine (1) the linear velocity and (2) the centripetal acceleration.

Solution

(1) $v = 1570 \, \text{m s}^{-1}$; (2) $a = 9.86 \times 10^6 \, \text{m s}^{-2}$

8.9. What is the maximum distance that a coin can be placed from the center of a record rotating at 78 rev min^{-1} without sliding off? (Assume the coefficient of static friction is 0.35.)

Solution

$r = 5.14 \, \text{cm}$

8.10. A record turntable is playing a record, rotating with a constant speed of 45 rev min^{-1}. Determine the time taken for the turntable to rotate through an angle of 180°.

Solution

$t = 0.67 \, \text{s}$

8.11. A certain record requires an angular speed of 78 rev min^{-1} in order to be played properly. If the turntable requires 1.75 s to attain this speed from rest, determine the (1) average acceleration of the turntable and (2) angular displacement after the first 1.75 s.

Solution

(1) $4.67 \, \text{rad s}^{-2}$; (2) 409.8°

8.12. For the record player in Problem 8.11, determine the tangential acceleration of a point 18 cm from the center.

Solution

$a_t = 84.1 \, \text{cm s}^{-2}$

8.13. A record turntable reaches its correct operating speed of 33 rev min^{-1} within 1.5 s after being turned on. Determine the magnitude of the total acceleration for a point on the edge of a record 20 cm in radius.

Solution

$a = 242.5 \, \text{cm s}^{-2}$

8.14. A car is traveling on an unbanked curve with a radius of 50 m. Determine the maximum speed that the car can accommodate without rolling over, assuming that the coefficient of static friction $\mu_s = 0.75$.

Solution

$v = 19.17 \, \text{m s}^{-2}$

8.15. A car traveling at a constant speed of 50 mi h^{-1} rounds a curve with a radius of curvature equal to 375 ft. What is the acceleration of the car?

Solution

$a = 14.3 \, \text{ft s}^{-2}$

8.16.   An object undergoes uniform circular motion within a radius of 1.5 m and a period of 4 s. Assuming that the object begins along the positive x axis and rotates in a counterclockwise direction, determine the average velocity vector in the time interval from $t = 1$ s to $t = 2$ s.

Solution

2.13 m s^{-2}; 135° from positive x axis

Chapter 9

Rotational Dynamics

In continuation of our comparison of translational motion and rotational motion, it was shown in Chapters 5 and 6 that translational forces exerted on objects can result in acceleration, which in turn causes work to be done on the object and energy to be expended by the object. Similar physical phenomena and interactions occur for objects in rotational motion. This chapter summarizes these interactions and the physics involved.

9.1 DEFINITIONS OF ROTATIONAL DYNAMICS

The *torque* τ, or moment of force, represents the rotational force generated by an external force acting perpendicular to an object with respect to a fixed point or pivot and is defined by

$$\tau = r \times F = |r||F| \sin \theta$$

where r is the distance between the pivot and the point where the force is exerted, also known as the *lever arm*, F is the external force, and θ is the angle between r and F. Since a body can rotate either clockwise or counterclockwise in two dimensions, there are accordingly two types of torques: clockwise torques τ_{cw} and counterclockwise torques τ_{ccw}. Torque is a vector quantity with dimensions of force multiplied by distance, or ML^2T^{-2}, and units of newton-meter (N m).

The *moment of inertia I*, or rotational inertia, represents the resistance of an object to rotational motion just as inertia represents the resistance of an object to translational or linear motion. The moment of inertia is dependent on the number of particles of mass m and a distance r from the pivot of a given body and can be expressed for a small body according to

$$I = mr^2$$

For large or irregular objects composed of several smaller, well-defined objects, the moment of inertia can be expressed as the total or sum of the moments of inertia for the smaller objects

$$I = \Sigma mr^2 = m_1 r_1^2 + m_2 r_2^2 + m_3 r_3^2 + \ldots + m_n r_n^2$$

The moment of inertia is a scalar quantity with dimensions of ML^2 and units of kg m^2. The moments of inertia for common rigid-body geometries are displayed in Figure 9-1.

The *radius of gyration k* of a rigid body represents the distance between the axis of rotation and the point where the weight is considered to be concentrated without changing the moment of inertia of the body. The radius of gyration k of a rigid body is related to its moment of inertia I by

$$I = mk^2$$

The radius of gyration is a scalar quantity with dimensions of L and units of meters.

Solved Problem 9.1. Alicia exerts a torque of 35 N m in turning the steering wheel of her car. If the radius of the steering wheel is 28 cm, determine the required force exerted by Alicia.

Solution

The force F is applied in a direction tangential to the motion of the steering wheel and perpendicular to its radius r. Therefore, from the definition of a torque,

$$\tau = rF$$

Solving for F gives

$$F = \frac{\tau}{r} = \frac{35\,\text{N m}}{0.28\,\text{m}} = 125\,\text{N}$$

139

RIGID BODY	DESCRIPTION	MOMENT OF INERTIA

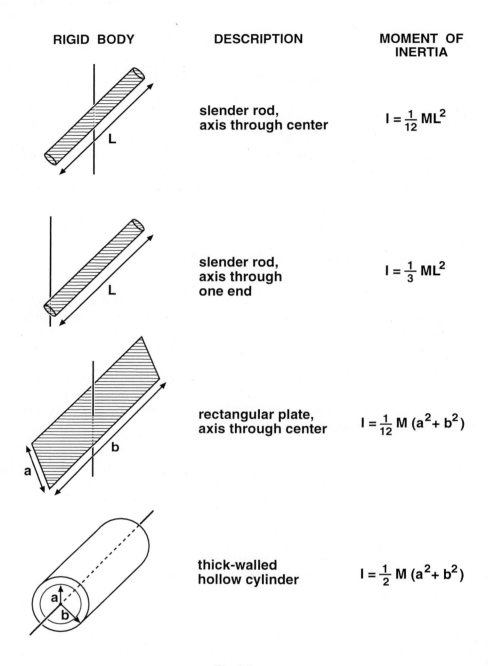

slender rod, axis through center — $I = \frac{1}{12} ML^2$

slender rod, axis through one end — $I = \frac{1}{3} ML^2$

rectangular plate, axis through center — $I = \frac{1}{12} M(a^2 + b^2)$

thick-walled hollow cylinder — $I = \frac{1}{2} M(a^2 + b^2)$

Fig. 9-1a

Solved Problem 9.2. Jerry, a contestant on Wheel of Fortune, is asked to spin the wheel of 2.5-m diameter. If he exerts a tangential force of 75 N for his spin, what torque is exerted on the wheel?

Solution

The tangential force applied to the wheel is perpendicular to the radius and is related to the torque by

$$\tau = rF = (1.25 \text{ m})(75 \text{ N}) = 93.8 \text{ N m}$$

RIGID BODY	DESCRIPTION	MOMENT OF INERTIA

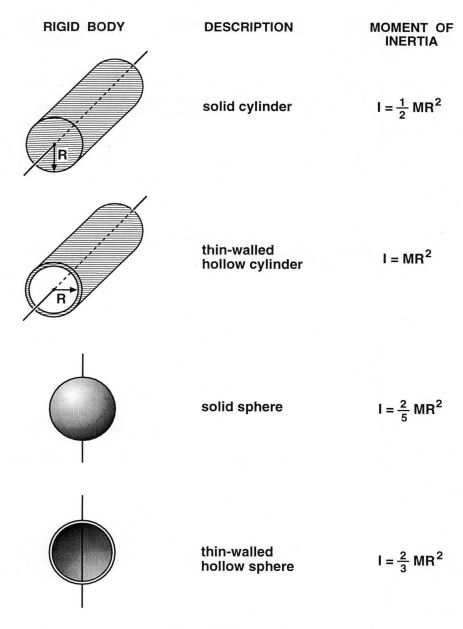

solid cylinder — $I = \frac{1}{2}MR^2$

thin-walled hollow cylinder — $I = MR^2$

solid sphere — $I = \frac{2}{5}MR^2$

thin-walled hollow sphere — $I = \frac{2}{3}MR^2$

Fig. 9-1b

Solved Problem 9.3. Lori of 35-kg mass is seated on one end of a seesaw, 2.5 m from the pivot. What torque is Lori exerting on the seesaw to make it rotate?

Solution

By definition, the torque is related to the force F exerted a distance r about a pivot:

$$\tau = rF$$

The force exerted by Lori on the seesaw is her weight, or

$$F = mg = (35\,\text{kg})(9.8\,\text{m s}^{-2}) = 343\,\text{N}$$

RIGID BODY	DESCRIPTION	MOMENT OF INERTIA

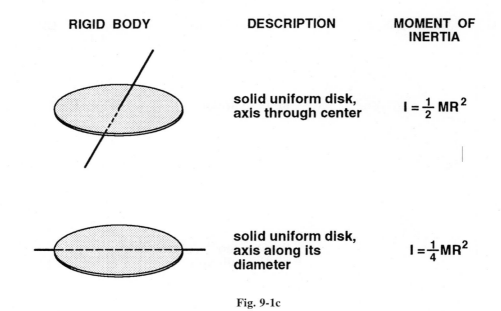

Fig. 9-1c

Thus, the torque is

$$\tau = (343 \text{ N})(2.5 \text{ m}) = 857.5 \text{ N m}$$

Solved Problem 9.4.   A thin hoop of mass of 0.40 kg has a radius of gyration of 0.25 m. Determine its moment of inertia.

Solution

The moment of inertia is related to the radius of gyration k by

$$I = mk^2$$

Thus,

$$I = (0.40 \text{ kg})(0.25 \text{ m})^2 = 0.025 \text{ kg m}^2$$

Solved Problem 9.5.   Determine the moment of inertia of the following: (1) thick-walled artery of 0.5-g mass, an inner radius of 0.035 cm, and an outer radius of 0.05 cm; (2) thin-walled vein of 0.1-g mass and a radius of 0.15 cm; (3) bowling ball of 10-kg mass and a radius of 0.1 m; (4) basketball of 4-kg mass and a radius of 0.25 m.

Solution

Each of these objects can be represented by common geometries given in Figure 9-1.

(1)   A thick-walled artery can be approximated as a hollow cylinder where the moment of inertia is

$$I = \frac{1}{2} m(r_{inner}^2 + r_{outer}^2) = \frac{1}{2}(0.5 \text{ g})[(0.035 \text{ cm})^2 + (0.05 \text{ cm})^2] = 9.3 \times 10^{-4} \text{ g cm}^2$$

(2)   A thin-walled vein can be approximated as a thin-walled hollow cylinder where the moment of inertia is

$$I = mr^2 = (0.1 \text{ g})(0.15 \text{ cm})^2 = 2.25 \times 10^{-3} \text{ g cm}^2$$

(3)   A bowling ball can be approximated as a solid sphere where the moment of inertia is

$$I = \frac{2}{5}mr^2 = \frac{2}{5}(10\text{ kg})(0.1\text{ m})^2 = 0.04\text{ kg m}^2$$

(4)   A basketball can be approximated as a hollow sphere where the moment of inertia is

$$I = \frac{2}{3}mr^2 = \frac{2}{3}(4\text{ kg})(0.25\text{ m})^2 = 0.17\text{ kg m}^2$$

9.2   PARALLEL-AXIS THEOREM

The moment of inertia values presented in Figure 9-1 for the various geometries holds only for the defined axis. However, if the problem at hand dictates rotation about an axis parallel to the original axis, you can use the parallel-axis theorem to compute the new moment of inertia, given by

$$I_{new} = I_{original} + md^2$$

where m is the total mass of the object and d is the perpendicular distance between the original and new axes.

Solved Problem 9.6.   Consider a cylinder of 6.0-kg mass and 25-cm radius. Determine the moment of inertia about an axis (1) through its center and (2) about a parallel axis 12 cm from the center.

   Solution

(1)   The moment of inertia of a cylinder about its axis is

$$I = \frac{1}{2}mr^2 = \frac{1}{2}(6.0\text{ kg})(0.25\text{ m})^2 = 0.125\text{ kg m}^2$$

(2)   The moment of inertia of a cylinder about its new axis is

$$I_{new} = I_{original} + md^2 = (0.125\text{ kg m}^2) + (6.0\text{ kg})(0.12\text{ m})^2 = 0.211\text{ kg m}^2$$

Solved Problem 9.7.   Consider a solid steel ball of mass 18.0 kg and 1.2-m radius. Determine the moment of inertia (1) about an axis through its center and (2) about a parallel axis 0.5 m from the center.

   Solution

(1)   The moment of inertia of a solid sphere about its axis is

$$I = \frac{2}{5}mr^2 = \frac{2}{5}(18.0\text{ kg})(1.2\text{ m})^2 = 10.4\text{ kg m}^2$$

(2)   The moment of inertia of the solid sphere about its new axis is

$$I_{new} = I_{original} + md^2 = (10.4\text{ kg m}^2) + (18.0\text{ kg})(0.5\text{ m})^2 = 14.9\text{ kg m}^2$$

9.3   ROTATIONAL EQUILIBRIUM

In Chapter 5, it was stated that for an object in translational equilibrium, the vector sum of forces must be zero, or $\Sigma F = 0$. However, for an object to be in static equilibrium, it must also be in rotational equilibrium, where the sum of torques or rotational forces must be zero, or

$$\Sigma \tau = \Sigma \tau_{cw} + \Sigma \tau_{ccw} = 0$$

9.4 ROTATIONAL DYNAMICS

The *kinetic energy of rotation*, denoted by KE_{rot}, is defined as

$$KE_{rot} = \frac{1}{2} I\omega^2 = \frac{1}{2}\left(mr^2\right)\omega^2$$

and is directed along the axis of rotation. Kinetic energy is in joules, and ω is in rad s^{-1}. The total kinetic energy of an object in rotational motion is the sum of (1) its rotational kinetic energy determined about an axis through its center of mass

$$KE_{rot} = \frac{1}{2} I\omega^2$$

and (2) its translational kinetic energy of the motion of the object's center of mass

$$KE_{trans} = \frac{1}{2} mv_{cm}^2$$

or

$$KE_{tot} = \frac{1}{2} I\omega^2 + \frac{1}{2} mv_{cm}^2$$

The rotational kinetic energy of a large object composed of several smaller masses, each possessing its own moment of inertia, can be expressed as

$$KE_{rot} = \frac{1}{2}(\Sigma m_n r_n^2)\,\omega^2 = \frac{1}{2}(m_1 r_1^2 + m_2 r_2^2 + m_3 r_3^2 + \cdots + m_n r_n^2)\,\omega^2 = \frac{1}{2} I\omega^2$$

The *rotational work* W_{rot} is the work done by an object subjected to a constant torque τ resulting in an angular displacement from θ_1 to θ_2, and it is given by

$$W_{rot} = \tau\theta = \tau(\theta_2 - \theta_1)$$

Rotational work is expressed in units of joules.

The *rotational power* P_{rot} is the power output of an object in rotational motion performing work over a defined time interval and is given by

$$P_{rot} = \tau\omega$$

where ω is the angular velocity. Rotational power is expressed in units of watts.

The *torque* τ for a rigid body can also be expressed as

$$\tau = I\alpha$$

where I is the moment of inertia and α is the angular acceleration.

The *angular momentum L* is the product of the linear momentum p and the radius r, or distance from the pivot point where the momentum is applied:

$$L = mvr$$

Angular momentum is a vector quantity and is given in dimensions of $ML^2 T^{-1}$ and in units of kg m^2 s^{-1}.

9.5 CONSERVATION OF ANGULAR MOMENTUM

The angular momentum of a system will remain constant, that is, will be conserved, if the net torque acting on the system is zero.

9.6 NEWTON'S LAWS OF ROTATIONAL MOTION

Analogous to translational motion, Newton's laws of motion can be extended to rotational motion.

Newton's first law: A body in rotational motion will continue its rotation until acted on by an external torque.

Newton's second law: The magnitude of a torque is equal to the product of the moment of inertia and the angular acceleration, or

$$\tau = I\alpha$$
$$F = ma$$

Newton's third law: For every torque acting on an object, there is another torque acting on the object equal in magnitude and opposite in direction to the original torque.

Solved Problem 9.8.   A crowbar 2.8 m in length and pivoted at 0.45 m from the end is used to pry a 550-N rock from the ground. Determine the force required to successfully pry the rock from the ground.

   Solution

   Applying the conservation of rotational equilibrium

$$\Sigma\tau = 0$$

we see that there are two counteracting torques exerted on the crowbar. The rock, positioned 0.45 m from the end of the crowbar, is exerting a counterclockwise torque of magnitude

$$\tau_{\text{ccw}} = Fr = (550\text{ N})(0.45\text{ m}) = 247.5\text{ N m}$$

This torque is counteracted by a clockwise torque generated by the person, given by

$$\tau_{\text{cw}} = Fr = F(2.35\text{ m})$$

where the radius is the length from the end to the pivot which is 0.45 m from the opposite end. This radius is

$$r = 2.8\text{ m} - 0.45\text{ m} = 2.35\text{ m}$$

Equating the two torques, we can solve for the force:

$$\Sigma\tau_{\text{ccw}} = \Sigma\tau_{\text{cw}}$$
$$247.5\text{ N} = F(2.35\text{ m})$$
$$F = 105.3\text{ N}$$

Solved Problem 9.9.   Suppose Tori of mass 42 kg and Cori of mass 70 kg join their sister Lori of mass 35 kg on the seesaw. If Tori and Lori are seated 2.0 and 3.5 m from the pivot, respectively, on the same side of a seesaw, where must Cori sit on the other side of the seesaw in order to balance it?

   Solution

   By the conservation of rotational motion,

$$\Sigma\tau = 0$$

or

$$\tau_{\text{Tori}} + \tau_{\text{Lori}} = \tau_{\text{Cori}}$$
$$(Fr)_{\text{Tori}} + (Fr)_{\text{Lori}} = (Fr)_{\text{Cori}}$$
$$(42\text{ kg})(9.8\text{ m s}^{-2})(2.0\text{ m}) + (35\text{ kg})(9.8\text{ m s}^{-2})(3.5\text{ m}) = (70\text{ kg})(9.8\text{ m s}^{-2})(r)$$

Solving for r gives

$$r = 2.95\text{ m}\qquad\text{from pivot}$$

Solved Problem 9.10. A Marine recruit of mass $m = 65$ kg is ordered to perform the daily exercise regimen of 20 push-ups, as illustrated in Figure 9-2. Determine the force exerted by his hands on the ground in order to successfully accomplish the push-ups.

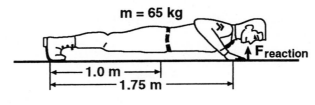

Fig, 9-2

Solution

From the conservation of rotational equilibrium,

$$\Sigma\tau = 0$$

there are two counteracting torques exerted by the Marine in performing the push-ups. The center of gravity of the Marine acting at a point 1.0 m from the pivot, that is, the feet, generates a clockwise torque of magnitude

$$\tau_{cw} = Fr = (65 \text{ kg})(9.8 \text{ m s}^{-2})(1.0 \text{ m}) = 637 \text{ N m}$$

This torque is counteracted by a counterclockwise torque generated by the ground on the Marine given by

$$\tau_{ccw} = Fr = F(1.75 \text{ m})$$

where the radius is the length from the hands to the pivot, which is 1.75 m from the hands. Equating the two torques, we can solve for the force:

$$\Sigma\tau_{ccw} = \Sigma\tau_{cw}$$
$$637 \text{ N m} = F(1.75 \text{ m})$$
$$F = 364 \text{ N}$$

Solved Problem 9.11. One of the muscles involved in the process of mastication, or chewing, is the masseter muscle. Assume that the masseter muscle exerts an upward force of 250 N at a distance of 2.5 cm from the pivot or temporomandibular joint, as shown in Figure 9-3. If the linear distance from the pivot to the molars is 4.8 cm, determine the force exerted on the molars or back teeth during the chewing process.

Solution

From the conservation of rotational equilibrium,

$$\Sigma\tau_{ccw} = \Sigma\tau_{cw}$$
$$(250 \text{ N})(2.5 \text{ cm}) = F_{molars}(4.8 \text{ cm})$$

Solving for F_{molars} gives

$$F_{molars} = 130.2 \text{ N}$$

Solved Problem 9.12. A ladder of weight $W = 150$ N and length $L = 6$ m rests against a smooth vertical wall at an angle of 30°, as shown in Figure 9-4. The center of gravity of the ladder is at a height $h = 3.5$ m above the floor. A man pulls the ladder at its midpoint in a horizontal direction with a force F. What is the minimum force F required for the man to pull the top of the ladder away from the wall? The friction on the floor is sufficient to prevent the bottom of the ladder from slipping.

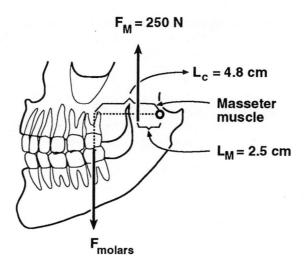

Fig. 9-3

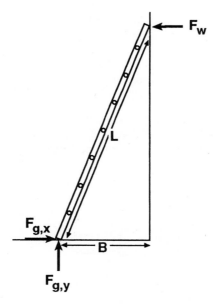

Fig. 9-4

Solution

By implementing the conservation of rotational equilibrium, the magnitude of the force F is determined by taking the moments and determining the torques about the lower end of the ladder:

$$Wh \tan \theta = \frac{1}{2} FL \cos \theta$$

Solving for F gives

$$F = \frac{2Wh \sin \theta}{L \cos^2 \theta}$$

Substituting appropriate values yields $F = 350 \text{ N}$.

Solved Problem 9.13. A uniform cylinder of mass 50 g and radius 3.0 cm is placed at the top of an inclined plane 8 m in length and is oriented 35° from the horizontal and is allowed to roll from rest. Determine the final velocity of the cylinder at the bottom of the inclined plane.

Solution

The vertical fall, or height, of the cylinder is $(8 \text{ m})(\sin 35°) = 4.6 \text{ m}$. Thus, in rolling down the inclined plane, the loss of potential energy due to rolling, given by PE $= mgh = (0.50 \text{ kg})(9.8 \text{ m s}^{-2})(4.6 \text{ m}) = 22.5 \text{ J}$, will be converted to kinetic energy at the bottom of the inclined plane, due to both translational and rotational motion. Assuming that the translational velocity is $v \text{ m s}^{-1}$ and rotational velocity is $\omega \text{ rad s}^{-1}$, then

$$KE = \frac{1}{2}mv^2 + \frac{1}{2}I\omega^2$$

The moment of inertia of a cylinder is

$$I = \frac{mr^2}{2}$$

Thus, the kinetic energy can be rewritten as

$$KE = \frac{1}{2}mv^2 + \frac{1}{2}\left(\frac{mr^2}{2}\right)\frac{v^2}{r^2} = 0.25v^2 + 0.125v^2 = 0.375v^2$$

However, since PE loss is equal to KE gain:

$$PE_{loss} = KE_{gain}$$
$$22.5 \text{ J} = 0.375v^2$$

Solving for v^2 gives

$$v^2 = 60 \text{ m}^2 \text{ s}^{-2} \Rightarrow v = 7.7 \text{ m s}^{-1}$$

Solved Problem 9.14. In a primitive Yo-Yo, the string is wrapped around a solid cylinder of mass M and radius R, as shown in Figure 9-5. Upon release, the Yo-Yo proceeds in a downward motion, assuming the string unwinds without slipping. Determine the downward acceleration and tension in the string.

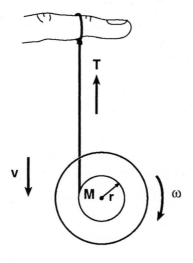

Fig. 9-5

Solution

We can apply conditions of translational and rotational equilibrium to the motion of the Yo-Yo:

Translational equilibrium

$$\Sigma F_x = 0 \tag{1}$$

$$\Sigma F_y = Mg - T = Ma_{cm} \tag{2}$$

Rotational equilibrium

$$\Sigma \tau = TR = I_{cm}\,\alpha = \left(\frac{1}{2}MR^2\right)\alpha \tag{3}$$

where I_{cm} is the moment of inertia for a cylinder with the axis through its center of mass. Since no slipping by the string is assumed, the following equations of rotational motion can be applied:

$$v_{cm} = R\omega \tag{4}$$

$$a_{cm} = R\alpha \tag{5}$$

Equation (5) can be substituted into Eq. (3) to eliminate α:

$$TR = \left(\frac{1}{2}MR^2\right)\alpha = \left(\frac{1}{2}MR^2\right)\frac{a_{cm}}{R}$$

$$T = \frac{1}{2}Ma_{cm} \tag{6}$$

Substituting Eq. (6) into Eq. (2) yields

$$Mg - \frac{1}{2}Ma_{cm} = Ma_{cm}$$

$$g = \frac{3}{2}a_{cm}$$

Solving for a_{cm} gives

$$a_{cm} = \frac{2}{3}g \tag{7}$$

Substituting Eq. (7) into Eq. (6) yields the tension in the string:

$$T = \frac{1}{2}Ma_{cm} = \frac{1}{2}M\left(\frac{2}{3}g\right) = \frac{1}{3}Mg$$

Solved Problem 9.15.   What is the speed of the Yo-Yo after it has dropped a distance h?

Solution

While it is in the hand, the potential energy PE_1 is

$$PE_1 = Mgh$$

and the kinetic energy KE_1 is

$$KE_1 = 0$$

Once the Yo-Yo has traversed a distance h, the potential energy PE_2 is

$$PE_2 = 0$$

and the kinetic energy KE_2 is

$$KE_2 = \frac{1}{2}Mv_{cm}^2 + \frac{1}{2}I\omega^2$$

The angular velocity ω is related to the linear velocity by

$$\omega = \frac{v_{cm}}{R}$$

and the moment of inertia I is given by

$$I = \frac{1}{2}MR^2$$

Substituting into the expression for the kinetic energy gives

$$KE_2 = \frac{1}{2}\left(\frac{1}{2}MR^2\right)\left(\frac{v_{cm}}{R}\right)^2 + \frac{1}{2}Mv_{cm}^2 = \frac{3}{4}Mv_{cm}^2$$

Application of the conservation of energy yields

$$\text{Total energy}_{before} = \text{total energy}_{after}$$
$$PE_1 + KE_1 = PE_2 + KE_2$$
$$mgh + 0 = \frac{3}{4}Mv_{cm}^2 + 0$$

Solving for v_{cm} gives

$$v_{cm} = \sqrt{\frac{4}{3}gh}$$

Solved Problem 9.16.   In pondering novel marketing schemes for her cola products, the chief executive of Poopsi Cola rolls a cylindrical soda container of radius $R = 4.5$ cm on a flat surface back and forth with a speed of 10 cm s^{-1}. Determine the kinetic energy of the soda can, assuming (1) the can is full of soda ($M = 10$ g) and (2) the can is empty ($M = 2$ g).

Solution

The kinetic energy of the soda can is given by

$$KE_{tot} = \frac{1}{2}Mv_{cm}^2 + \frac{1}{2}I\omega^2$$

where

$$\omega = \frac{v_{cm}}{R}$$

(1)   For a full can of soda, we have a solid cylinder with a moment of inertia of

$$I = MR^2$$

and the kinetic energy can be written as

$$KE_{tot} = \frac{1}{2}Mv_{cm}^2 + \frac{1}{2}(MR^2)\left(\frac{v_{cm}}{R}\right)^2 = Mv_{cm}^2 = (0.01 \text{ kg})(0.1 \text{ m s}^{-1})^2 = 1 \times 10^{-4}\,\text{J}$$

(2)   For an empty can of soda, we have a hollow cylindrical shell with a moment of inertia of

$$I = \frac{1}{2}MR^2$$

and the kinetic energy can be written as

$$KE_{tot} = \frac{1}{2}Mv_{cm}^2 + \frac{1}{2}\left(\frac{1}{2}MR^2\right)\left(\frac{v_{cm}}{R}\right)^2 = \frac{3}{4}Mv_{cm}^2 = \frac{3}{4}(0.01 \text{ kg})(0.1 \text{ m s}^{-1})^2 = 7.5 \times 10^{-5}\,\text{J}$$

Solved Problem 9.17.   A solid disk of 1.2-m diameter and 150-kg mass is subjected to a torque of 135 N m. Determine the angular acceleration of the disk.

Solution

The torque is related to the angular acceleration by $\tau = I\alpha$. The moment of inertia of a solid disk with an axis perpendicular and through its center is

$$I = \frac{1}{2}mr^2 = \frac{1}{2}(150 \text{ kg})(0.6 \text{ m})^2 = 27 \text{ kg m}^2$$

Substituting into the above expression yields

$$\alpha = \frac{\tau}{I} = \frac{135 \text{ N m}}{27 \text{ kg m}^2} = 5.0 \frac{\text{rad}}{\text{s}^2}$$

Solved Problem 9.18.   In playing a long-playing (LP) album, a record turntable of mass 1.2 kg and a 40-cm diameter accelerates from rest to 33 rev min^{-1} in 4.5 s. Determine the (1) angular acceleration, (2) angular momentum, and (3) torque exerted by the turntable.

Solution

From this problem

$$\omega_o = 0 \qquad \omega_f = 33 \text{ rev min}^{-1} \qquad t = 4.5 \text{ s}$$

Converting the angular speed from rev min^{-1} to rad s^{-1} gives

$$33 \frac{\text{rev}}{\text{min}} \cdot \frac{2\pi \text{ rad}}{1 \text{ rev}} \cdot \frac{1 \text{ min}}{60 \text{ s}} = 3.45 \text{ rad s}^{-1}$$

(1)   The angular acceleration is

$$\alpha = \frac{\omega_f - \omega_o}{t} = \frac{3.45 \text{ rad s}^{-1} - 0}{4.5 \text{ s}} = 0.77 \text{ rad s}^{-2}$$

(2)   Angular momentum L is given by

$$L = I\omega$$

The moment of inertia for a circular disk rotating about a perpendicular axis is

$$I = \frac{1}{2}MR^2 = \frac{1}{2}(1.2 \text{ kg})(0.2 \text{ m})^2 = 0.024 \text{ kg m}^2$$

(3)   The torque exerted by the turntable is given by

$$\tau = I\alpha = (0.024 \text{ kg m}^2)(0.77 \text{ rad s}^{-2}) = 0.018 \text{ N m}$$

In Chapter 5, the forces exerted on skeletal joints such as the hip, elbow, and ankle were analyzed in translational equilibrium. Since joints represent fixed points at which rotational motion can occur, all torques must be analyzed in rotational equilibrium, that is, $\Sigma\tau = 0$. We now consider rotational equilibrium as applied to the elbow, hip, and ankle.

ILLUSTRATIVE EXAMPLE 9.1.   Skeletal Mechanics of the Joint: Elbow

Consider the arm of a person lifting an object of weight $W_{\text{ob}} = 490$ N, as shown in Figure 9-6. We wish to analyze and obtain an expression for all forces (magnitude and direction) acting on the elbow. The elbow is chosen as the pivot point. Approximate measurements include 40 cm from the elbow to the hand, 15 cm from the elbow to the arm's center of gravity, and 4 cm from the elbow to the biceps muscle. The weight of the lower arm W_a is taken as 24.5 N. Lifting a weight introduces four forces and three unknown physical quantities:

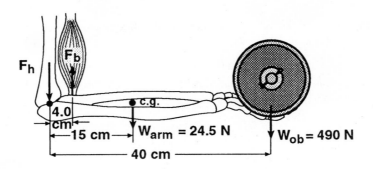

Fig. 9-6

F_{bm}—the pulling force exerted by the biceps muscle. The direction of this force is the angle θ_{bm} between the biceps muscle and the lower arm and is approximately 73°.

F_{ej}, θ_{ej}—the magnitude and direction of the force exerted on the elbow joint by the humerus or upper arm.

W_{ob}—the weight of the object acting downward, given in the example as 490 N.

W_a—the weight of the lower arm acting downward at its center of gravity, given in the example as 24.5 N.

From translational equilibrium, that is, $\Sigma F = 0$:

ΣF_x: $\qquad\qquad\qquad\qquad\qquad F_{bm} \cos 73° = F_{ej} \cos \theta_{ej}$

ΣF_y: $\qquad\qquad\qquad\qquad\qquad F_{bm} \sin 73° = W_{ob} + W_a + F_{ej} \sin \theta_{ej}$

where W_{ob} is the weight of the object. Solution of the three unknowns requires an additional equation which can be derived from the condition of rotational equilibrium, or $\Sigma \tau = 0$. Three torques acting at the elbow or pivot can be identified from this problem:

τ_{cw}: $\qquad$ (0.40 m) (W_{ob}) due to weight of object

$\qquad\qquad$ (0.15 m) (W_a) due to weight of arm

τ_{ccw}: $\qquad$ (0.04 m) ($F_{bm} \sin 73°$) due to vertical component of biceps muscle

From rotational equilibrium,

$$\Sigma \tau_{cw} = -\Sigma \tau_{ccw}$$

Therefore,

$$(0.40 \text{ m}) (W_{ob}) + (0.15 \text{ m}) (W_a) = -(0.04 \text{ m}) (F_{bm} \sin 73°)$$

Solving for F_{bm} gives

$$F_{bm} = -5255 \text{ N}$$

The minus sign corresponds to the pulling force by the biceps muscle in a counterclockwise direction. The equations obtained by invoking the condition of translational equilibrium provide the two other unknowns.

$$F_{ej} \cos \theta_{ej} = F_{bm} \cos 73° = -1536 \text{ N}$$

$$F_{ej} \sin \theta_{ej} + W_{ob} = F_{bm} \sin 73° \qquad \text{or} \qquad F_{ej} \sin \theta_{ej} = -4535 \text{ N}$$

The magnitude and direction of F_{ej} can be determined by

$$F_{ej} = \overline{)(F_{ej} \cos \theta_{ej})^2 + (F_{ej} \sin \theta_{ej})^2} = \overline{)(-1536 \text{ N})^2 + (-4535 \text{ N})^2} = 4788 \text{ N}$$

$$\theta_{ej} = \tan^{-1} \frac{F_{ej} \sin \theta_{ej}}{F_{ej} \cos \theta_{ej}} = \tan^{-1} \left(\frac{-4535 \text{ N}}{-1536 \text{ N}} \right) = 71.3°$$

ILLUSTRATIVE EXAMPLE 9.2.   Skeletal Mechanics of the Joint: Hip

At the hip, the femur or thigh bone rotates in a socket embedded in the pelvis and is stabilized by a group of three independent muscles, known as the *hip abductor muscle*, as shown in Figure 9-7. In this problem involving a standing person, we are confronted with four forces and three unknown physical quantities:

F_{hj}, θ_{hj}—the magnitude and direction of the force acting at the hip joint.

F_{am}—the magnitude of the force exerted by the hip abductor muscle. The direction of the hip abductor muscle is 71°.

W—the reaction force of the ground which is equal in magnitude and opposite in direction to the weight of the person.

W_L—the weight of the leg acting vertically downward at the center of gravity of the leg.

Let's assume that the weight of the person is 700 N corresponding to a mass of 71 kg. Employing conditions of translational equilibrium, we find

ΣF_x: $F_{am} \cos 71° = F_{hj} \cos \theta_{hj}$

ΣF_y: $F_{am} \sin 71° + W = W_L - F_{hj} \sin \theta_{hj}$

The weight of the leg can be approximated according to $W_L \approx W/7 = 0.143W$. Also, the hip joint is chosen as the pivot point, and the hip abductor muscle is attached to the femur 7 cm from the pivot or point of rotation. Since we have only two equations for three unknowns, the other equation can be obtained from rotational equilibrium, or $\Sigma \tau = 0$:

$$F_{am} \sin \theta_{am} \cdot 0.07\,\text{m} + W_L \cdot 0.03\,\text{m} - W \cdot 0.11\,\text{m} = 0$$
$$F_{am} \sin 71° \cdot 0.07\,\text{m} + 100\,\text{N} \cdot 0.03\,\text{m} - 700\,\text{N} \cdot 0.11\,\text{m} = 0$$
$$F_{am}(0.06\,\text{m}) + 3\,\text{N}\,\text{m} - 77\,\text{N}\,\text{m} = 0$$
$$F_{am}(0.06\,\text{m}) - 74\,\text{N}\,\text{m} = 0$$
$$F_{am} = \frac{74\,\text{N}\,\text{m}}{0.06\,\text{m}} = 1233\,\text{N}$$

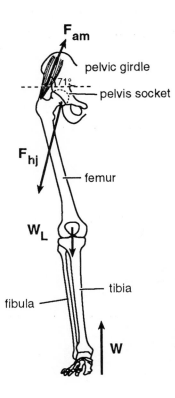

$\mathbf{F_{am}}$ = force exerted by the hip abductor muscle

$\mathbf{F_{hj}}$ = force acting at the hip joint

$\mathbf{W}$ = upward reaction force of the ground

$\mathbf{W_L}$ = weight of the leg acting downward at the center of gravity
$W_L \cong W/7$

Fig. 9-7

Substituting into ΣF_y gives

$$(1233\text{ N})(0.946) + 700\text{ N} = 100\text{ N} - F_{hj}\sin\theta_{hj}$$
$$F_{hj}\sin\theta_{hj} = -1766.4\text{ N}$$

From ΣF_x:

$$(1233\text{ N})(0.33) = F_{hj}\cos\theta_{hj}$$
$$F_{hj}\cos\theta_{hj} = 406.9\text{ N}$$
$$F_{hj} = \sqrt{(F_{hj}\cos\theta_{hj})^2 + (F_{hj}\sin\theta_{hj})^2} = \sqrt{(406.9\text{ N})^2 + (-1766.4\text{ N})^2} = 1812.6\text{ N}$$
$$\theta_{hj} = \tan^{-1}\frac{F_{hj}\sin\theta}{F_{hj}\sin\theta} = -77° \text{ or } 283° \qquad \text{from positive } x \text{ axis}$$

ILLUSTRATIVE EXAMPLE 9.3.   Skeletal Mechanics of the Joint: Ankle

In a walking person, three forces acting on the foot can be identified: (1) a reaction force by the ground acting upward at the toes, or more precisely the ball of the foot, that is equal to the person's weight but opposite in direction; (2) a pulling force by the Achilles tendon acting upward and at an angle to the right; and (3) a force corresponding to the weight of the leg acting downward on the ankle, as shown in Figure 9-8. Let's suppose that we want to determine only the magnitude of the force exerted by the Achilles tendon. Its direction is approximately 25° from the vertical or y axis. Assuming the ankle is the pivot point and applying conditions of rotational equilibrium, we see there are two torques:

τ_{cw}—F_{AT} (0.06 m) due to the force of the Achilles tendon acting at a distance of 6 cm from the ankle

τ_{ccw}—W_p (0.11 m) due to the weight of the person acting on the ball of the foot, which is a distance of approximately 11 cm from the ankle. We will assume for the problem that the weight of the person is 800 N.

According to rotational equilibrium,

$$\Sigma\tau_{cw} = \Sigma\tau_{ccw}$$
$$F_{AT}\ (0.06\text{ m}) = (800\text{ N})\ (0.11\text{ m})$$

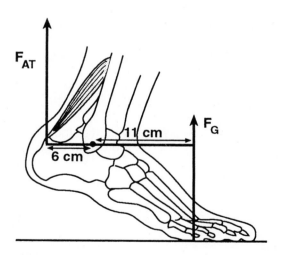

F_{AT} = force of the Achilles tendon

F_G = force of the ground (= weight of the body)

Fig. 9-8

Solving for F_{AT} gives

$$F_{AT} = 1467\,\text{N}$$

In summary, the analogies between translational and rotational quantities can be seen by the table below:

Translational Quantities	Rotational Quantities
Kinematics	
Linear displacement x (m)	Angular displacement θ (rad)
Linear velocity v (m s^{-1})	Angular velocity ω (rad s^{-1})
Linear acceleration a (m s^{-2})	Angular acceleration α (rad s^{-2})
Equations of Motion	
$x = vt$	$\theta = \omega t$
$x = v_o t + \dfrac{1}{2} a t^2$	$\theta = \omega_o t + \dfrac{1}{2} \alpha t^2$
$v_f^2 = v_o^2 + 2ax$	$\omega_f^2 = \omega_o^2 + 2\alpha\theta$
$a = \dfrac{v_f - v_o}{t}$	$\alpha = \dfrac{\omega_f - \omega_o}{t}$
$a = \dfrac{v_f + v_o}{2}$	$\omega = \dfrac{\omega_f + \omega_o}{2}$
Dynamics	
Mass (inertia) m (kg)	Moment of inertia I (kg m^2)
Force F (N)	Torque τ (N m)
Work W (J)	Rotational work W_{rot} (J)
Kinetic energy KE (J)	Rotational kinetic energy KE$_{\text{rot}}$ (J)
Potential energy PE (J)	Rotational potential energy PE$_{\text{rot}}$ (J)
Power P (W)	Rotational power P_{rot} (W)
Momentum p (kg m s^{-1})	Angular momentum L (kg m^2 s^{-2})
Conditions of Equilibrium	
$\Sigma F = 0$	$\Sigma \tau = 0$
$\Sigma F_x = 0 \qquad \Sigma F_y = 0$	$\Sigma \tau_{\text{cw}} + \Sigma \tau_{\text{ccw}} = 0$

Newton's Three Laws of Motion

1. An object at rest or in motion will remain at rest or in motion unless acted on by an external force.

2. The magnitude of a force is equal to the product of the object's mass and its resultant acceleration, or $F = ma$.

3. For every force acting on an object, there is another force acting on the object equal in magnitude and opposite in direction to the original force.

1. An object in rotational motion will remain in rotation unless acted on by an external torque.

2. The magnitude of a torque is equal to the product of the moment of inertia and the angular acceleration, or $\tau = I\alpha$.

3. For every torque acting on an object, there is another torque acting on the object equal in magnitude and opposite in direction to the original torque.

Conservation of Momentum

If the net force acting on a system is zero, then the total linear momentum of the system will remain constant.	If the net torque acting on a system is zero, then the total angular momentum of the system will remain constant.

Supplementary Problems

9.1. Determine the torque exerted by Jane in opening a door if she applies a 40-N force perpendicular to the door at a distance of (1) 0.9 m and (2) 0.5 m from the hinges.

Solution

(1) 36 N m; (2) 20 N m

9.2. Zeebo of mass 25 kg and Zeffer of mass 21 kg are playing on a seesaw. If Zeebo is sitting 1.5 m from the pivot point, where should Zeffer sit in order to balance the seesaw?

Solution

1.78 m

9.3. Wes the mechanic tells his apprentice Les that a torque of 80 N is required to tighten a nut on a bolt in a proper manner. If Les uses a wrench that is 0.14 m long, determine the force he must exert perpendicularly on a wrench at a distance 0.01 m from the end of the wrench.

Solution

615.4 N

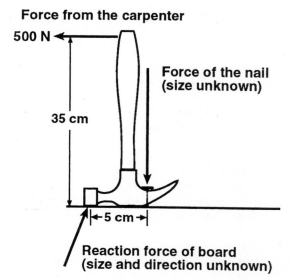

Force from the carpenter

500 N

Force of the nail (size unknown)

35 cm

9.4. Determine the radius of gyration of a 45-lb disk of diameter 21 in as it rotates about a diameter as an axis.

Solution

5.25 in

9.5. A carpenter exerts a force of 500 N on the end of the handle of a hammer of length 32 cm to remove a nail, as depicted in Figure 9-9. If the distance from the nail to the head of the hammer (assumed to be the pivot) is 9 cm, determine the magnitude of the force exerted by the nail on the hammer.

Solution

1777 N m in a counterclockwise direction

5 cm

Reaction force of board (size and direction unknown)

Fig. 9-9

9.6. Determine the kinetic energy of a rotating wheel with a moment of inertia of $15 \, \text{kg m}^2$ revolving at a speed of $120 \, \text{r min}^{-1}$.

Solution

1191 J

9.7. Determine the kinetic energy of a 7.5-kg bowling ball 0.12 m in diameter as it rolls along a level surface with a speed of $5 \, \text{m s}^{-1}$.

Solution

131.4 J

9.8. In driving her toy car at the park, Michelle exerts a force of 5 N in turning the steering wheel of diameter 30 cm. Find the torque exerted by Michelle in turning the steering wheel.

Solution

1.5 N m

9.9. A 5-kg bicycle wheel with a radius of gyration of 60 cm is rotating with a constant speed of $240 \, \text{rev min}^{-1}$. Determine the (1) moment of inertia and (2) rotational kinetic energy.

Solution

(1) $1.8 \, \text{kg m}^2$; (2) 567 J

9.10. The propeller of an airplane has a mass of 110 kg and a radius of gyration of 90 cm. Determine (1) its moment of inertia and (2) the torque required to impart an angular acceleration of $12 \, \text{rev s}^{-2}$.

Solution

(1) $89.1 \, \text{kg m}^2$; (2) 6718 N m

9.11. A record turntable of mass 1.4 kg and radius 0.15 m accelerates from rest to $78 \, \text{rev min}^{-1}$ in 4.5 s. Calculate (1) the moment of inertia of the turntable, (2) the change in angular momentum of the record turntable, and (3) torque required to achieve the acceleration of the turntable.

Solution

(1) $0.016 \, \text{kg m}^2$; (2) $0.13 \, \text{kg m}^2 \, \text{s}^{-1}$; (3) $0.029 \, \text{kg m}^2 \, \text{s}^{-2}$

9.12. Consider a primitive Yo-Yo consisting of a string wrapped around a cylinder of mass = 8 kg and radius = 0.14 m. Assuming an angular speed of $3.6 \, \text{rad s}^{-1}$ during its trip, determine the work done on the cylinder.

Solution

$0.51 \, \text{kg m}^2 \, \text{s}^{-2}$

9.13. Consider a metal sphere of mass 13.5 kg and a radius 0.85 m. Determine the moment of inertia (1) about an axis through its center and (2) about a parallel axis 0.35 m from the center.

Solution

(1) $3.9 \, \text{kg m}^2$; (2) $5.5 \, \text{kg m}^2$

9.14. A wheel with a moment of inertia of 2.5 kg m^2 increases its speed from 4 to 8 rev s^{-1} in 10 rev. Determine the unbalanced torque required to maintain the increase in speed.

Solution

5 kg m^2 s^{-1}

9.15. A rotating disk of mass 30 kg has a radius of gyration of 1.5 m. Determine the torque required to impart an angular acceleration of 5.5 rad s^{-2}.

Solution

371.3 N m

9.16. A rotating object has a radius of gyration k. If the object is acted on by a rotational power P over a time interval t, what is the angular speed?

Solution

$$\omega = \sqrt{\frac{2Pt}{mk^2}}$$

9.17. Determine the work done in turning a steering wheel of radius 0.30 m through 80° if a force of 400 N is applied tangentially.

Solution

165.8 J

9.18. A Yo-Yo consisting of string wrapped around a cylinder of mass 1.5 kg and radius 0.02 m falls from rest. Determine the velocity of the Yo-Yo after it falls 30 cm.

Solution

1.98 m s^{-1}

9.19. Determine the torque required to stop a disk of mass 5 kg and radius 0.45 m, originally rotating about an axis through its center at 1500 rev min^{-1}, in 5.0 min.

Solution

-0.27 N m

Chapter 10

Oscillatory Motion

Oscillations describe processes that are periodic in time such as birthdays, the seasons, a bouncing ball, and alternating-current (ac) voltage at a typical wall outlet. Oscillatory motion describes any physical process that moves in a repetitive or cyclical manner. The simplest and probably the most important type of oscillatory motion is simple harmonic motion. This chapter addresses the physical characteristics and concepts of oscillatory motion, particularly simple harmonic motion.

10.1 DEFINITIONS OF OSCILLATORY MOTION

Simple harmonic motion is the simplest type of oscillatory motion. An example of simple harmonic motion is an object attached to a spring. When the object is pulled or displaced a distance $-A$ from the point of equilibrium and released, the restoring force of the spring pulls the object a distance $+A$ above the point of equilibrium, causing the spring to become compressed. The object then oscillates between the limits of $-A$ and $+A$, resembling a sinusoidal wave, as depicted in Figure 10-1.

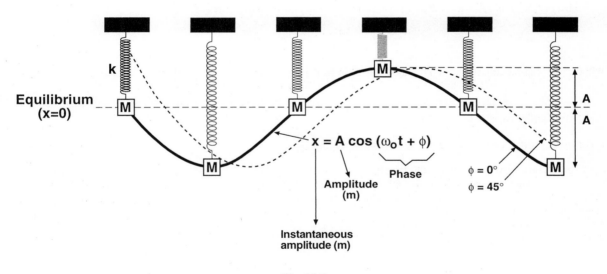

Fig. 10-1

10.2 PHYSICAL CHARACTERISTICS OF SIMPLE HARMONIC MOTION

The motion of the object, consisting of a continuous cycle of crests or peaks and valleys or troughs, is graphically shown in Figure 10-1. Four physical characteristics which represent the simple harmonic motion of an object are amplitude, frequency, period, and phase; and they are related according to the following equation describing the displacement of the oscillating object from its point of equilibrium:

$$x = A \cos(\omega t + \phi)$$

where x is the position as a function of time in meters, A is the amplitude of the oscillatory motion in meters, ω is the angular velocity in rad s^{-1}, t is time in seconds, and ϕ is the phase angle in radians.

Amplitude A is the maximum magnitude of displacement from equilibrium. Units of amplitude are typically in length or *L* but can be represented by any physical quantity whose magnitude varies in an oscillatory manner.

Frequency ν is the number of cycles in a time interval and is given by

$$\nu = \frac{1}{T}$$

where *T* is the period of the oscillation. For the mass-spring system presented in Figure 10-1, the frequency is

$$\nu = \frac{1}{2\pi}\sqrt{\frac{k}{m}}$$

where *k* is the force constant of the spring and *m* is the mass of the oscillating object. Units of frequency are hertz (Hz) or cycles s^{-1}. Cycles are in essence unitless, and therefore frequency is actually given in units of s^{-1}.

The frequency is related to angular frequency by

$$\omega = 2\pi\nu = \frac{2\pi}{T} = \sqrt{\frac{k}{m}}$$

Units of angular frequency are rad s^{-1}.

The *period T* is the time taken to complete 1 cycle and is defined as

$$T = \frac{2\pi}{\omega} = 2\pi\sqrt{\frac{x}{a}} = 2\pi\sqrt{\frac{m}{k}}$$

where *x* is displacement and *a* is acceleration. Period is in units of *T* and is often expressed as seconds per cycle.

The *phase φ* represents a constant that can be adjusted to conform with the value of the displacement at *t* = 0 and is presented as an angle, in radians.

ILLUSTRATIVE EXAMPLE 10.1.   Blood Flow Waveforms

The beating heart ejects blood into the circulatory system in a cyclical or periodic manner with a complete cycle or heartbeat lasting approximately 1.1 s. The pulsatile nature of the heart imparts oscillatory motion to the blood flowing through the circulatory system, where the blood flow velocity represents the amplitude. More specifically, there is a dramatic surge or impulsive rise in blood flow velocity from the heart corresponding to the amplitude at the onset of systole or followed by a sharp but continual decrease in blood flow velocity while the heart is filling for its next heartbeat. A typical blood flow waveform is shown in Figure 10-2.

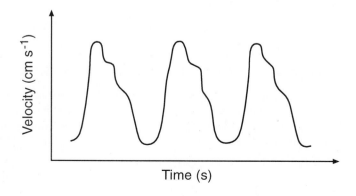

Fig. 10-2

ILLUSTRATIVE EXAMPLE 10.2.   Electrocardiogram

In the circulatory system, the initial contraction and subsequent surge of pressure generated by the heart at the onset of a heartbeat are caused by an electric signal. The electrical activity of the heart can be measured, recorded, and displayed in an electrocardiogram (ECG), as shown in Figure 10-3. An ECG consists of a continuous series or cycles of electric signals in which the large spike corresponds to the primary ventricular contraction. The cycles of electrical activity reveal a large amount of information regarding normal and abnormal cardiac physiology.

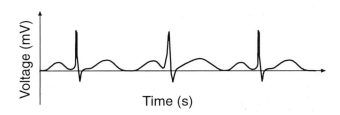

Fig. 10-3

Solved Problem 10.1.   A vibrating object oscillates at a rate of 3 complete vibrations per second. Determine the (1) angular frequency and (2) period of the motion.

Solution

(1)   The angular frequency of the vibrating object can be determined from

$$\omega = 2\pi\nu = 2\cdot3.14\cdot3 \text{ cycles s}^{-1} = 18.8 \text{ cycles s}^{-1} = 18.8 \text{ Hz}$$

(2)   The period of motion is

$$T = \frac{2\pi}{\omega} = \frac{2\cdot3.14}{18.8 \text{ cycles s}^{-1}} = 0.33 \text{ s}$$

Solved Problem 10.2.   A marble attached to a spring is displaced 4.0 cm from its point of equilibrium and is released. If the marble reaches 4.0 cm on the opposite side of the point of equilibrium in 0.85 s, determine the (1) amplitude, (2) period of motion, and (3) frequency.

Solution

(1)   The amplitude is the maximum distance from the point of equilibrium attained by the oscillating marble, or $A = 4.0$ cm.

(2)   The period is the time required for the marble to complete 1 entire cycle. The time taken to traverse from -4.0 cm below the equilibrium point to 4.0 cm above the equilibrium point was 0.85 s and constituted one-half of a cycle. The time taken to complete the cycle is the period, or $T = 2 \times 0.85$ s $= 1.7$ s.

(3)   The frequency of the oscillating motion of the marble is

$$\nu = \frac{1}{T}$$

where T is the period of motion. Therefore,

$$\nu = \frac{1}{1.7 \text{ s}} = 0.59 \text{ cycles s}^{-1}$$

Solved Problem 10.3. For the oscillating marble in the previous problem, derive the equation that describes the displacement of the marble.

Solution

In this problem, we need to retrieve all pertinent information concerning the oscillatory behavior of the marble and display this information in the form

$$x = A \cos \omega t$$

The amplitude of the oscillating marble is 4.0 cm. The angular frequency can be determined from the period by

$$\omega = \frac{2\pi}{T} = \frac{2\pi}{1.7 \text{ s}} = 1.18 \text{ rad s}^{-1}$$

The equation can thus be written

$$x = 4.0 \cos 1.18t \text{ (cm)}$$

Solved Problem 10.4. A force of 75 N is applied to the end of a spring, resulting in a displacement of 15.0 cm. Determine the force constant k of the spring.

Solution

Using Hooke's law for a spring

$$F = -kx$$

and solving for k, we have

$$k = -\frac{F}{x} = -\frac{75 \text{ N}}{0.15 \text{ m}} = -500 \text{ N m}^{-1}$$

The minus sign implies that the restoring force acts in a direction opposite to the external force responsible for the displacement.

Solved Problem 10.5. The force constant of a spring is 2500 N m^{-1}. Determine the force required to displace the spring 4.0 cm.

Solution

Using Hooke's law for a spring

$$F = -kx$$

and solving for F, we find

$$F = -(2500 \text{ N m}^{-1})(0.04 \text{ m}) = -100 \text{ N}$$

Solved Problem 10.6. Express the wave $x = A \cos \omega t$ in terms of (1) frequency and (2) period.

Solution

(1) The angular velocity is related to the frequency by

$$\nu = \frac{\omega}{2\pi} \qquad \text{or} \qquad \omega = 2\pi\nu$$

Substituting into the expression for the wave position gives

$$x = A \cos 2\pi\nu t$$

(2) The angular velocity is related to the period by

$$\omega = \frac{2\pi}{T}$$

Substituting into the expression for the wave position yields

$$x = A \cos \frac{2\pi t}{T}$$

Solved Problem 10.7.   The wave described by

$$x = A \cos{(\omega t + \phi)}$$

can also be expressed as

$$x = a \sin{\omega t} + b \cos{\omega t}$$

What is the amplitude of this wave?

Solution

We begin by using the following trigonometric identity:

$$\cos{(x + y)} = \cos{x} \cos{y} - \sin{x} \sin{y}$$

where $x = \omega t$ and $y = \phi$. Thus,

$$x = A \cos(\omega t + \phi) = A \cos{\omega t} \cos{\phi} - A \sin{\omega t} \sin{\phi}$$

By defining

$$a = -A \sin{\phi} \qquad \text{and} \qquad b = A \cos{\omega t}$$

then

$$x = a \sin{\omega t} + b \cos{\omega t}$$

Squaring a and b yields

$$a^2 = (-A \sin{\phi})^2 = A^2 \sin^2{\phi}$$
$$b^2 = (A \cos{\phi})^2 = A^2 \cos^2{\phi}$$

Adding a^2 and b^2 gives

$$a^2 + b^2 = A^2 \sin^2{\phi} + A^2 \cos^2{\phi} = A^2(\sin^2{\phi} + \cos^2{\phi}) = A^2$$

Thus,

$$A = \sqrt{a^2 + b^2}$$

Solved Problem 10.8.   Show that the relation

$$x = A \cos{(\omega t + \phi)}$$

satisfies Newton's second law of motion as applied to the oscillatory motion of a spring.

Solution

Newton's second law of motion is given by

$$F = ma = m\frac{d^2x}{dt^2} = -kx$$

To show that the equation for oscillatory displacement is indeed a solution to the differential equation of Newton's second law, we must differentiate the function twice and substitute the result into that for Newton's second law, at which point both sides must be equal. Taking the two derivatives, we get

$$\frac{dx}{dt} = -A\omega \sin{(\omega t + \phi)}$$

$$\frac{d^2x}{dt^2} = -A\omega^2 \cos{(\omega t + \phi)}$$

Substituting into the expression for F yields

$$m[-A\omega^2 \cos{(\omega t + \phi)}] = -k[A \cos{(\omega t + \phi)}]$$
$$m\omega^2 = k$$

or

$$\omega = \sqrt{\frac{k}{m}}$$

Thus, the displacement equation is indeed a solution to Newton's second law as it pertains to oscillatory motion.

Solved Problem 10.9.   From the equation describing the displacement of an oscillating object

$$x = A \cos(\omega t + \phi)$$

determine the (1) maximum velocity and (2) maximum acceleration for an oscillating object.

Solution

The velocity and acceleration of an oscillating object can be determined by differentiating the equation for the displacement:

$$v = \frac{dx}{dt} = -A\omega \sin(\omega t + \phi)$$

$$a = \frac{d^2 x}{dt^2} = -A\omega^2 \cos(\omega t + \phi)$$

The maximum corresponds to the coefficient of the trigonometric function. Thus,

(1) $|v_{max}| = A\omega$

(2) $|a_{max}| = A\omega^2$

10.3   DYNAMICS OF A SPRING

When displaced from equilibrium, an object connected to a spring is performing work and, depending on its position, possesses either potential or kinetic energy. At the equilibrium point ($x = 0$), the object possesses only kinetic energy. At the two extremes—maximum compression ($x = A$) and maximum elongation ($x = -A$)—the object possesses only potential energy. At all points during the oscillation, the object is continuously alternating between possessing potential energy and kinetic energy.

10.3.1   Potential Energy of a Spring

The potential energy U of a spring is given by

$$U = \frac{1}{2} kx^2$$

where k is the spring constant and x is the displacement of harmonic motion. Potential energy is maximum when $x = \pm A$ and is minimum at $x = 0$ (equilibrium point). This can be rewritten in terms of the physical characteristics of frequency and mass by

$$k = m\omega^2 = 4\pi^2 \nu^2 m$$

The potential energy now becomes, following substitution,

$$U = 2\pi^2 \nu^2 x^2 m$$

10.3.2   Kinetic Energy of a Spring

The kinetic energy (KE) of a spring is given by

$$KE = \frac{1}{2}mv^2$$

Kinetic energy is maximum at $x = 0$ (equilibrium point) and minimum (zero) when $x = \pm A$. The gain in kinetic energy is equal to the potential energy lost by the spring.

10.3.3   Total Mechanical Energy of a Spring

The total mechanical energy (TME) of a harmonic oscillator is the sum of the potential and kinetic energies of the spring and is given by

$$TME = U + KE = \frac{1}{2}kx^2 + \frac{1}{2}mv^2 = \text{constant}$$

The TME of an oscillating system is constant and depends solely on object mass, the stiffness of the spring and the vibration displacement.

Solved Problem 10.10.   Given an expression for the displacement of a simple harmonic oscillator

$$x = A\cos(\omega t + \phi)$$

determine the (1) potential energy and (2) kinetic energy of the harmonic oscillator.

Solution

(1)   The potential energy U of a harmonic oscillator is

$$U = \frac{1}{2}kx^2 = \frac{1}{2}k[A\cos(\omega t + \phi)]^2 = \frac{1}{2}k(A^2\cos^2(\omega t + \phi))$$

(2)   The kinetic energy of a harmonic oscillator is

$$KE = \frac{1}{2}mv^2$$

From Solved Problem 10.9,

$$v = \frac{dx}{dt} = -A\omega\sin(\omega t + \phi)$$

Substituting into the expression for the kinetic energy yields

$$KE = \frac{1}{2}mA^2\omega^2\sin^2(\omega t + \phi)$$

Solved Problem 10.11.   A 3-kg object connected to a spring requires a force of 5 N for a displacement of 2 cm from equilibrium. When it is set into motion, the displacement varies with time according to

$$x = 2\,\text{cm}\cos\frac{2\pi}{3}t$$

Determine the (1) potential energy and (2) kinetic energy of the harmonic oscillator.

Solution

(1)   From the previous problem, the potential energy U is given by

$$U = \frac{1}{2}kx^2$$

The spring constant k can be determined from

$$F = -kx$$

$$k = -\frac{F}{-x} = \frac{5\,\text{N}}{0.02\,\text{m}} = 250\,\text{N m}^{-1}$$

The potential energy can now be determined:

$$U = \frac{1}{2}(250\,\text{N m}^{-1})\left(0.02\,\text{m}\cos\frac{2\pi}{3}t\right)^2 = 0.05\cos^2\left(\frac{2\pi}{3}t\right)\text{J}$$

(2)   The kinetic energy is given by

$$\text{KE} = \frac{1}{2}mv^2 = \frac{1}{2}mA^2\omega^2\sin^2\omega t$$

Making the appropriate substitutions yields

$$\text{KE} = \frac{1}{2}(3\,\text{kg})(0.02\,\text{m})^2\left(\frac{2\pi}{3}\right)^2\sin^2\frac{2\pi}{3}t = 2.6\times10^{-3}\sin^2\frac{2\pi}{3}t\,\text{J}$$

10.4   PENDULUM MOTION

10.4.1   Simple Pendulum

A simple pendulum consists of a mass or bob attached to the end of a massless cord, as shown in Figure 10-4. The pendulum bob is pulled to one side and released, causing the bob to move back and forth in oscillatory motion under the force of gravity. The period of a simple pendulum is given by

$$T = 2\pi\sqrt{\frac{L}{g}}$$

where L is the length of the cord and g is the gravitational acceleration constant.

Solved Problem 10.12.   A simple pendulum consists of a 3.5-kg mass attached to a cord 75 cm in length. Determine the period of oscillation of pendulum motion.

Solution

The period of the simple pendulum can be determined from the formula:

$$T = 2\pi\sqrt{\frac{L}{g}} = 2\pi\sqrt{\frac{0.75\,\text{m}}{9.8\,\text{m s}^{-2}}} = 1.74\,\text{s}$$

Solved Problem 10.13.   Harry, a hypnotist and part-time physicist, uses a makeshift simple pendulum to hypnotize

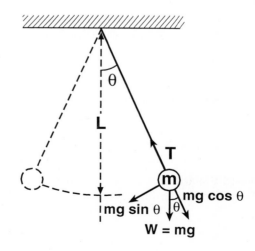

Fig. 10-4

members of the audience. He has found that he can successfully hypnotize a person by maintaining oscillations with a period of 2 s. Determine the length of the cord that will yield this period.

Solution

The period of an oscillation is related to the length of the string by

$$T = 2\pi \sqrt{\frac{L}{g}}$$

Squaring both sides of the relation and solving for L yields

$$T^2 = 4\pi^2 \left(\frac{L}{g}\right)$$

$$L = \frac{gT^2}{4\pi^2} = \frac{(9.8 \text{ m s}^{-2})(2 \text{ s})^2}{4\pi^2} = 0.99 \text{ m}$$

10.4.2   Conical Pendulum

The conical pendulum is similar to the simple pendulum in that it consists of a mass or bob attached to the end of a massless cord, as shown in Figure 10-5. However, when set into motion, the bob of the conical pendulum rotates around the vertical axis in a circular orbit, rather than being constrained to one dimension as in the simple pendulum. The period of a compound pendulum is given by

$$T = 2\pi \sqrt{\frac{L \cos \theta}{g}}$$

where L is the length of the cord and g is the gravitational acceleration constant.

Solved Problem 10.14.   Tetherball, an example of a conical pendulum, uses a ball of mass m attached to the top of an anchored pole by a long cord or string. When hit, the ball proceeds in circular motion about the vertical pole. For a tetherball of mass 2.5 kg suspended by a cord length 1.5 m in circular motion at an angle of 35° from

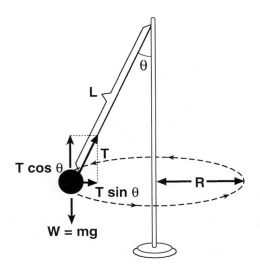

Fig. 10-5

the vertical axis, determine the (1) tension in the cord, (2) velocity of the tetherball, and (3) angular speed of the tetherball.

Solution

(1)   The two forces acting on the tetherball are the tension in the cord and acting upward at an angle θ and the weight of the ball acting downward. Applying the condition of translational equilibrium gives

$$\Sigma F_x = ma: \qquad T \sin \theta = \frac{mv^2}{r} \qquad\qquad\qquad (1)$$

$$\Sigma F_y = ma: \qquad T \cos \theta = mg \qquad\qquad\qquad (2)$$

From Eq. (2), we obtain an expression for the tension F:

$$T = \frac{mg}{\cos \theta} = \frac{(2.5 \text{ kg})(9.8 \text{ m s}^{-2})}{\cos 35°} = 29.9 \text{ N}$$

(2)   From the equations of rotational equilibrium, dividing Eq. (1) by Eq. (2) yields

$$\tan \theta = \frac{v^2}{gr}$$

$$v = \sqrt{gr \tan \theta}$$

where r is the radius of the circular orbit. The parameter r can be expressed in terms of the length of the string by elementary trigonometric identities:

$$r = L \sin \theta = (1.5 \text{ m})(\sin 35°) = 0.86 \text{ m}$$

Therefore, the velocity is

$$v = \sqrt{(9.8 \text{ m s}^{-2})(0.86 \text{ m})(\tan 35°)} = 2.43 \text{ m s}^{-1}$$

(3)   The centripetal force $F_c = m\omega^2 R$ for the motion of the ball around a circle of radius $R = L \sin \theta$ can be expressed in terms of the ball's weight:

$$mg \tan \theta = m\omega^2 L \sin \theta$$

Solving for the angular speed gives

$$\omega^2 = \frac{g}{L \cos \theta}$$

$$\omega = \sqrt{\frac{g}{L \cos \theta}} = \sqrt{\frac{9.8 \text{ m s}^{-1}}{(1.5 \text{ m})(\cos 35°)}} = 2.82 \text{ s}^{-1}$$

Solved Problem 10.15.   In Solved Problem 10.14, is the acceleration of the tetherball constant?

Solution

No. Although the magnitude of the centripetal acceleration is constant, the direction is continually changing.

10.4.3   Physical Pendulum

A physical pendulum consists of an arbitrarily shaped object with a distributed mass oscillating about a fixed pivot point, as shown in Figure 10-6. The period of a physical pendulum is given by

$$T = 2\pi \sqrt{\frac{I}{mgd}}$$

where I is the moment of inertia of the pendulum, m is the mass of the pendulum, g is the gravitational acceleration constant, and d is the distance from the center of gravity to the point of suspension or pivot.

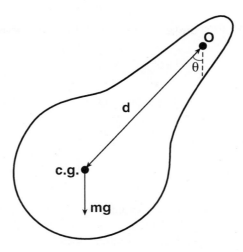

Fig. 10-6

Solved Problem 10.16. Determine the period of oscillation for a meter stick pivoted at one end.

Solution

The period of oscillation can be determined from the relation

$$T = 2\pi\sqrt{\frac{I}{mgd}}$$

The values for d, the distance from the pivot to the center of gravity of the meter stick, is one-half the length of the meter stick, or $d = 0.5$ m. The moment of inertia for a rod about an axis at its end is given by

$$I = \frac{1}{3}mL^2$$

Thus, the period of the meter stick can be rewritten as

$$T = 2\pi\sqrt{\frac{\frac{1}{3}mL^2}{mg(L/2)}} = 2\pi\sqrt{\frac{\frac{1}{3}L}{g(\frac{1}{2})}} = 2\pi\sqrt{\frac{2L}{3g}} = 2\pi\sqrt{\frac{2(1.0\text{ m})}{3(9.8\text{ m s}^{-2})}} = 1.63\text{ s}$$

ILLUSTRATIVE EXAMPLE 10.3. Physical Pendulum and the Process of Walking

The continual process of walking at a steady pace can be analyzed in terms of pendulum motion, with the leg representing a physical pendulum that is pivoted at the hip joint. The geometry of the leg can be adequately approximated as a thin rod rotating about an axis through one end. Thus, the moment of inertia for an object possessing such geometry is

$$I = \frac{mL^2}{3}$$

We now can determine the period, or the time required per stride. For a person of mass $m = 65$ kg with a leg 0.85 m in length, the period T is

$$T = 2\pi\sqrt{\frac{I}{mgd}}$$

where d is the distance between the pivot point and the center of gravity. For this discussion, $d = L/2$. The moment of inertia can now be calculated by

$$I = \frac{(65 \text{ kg})(0.85 \text{ m})^2}{3} = 15.7 \text{ kg m}^2$$

Making the appropriate substitutions, we see that the period of one stride in the walking process is

$$T = 2\pi \sqrt{\frac{15.7 \text{ kg m}^{-2}}{(65 \text{ kg})(9.8 \text{ m s}^{-2})(0.85 \text{ m}/2)}} = 1.51 \text{ s}$$

10.4.4 Torsional Pendulum

A torsional pendulum consists of a solid disk suspended by a wire attached to a fixed surface, as shown in Figure 10-7. The disk oscillates in a rotational motion about the vertical axis. Assuming a small angular displacement, the disk is subject to a restoring force or torque defined by

$$\tau = -\kappa\theta$$

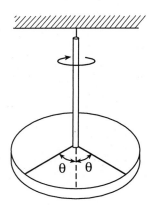

Fig. 10-7

where τ is the torque causing the angular displacement, κ is the torsional constant of the wire, and θ is the angular displacement in radians. The period of a torsional pendulum is given by the formula

$$T = 2\pi \sqrt{\frac{I}{L/\theta}}$$

where I is the moment of inertia.

Solved Problem 10.17. A metal cylinder, positioned horizontally, is suspended by a steel wire vertically attached to the top of the cylinder. A 10-N m torque is applied to the cylinder, twisting it horizontally through an angle of 35°. Once the torque is removed, the cylinder vibrates with a period of 3.2 s. Determine the moment of inertia of the cylinder.

Solution

The period of oscillation for a torsional pendulum is given by

$$T = 2\pi \sqrt{\frac{I}{L/\theta}}$$

Squaring both sides and solving for I yields

$$T^2 = 4\pi^2 \frac{I}{L/\theta}$$

$$I = \frac{T^2 L}{4\pi^2 \theta}$$

The angular displacement $\theta = 35°$ must be determined in radians.

$$\theta = (35°)\left(\frac{2\pi\,\text{rad}}{360°}\right) = \frac{\pi}{5}\,\text{rad}$$

The moment of inertia of the cylinder is determined as

$$I = \frac{(3.2\text{ s})^2 (10\text{ N m})}{4\pi^2 (\pi/5)\,\text{rad}} = 4.1\text{ kg m}^2$$

Supplementary Problems

10.1. A 0.5-kg mass, attached to a spring, is set into motion and oscillates about its point of equilibrium according to

$$x = 5\sin(6t + 4)\text{ m}$$

where x is displacement from the point of equilibrium, given in meters, and t is time, given in seconds. (*a*) What is the amplitude of the oscillatory motion? (*b*) What is the value of the angular speed? (*c*) What is the value of the initial phase angle? (*d*) What is the period of the oscillation? (*e*) What is the value of the frequency?

Solution

(*a*) 5.0 m; (*b*) 6 rad s^{-1}; (*c*) 4 rad; (*d*) $\pi/3$ s; (*e*) $3/\pi$ s^{-1}

10.2. For the mass in Problem 10.1, find the (1) maximum oscillation speed of the mass, (2) maximum oscillation acceleration of the mass, and (3) force exerted on the mass to initiate oscillation.

Solution

(1) 30 m s^{-1}; (2) 180 m s^{-2}; (3) $-90\sin(6t + 4)$ N

10.3. For the mass in Problem 10.1, find the total mechanical energy of the oscillating mass.

Solution

225 J

10.4. A mass m attached to a spring executes simple harmonic motion. Using choices of *maximum*, *minimum*, or *zero*, answer the following questions referring to the point when the spring is stretched the most:

Solution

The speed is: Zero
The acceleration is: Maximum
The potential energy is: Maximum
The kinetic energy is: Zero (because speed is zero)
The total energy is: Neither (it doesn't change with time; it's constant)

10.5. A 25-kg mass attached to a spring causes it to extend by 7×10^{-2} m. Determine the force constant of the spring.

Solution

3500 N m^{-1}

10.6. A spring with a force constant of 25 N m^{-1} oscillates with a frequency of 20 cycles min^{-1}. Determine the mass of the object in kilograms.

Solution

5.8 kg

10.7. A 2-N object attached to a spring is displaced initially 5 cm and is released, resulting in an acceleration of 2.8 m s^{-2}. Determine the period and frequency of the vibration.

Solution

0.84 s; 1.2 cycles s^{-1}

10.8. A 5-kg object exhibits simple harmonic motion with a frequency of 0.5 cycle s^{-1} when attached to a spring. Determine the frequency of vibration if a 20-kg object is attached to the spring.

Solution

0.25 cycle s^{-1}

10.9. An object of 3.2-kg mass attached to a spring experiences a force of 15 N as it is displaced 2 cm from its position of equilibrium. Determine the (1) force constant of the spring and (2) period of oscillation.

Solution

(1) 750 N m^{-1}; (2) 0.065 s

10.10. For a simple pendulum consisting of an object of mass $m = 0.5$ kg attached to a string of length $L = 0.75$ m, determine the torque τ exerted by the object at an angle of 15°.

Solution

-0.95 N m

10.11. Consider a simple pendulum consisting of an object of mass $m = 4$ kg attached to a massless string of length $L = 1.0$ m. A variable horizontal force F is exerted in pulling the object until it is displaced by an angle $\theta = 35°$ with the vertical. Determine the work done by the force F.

Solution

7.1 J

10.12. A conical pendulum consists of a string 0.8 m long attached to a mass of 3.5 kg. The pendulum rotates at an average speed of 1.2 r s^{-1}. Determine the (1) tension in the string and (2) angle between the string and the vertical axis.

Solution

(1) 159 N; (2) 77.6°

10.13. The displacement of an oscillating body that vibrates in simple harmonic motion is described by

$$x = 12 \cos\left(\frac{2\pi}{3}t + \frac{\pi}{4}\right) \text{ mm}$$

At $t = 0$, determine the (1) displacement, (2) velocity, and (3) acceleration.

Solution

(1) 0.85 cm; (2) 1.8 cm s^{-1}; (3) 3.7 cm s^{-2}

10.14. Assuming small angles, that is, $\sin \theta \approx \theta$, derive an expression for Newton's second law as it applies to the simple pendulum and suggest a solution.

Solution

$$\frac{d^2\theta}{dt^2} = -\frac{g}{L}\theta$$

where

$$\theta(t) = \theta_o \cos(\omega t + \phi)$$

10.15. A vibrating object completes 15 oscillations in 30 s. Determine the (1) period and (2) frequency of the oscillating object.

Solution

(1) 2 s; (2) 0.5 Hz

10.16. For a simple pendulum, derive an expression for the velocity of the bob in terms of its angular displacement.

Solution

$$v = L\omega\theta_o(-\sin \omega t)$$

10.17. An object oscillates in simple harmonic motion with a period of 2.5 s. At what times will the object pass the point of equilibrium?

Solution

0.63 and 1.87 s

10.18. Determine the period of a vibrating object that exhibits an acceleration of 2.4 m s^{-2} when its displacement is 0.35 m.

Solution

2.4 s

10.19. Determine the rotational kinetic energy expended by an 80-kg person with legs of length $L = 95$ cm running at a constant velocity of 12 m s^{-1}.

Solution

1913 J

Chapter 11

Elasticity

Regardless of its structural composition, an object acted on by external forces such as pushing, pulling, twisting, bending, or stretching will respond to the forces through a deformation or change in the size and/or shape of the object. This is guaranteed by Newton's third law, which states that for every force acting on an object, the object responds with an equal yet opposite reacting force. The magnitude of force required to elicit such a response depends on the object's internal structure and size. What is of interest is not how the object responds during the exertion of these forces, but how and why the object responds once these forces are removed. These questions form the physical basis for elasticity.

11.1 DEFINITIONS OF ELASTICITY

Elasticity is an inherent property of all materials describing their response to an external force. It is characterized by the presence and degree of deformation and depends primarily on the size and internal structure of the object, on the magnitude of the force, and on the time interval or rate over which the force is applied. The elasticity of a substance can be characterized by two physical parameters: stress and strain.

Stress S reflects the magnitude of an external force F required to induce the given deformation of an object, divided by the surface area A upon which the force is exerted perpendicularly, or

$$S = \frac{F}{A}$$

and is given in units of $N\,m^{-2}$, $dyn\,cm^{-2}$, or $lb\,in^{-2}$.

Strain ε is a quantitative measure of the fractional extent of deformation of an object produced as a result of the applied stresses, and it is a unitless quantity. Strain is defined as the ratio of the increase in a particular dimension in the deformed state to that dimension in its initial, undeformed state and can be expressed mathematically as

$$\varepsilon = \frac{x_d - x_n}{x_n}$$

where x_d and x_n represent the dimensional quantities of the elastic object in the deformed and normal states, respectively.

Hooke's law of elasticity describes, for any material, a linear relation between tensile stress and tensile strain with the proportionality constant defined as an elastic modulus:

$$\text{Elastic modulus} = \frac{\text{stress}}{\text{strain}}$$

Since the elastic modulus is linearly proportional to stress, a large value for the elastic modulus implies that a large stress is required to produce a given strain.

The *elastic modulus* is a physical constant which describes the elastic properties of a solid. Elastic modulus is similar to stress in that it is also expressed in units of $N\,m^{-2}$, $dyn\,cm^{-2}$, or $lb\,in^{-2}$. There are three types of elastic moduli, described below:

Young's modulus Y is given as the ratio of the tensile stress to the tensile strain and can be expressed in terms of physical parameters unique to an elastic solid by

$$\text{Young's modulus } Y = \frac{\text{tensile stress}}{\text{tensile strain}} = \frac{FL}{A\,\Delta L}$$

where F is the force applied to the solid, A is the cross-sectional area of the solid, L is the original length of the solid, and ΔL is the fractional amount of deformation as a result of the applied force.

Shear modulus G relates the shear stress to the shear or angular strain and is given by

$$\text{Shear modulus } G = \frac{\text{shear stress}}{\text{shear strain}} = \frac{FL}{A\,\Delta L}$$

However, note that the strain or $\Delta L/L$ is generally small, and hence the elastic modulus can be simplified by a small-angle approximation

$$\frac{\Delta L}{L} \approx \phi$$

where ϕ is the angle by which the elastic solid is displaced by the shearing stress. Thus, the elastic modulus G becomes

$$G = \frac{F}{A\phi}$$

Bulk modulus B is the elastic modulus due to volumetric compression of an elastic solid and is given by

$$B = -\frac{V\Delta P}{\Delta V}$$

where V is volume and P is pressure. The minus sign is used to negate the negative effects of volumetric change, making the bulk modulus positive. Because the volumetric units cancel, the bulk modulus has units of pressure, or $N\,m^{-2}$, $dyn\,cm^{-2}$, or $lb\,in^{-2}$.

The physical interactions of elasticity and their corresponding equations described above are summarized in Figure 11-1.

Solved Problem 11.1. Given that the Young's modulus for steel is $2 \times 10^{11}\,dyn\,cm^{-2}$, determine the force required to stretch a wire of 1-mm^2 cross-sectional area by 10 percent of its original length.

Solution

Young's modulus is defined as

$$Y = \frac{\text{stress}}{\text{strain}} = \frac{F/A}{\Delta L/L}$$

From the text of the problem, $\Delta L = 0.1L$, $Y = 2 \times 10^{11}\,dyn\,cm^{-2}$, and

$$A = 1\,mm^2 \cdot \frac{1\,cm^2}{100\,mm^2} = 0.01\,cm^2$$

The force can now be found:

$$F = (2 \times 10^{11}\,dyn\,cm^{-2})(0.01\,cm^2)(0.1) = 2 \times 10^8\,dyn$$

Solved Problem 11.2. A 2.5-kg mass is attached to one end of a sample of artery with 0.05-cm radius and 15-mm length. The other end of the artery sample is secured to a rigid support. Determine the elongation of the artery sample if $Y = 1 \times 10^8\,N\,m^{-2}$.

TYPE OF ELASTICITY	VISUAL DESCRIPTION	ELASTICITY RELATIONS

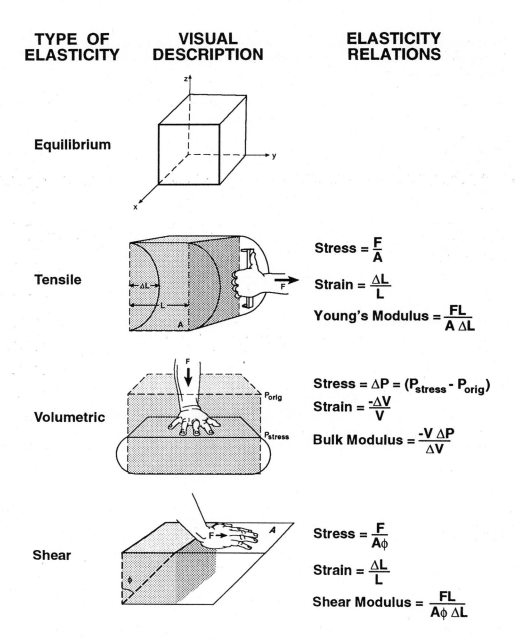

Equilibrium

Tensile

$$\text{Stress} = \frac{F}{A}$$

$$\text{Strain} = \frac{\Delta L}{L}$$

$$\text{Young's Modulus} = \frac{FL}{A \, \Delta L}$$

Volumetric

$$\text{Stress} = \Delta P = (P_{stress} - P_{orig})$$

$$\text{Strain} = \frac{-\Delta V}{V}$$

$$\text{Bulk Modulus} = \frac{-V \, \Delta P}{\Delta V}$$

Shear

$$\text{Stress} = \frac{F}{A\phi}$$

$$\text{Strain} = \frac{\Delta L}{L}$$

$$\text{Shear Modulus} = \frac{FL}{A\phi \, \Delta L}$$

Fig. 11-1

Solution

The mass attached to the artery sample exerts a force $F = mg$. The strain of the elastic object is given by

$$\varepsilon = \frac{\Delta L}{L_o} = \frac{S}{Y} = \frac{F}{AY}$$

and it can be used to derive an expression for the change in length as a result of the mass:

$$\Delta L = \frac{FL_o}{A_o Y} = \frac{mgL_o}{\pi r^2 Y} = \frac{(2.5 \text{ kg})(9.8 \text{ m s}^{-2})(15 \times 10^{-3} \text{ m})}{(3.14)(0.5 \times 10^{-3} \text{ m})^2(1 \times 10^8 \text{ N m}^{-2})} = 4.6 \text{ mm}$$

Solved Problem 11.3.   A sample of muscle in a relaxed state elongates 4.5 cm when subjected to a force of 40 N. The same muscle sample under conditions of maximum tension requires a force of 675 N to achieve the same elongation. Assuming that the muscle sample can be approximated as a uniform cylinder of 0.35-m length and 80-cm^2 cross-sectional area, determine (1) the stress, (2) the strain, and (3) Young's modulus for the muscle sample in both cases.

Solution

For the muscle in the relaxed state,

$$\text{Stress} = \frac{F}{A} = \frac{40\,\text{N}}{80 \times 10^{-4}\,\text{m}^2} = 5.0 \times 10^3\,\text{N m}^{-2}$$

$$\text{Strain} = \frac{\Delta L}{L} = \frac{0.045\,\text{m}}{0.35\,\text{m}} = 0.13$$

$$Y = \frac{\text{stress}}{\text{strain}} = \frac{5.0 \times 10^3\,\text{N m}^{-2}}{0.13} 3.85 \times 10^4\,\text{N m}^{-2}$$

For the muscle under maximum tension,

$$\text{Stress} = \frac{F}{A} = \frac{675\,\text{N}}{80 \times 10^{-4}\,\text{m}^2} = 8.4 \times 10^4\,\text{N m}^{-2}$$

$$\text{Strain} = \frac{\Delta L}{L} = \frac{0.045\,\text{m}}{0.35\,\text{m}} = 0.13$$

$$Y = \frac{\text{stress}}{\text{strain}} = \frac{8.4 \times 10^4\,\text{N m}^{-1}}{0.13} = 6.46 \times 10^5\,\text{N m}^{-2}$$

Solved Problem 11.4.   A steel rod 0.25 m in diameter and 1.5 m long can support a total load of 10,000 kg. Given that Young's modulus for steel is $2 \times 10^{11}\,\text{N m}^{-2}$, determine the (1) stress, (2) strain, and (3) change in length of the rod.

Solution

(1)   The cross-sectional area of the rod is

$$A = \pi r^2 = (3.14)(0.125\,\text{m})^2 = 0.05\,\text{m}^2$$

$$\text{Stress} = \frac{F}{A} = \frac{(10,000\,\text{kg})(9.8\,\text{m s}^{-2})}{0.05\,\text{m}^2} = 1.96 \times 10^6\,\text{N m}^{-2}$$

(2)   The strain can be determined by the definition for Young's modulus:

$$Y = \frac{\text{stress}}{\text{strain}} \Rightarrow \text{strain} = \frac{\text{stress}}{Y}$$

$$\text{Strain} = \frac{1.96 \times 10^6\,\text{N m}^{-2}}{2.0 \times 10^{11}\,\text{N m}^{-2}} = 9.8 \times 10^{-6}$$

(3)   To determine the change in length of the rod,

$$\text{Strain} = \frac{\Delta L}{L} \Rightarrow \Delta L = \text{strain} \cdot L = (9.8 \times 10^{-6})(1.5\,\text{m}) = 1.48 \times 10^{-5}\,\text{m} = 0.015\,\text{mm}$$

Solved Problem 11.5.   A force of 50 pN is required to stretch a sample of DNA of length 50 nm and Young's modulus $Y = 1 \times 10^8\,\text{N m}^2$ by 10 percent of its original length. Determine the cross-sectional area of the DNA molecule.

Solution

We begin by first calculating the strain on the DNA molecule:

$$\text{Strain} = \frac{\Delta L}{L} = \frac{5\,\text{nm}}{50\,\text{nm}} = 0.1$$

Knowing Young's modulus, we can calculate the stress:

$$\text{Stress} = Y \cdot \text{strain} = (1 \times 10^8\,\text{N m}^{-2})(0.1) = 1 \times 10^7\,\text{N m}^{-2} = \frac{F}{A}$$

Solving for A gives

$$A = \frac{50 \times 10^{-12}\,\text{N}}{1 \times 10^7\,\text{N m}^{-2}} = 5.0 \times 10^{-18}\,\text{m}^2 = 5.0\,\text{nm}^2$$

Solved Problem 11.6.   In creating his dessert specialty, Chef Pierre Flambé must mold a block of gelatin with a 20-cm^2 surface area of the top and a height of 6 cm. In molding the gelatin, Chef Pierre applies a shearing force of 0.75 N to the top surface, causing a net displacement of 2.5 mm relative to the bottom surface. Determine the (1) shear stress, (2) shear strain, and (3) shear modulus.

Solution

(1)   The shear stress can be calculated by

$$\text{Shear stress} = \frac{\text{tangential force}}{\text{surface area}} = \frac{0.75\,\text{N}}{20 \times 10^{-4}\,\text{m}^2} = 375\,\text{N m}^{-2}$$

(2)   The shear strain is determined by

$$\text{Shear strain} = \frac{\text{displacement}}{\text{height}} = \frac{0.25\,\text{cm}}{6\,\text{cm}} = 0.042$$

(3)   The shear modulus is

$$\text{Shear modulus} = \frac{\text{shear stress}}{\text{shear strain}} = \frac{375\,\text{N m}^{-2}}{0.042} = 8928\,\text{N m}^{-2}$$

Solved Problem 11.7.   The top surface of a lead cube of side length $L = 0.05$ m is subjected to a shearing force of 500 N. Given the shear modulus for lead $G = 5.5 \times 10^9\,\text{N m}^{-2}$, determine the (1) shear stress, (2) shear strain, and (3) lateral displacement of the cube.

Solution

(1)   To determine the shear stress, we must determine the surface area of a face of the cube:

$$A = L^2 = (0.05\,\text{m})^2 = 0.0025\,\text{m}^3$$

The shear stress is

$$\text{Shear stress} = \frac{\text{tangential force}}{\text{surface area}} = \frac{500\,\text{N}}{2.5 \times 10^{-3}\,\text{m}^2} = 2.0 \times 10^5\,\text{N m}^{-2}$$

(2)   The shear strain is related to the shear stress by the shear modulus, or

$$G = \frac{\text{shear stress}}{\text{shear strain}} \Rightarrow \text{shear strain} = \frac{\text{shear stress}}{G} = \frac{2.0 \times 10^5\,\text{N m}^{-2}}{5.5 \times 10^9\,\text{N m}^{-2}} = 3.6 \times 10^{-5}$$

(3)   The lateral displacement is related to shear strain by

$$\text{Shear strain} = \frac{\text{displacement}}{\text{height}}$$

$$\text{Displacement} = \text{shear strain} \cdot \text{height} = (3.6 \times 10^{-5})(0.05 \text{ m}) = 1.8 \times 10^{-6} \text{ m}$$

Solved Problem 11.8.   A pressure of 30×10^5 N m^{-2} is applied to a volume of 1×10^{-4} m^3 of mercury. Given that the bulk modulus for mercury is $B = 2.8 \times 10^{10}$ N m^{-2}, determine the decrease in volume.

Solution

By definition, the bulk modulus is related to the volume by

$$B = -\frac{PV}{\Delta V}$$

Solving for ΔV gives

$$\Delta V = -\frac{PV}{B} = -\frac{(30 \times 10^5 \text{ N m}^{-2})(1.0 \times 10^{-4} \text{ m}^3)}{2.8 \times 10^{10} \text{ N m}^{-2}} = -1.1 \times 10^{-8} \text{ m}^3$$

Solved Problem 11.9.   To reduce the volume of 500 in^3 of water by 2.0 percent, a pressure of 7000 lb in^{-2} must be applied. Determine the bulk modulus for water.

Solution

Computing the change in volume as a result of the pressure, we find

$$\Delta V = (0.02)(500 \text{ in}^3) = 10.0 \text{ in}^3$$

and it is negative since the volume decreases or contracts. By definition,

$$B = -\frac{PV}{\Delta V} = \frac{-(7000 \text{ lb in}^{-2})(500 \text{ in}^3)}{-10.0 \text{ in}^3} = 3.5 \times 10^5 \text{ lb in}^{-2}$$

ILLUSTRATIVE EXAMPLE 11.1.   Elastic Properties of Blood Vessels

The circulatory system is composed primarily of three types of blood vessels: arteries, veins, and capillaries. Arteries transport oxygen-rich blood from the heart under high pressure to beds of capillary vessels embedded in tissues and organs of the human body. The role of the veins is the exact opposite of that of arteries in that they transport oxygen-poor blood away from the capillary bed under low pressure and return it to the heart. Each blood vessel consists of four types of structural components in varying proportions: endothelial cell lining, smooth muscle, elastin fibers, and collagen fibers.

Endothelial cell lining provides the smooth surface of the inner wall of the vessel and permits selective permeability to the various substances transported in the bloodstream. It does not, however, contribute to the elasticity of the vessel wall.

Smooth muscle performs an important physiological role by contracting and relaxing the vessel wall to control the flow of proper volume of blood through the vessel. Its elastic modulus ranges from 6×10^4 to 6×10^7 dyn cm^{-2}.

Elastin and collagen are structural proteins that contribute dominantly to the elasticity of the blood vessel. Elastin fibers are easily stretched and possess an elastic modulus of 5×10^6 dyn cm^{-2} while collagen fibers are much stronger and more resistant to stretching with an elastic modulus of 1×10^8 dyn cm^{-2}.

ILLUSTRATIVE EXAMPLE 11.2.   Stresses of the Leg during Movement

The majority of bones of the skeletal system are continually subjected to stresses as a result of everyday activity and movement, some bones more than others. Consider a 75-kg person positioned on the balls of one foot, as would commonly be experienced during walking or running. We want to determine the stress exerted on the shinbone or tibia. The weight of the person $W = 735$ N is acting downward on the leg. Added to this weight is the contribution of tension from the Achilles tendon, connected at the heel on one end and the thigh bone just above the knee cap at the other end. The Achilles tendon pulls downward with a force (tension) of 1350 N. Thus, the combined weight exerted on the ankle is the sum of these two forces, or 2085 N. This force is counteracted by a force equal in magnitude and opposite in direction acting upward on the tibia. Therefore, a compressional force of 2085 N is exerted on the tibia. Assuming that the cross-sectional area of the tibia is 4.0 cm^2, the stress exerted on the tibia is

$$S = \frac{F}{A} = \frac{2085\,\text{N}}{4 \times 10^{-4}\,\text{m}^2} = 5.2 \times 10^6\,\text{N m}^{-2}$$

ILLUSTRATIVE EXAMPLE 11.3.   Bone Fracture from a Fall

All solid objects, regardless of their material structure and composition, can absorb energy and safely avoid permanent distortion or deformation. When we discuss human bones, this becomes of particular importance since the amount of energy imparted to a bone dictates the severity of injury inflicted to the limb. Consider two bones of the leg—the femur and tibia or fibula, with a total length $L = 1.0$ m and an average cross-sectional area $A = 5$ cm^2. Since the elastic modulus for bone is $Y = 1 \times 10^{10}$ dyn cm^{-2} and the breaking stress for bone is $S_b = 1 \times 10^9$ dyn cm^{-2}, the force required to fracture the bone is

$$F = SA = (1 \times 10^9\,\text{dyn cm}^{-2})(5\,\text{cm}^2) = 5 \times 10^9\,\text{dyn}$$

From this force, we can determine the amount of energy capable of being stored within the bone:

$$E = \frac{1}{2}\frac{ALS_b^2}{Y} = \frac{1}{2}\frac{(5\,\text{cm}^2)(100\,\text{cm})(1 \times 10^9\,\text{dyn cm}^{-2})^2}{(1 \times 10^{10}\,\text{dyn cm}^{-2})} = 2.5 \times 10^{10}\,\text{erg}$$

Converting the energy from ergs to newton meters or joules, we have

$$(2.5 \times 10^{10}\,\text{erg})\left(\frac{1\,\text{N m}}{10^7\,\text{erg}}\right) = 2500\,\text{J}$$

Assuming that the person lands on both legs, the energy is doubled, yielding

$$E = 5000\,\text{J}$$

We now want to find out, if this energy corresponded to the potential energy of a person, from what height the person must jump to attain this energy upon impact with the ground. Assuming the person has a mass of 70 kg,

$$\text{Potential energy} = mgh$$
$$5000\,\text{J} = (70\,\text{kg})\,(9.8\,\text{m s}^{-2})\,(h)$$

Solving for h gives

$$h = 7.28\,\text{m}$$

It should be emphasized that there are a number of other factors, such as surface texture and weather conditions, which could impact the energy transferred to the leg bones and hence alter their likelihood of fracture upon impact.

11.2   ELASTIC LIMIT

The elastic limit of a body represents the minimum value of stress required to produce a permanent deformation. If the body has been exposed to a stress that exceeds its elastic limit, then the body will not revert to its original state following removal of the stress.

Solved Problem 11.10.   An aluminum wire 5 mm in length and 4.0 mm in diameter can support a mass of 80 kg. Determine the (1) elongation of the wire and (2) maximum mass that the wire can support without exceeding its elastic limit. Young's modulus for aluminum is 7×10^{10} N m^{-2}, and the elastic limit for aluminum is 1.3×10^8 N m^{-2}.

Solution

(1) Before we can calculate the stress placed on the wire, we must determine the force exerted on the wire by the load and the cross-sectional area of the wire. The force exerted on the wire as a result of the load is

$$F = mg = (80 \text{ kg})(9.8 \text{ m s}^{-2}) = 784 \text{ N}$$

The cross-sectional area of the aluminum wire is

$$A = \pi r^2 = (3.14)(2 \times 10^{-3} \text{ m})^2 = 12.5 \times 10^{-6} \text{ m}^2$$

From the definition for the elastic modulus,

$$Y = \frac{\text{stress}}{\text{strain}} = \frac{F/A}{\Delta L/L}$$

Solving for ΔL gives

$$\Delta L = \frac{L_o F}{YA} = \frac{(5.0 \text{ m})(784 \text{ N})}{(7 \times 10^{10} \text{ N m}^{-2})(12.5 \times 10^{-6} \text{ m}^2)} = 4.5 \text{ mm}$$

(2) For this problem, the elastic limit becomes the stress, and given the cross-sectional area of the wire, we want to determine the applied force F.

$$\text{Stress} = \frac{F}{A} \Rightarrow F = \text{stress} \cdot A = (1.3 \times 10^8 \text{ N m}^{-2})(12.5 \times 10^{-6} \text{ m}^2) = 1625 \text{ N}$$

From the force, we can determine the mass:

$$m = \frac{F}{g} = \frac{1625 \text{ N}}{9.8 \text{ m s}^{-2}} = 165.8 \text{ kg}$$

11.3 LAPLACE'S LAW

Laplace's law describes the relation between the circumferential tension and the radius for any curved elastic surface. Laplace's law, in effect, is based on Newton's third law and equates a force F_P produced by the transmural blood pressure over the cross-sectional area of the structure to a circumferential force (tension) T that compensates for the distension and is required to maintain equilibrium. Laplace's law can be applied to various organs and structures of the human body including the heart, lungs, stomach, blood vessels, eye, diaphragm, colon, bowel, ureter, bladder, and uterus. The dimensions of tension are MT^{-2}, and the units of tension are force per unit length, that is, N m^{-1}, dyn cm^{-1}, or lb in^{-1}.

ILLUSTRATIVE EXAMPLE 11.4. Application of Laplace's Law for an Elastic Cylinder: Forces Acting within a Blood Vessel

Blood vessels are complex elastic tubes through which blood circulates in the human body and can be approximated as long circular cylinders. Consider the diagram of the cylindrical vessel with radius R and length L in Figure 11-2. In static equilibrium, two forces are exerted along the vessel wall in opposing directions. The first force results from a circumferential tension T acting along both sides of the curved portion of the vessel and is defined mathematically by

$$F_{\text{down}} = (T \sin \theta)(2L)$$

while the second opposing force is the result of the intraluminal pressure P acting outward against an area of the vessel in a radial direction. The affected vessel area A has length L, multiplied by width $W = 2 \cdot r\theta$, yielding the outward force F_{out}:

$$F_{\text{out}} = P \cdot L \cdot 2r\theta$$

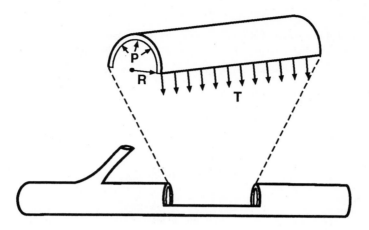

Fig. 11-2

Since these are the only two forces acting on the cylinder, they may now be equated:

$$F_{\text{down}} = F_{\text{out}}$$
$$(T\sin\theta)(2L) = P \cdot L \cdot 2r\theta$$

The angle θ is small, allowing for the small-angle approximation:

$$\sin\theta \approx \theta$$

The force equation may be rewritten as

$$T \cdot \theta \cdot 2L = P \cdot L \cdot 2r\theta$$

Solving for T yields the equation for tension in the wall of an elastic cylinder:

$$T = Pr$$

ILLUSTRATIVE EXAMPLE 11.5.   Application of Laplace's Law for an Elastic Sphere: Forces Acting within a Brain Aneurysm

A brain aneurysm is a ballooning or saclike distension of a diseased region of a blood vessel wall and can be approximated as a sphere. Consider the free-body diagram shown in Figure 11-3 applied to the cross-sectional geometry of a sphere of radius r. There are two forces acting on the spherical surface. One force, which acts outward, against the spherical surface, is that produced by the pressure, denoted by F_{out}. Pressure is a radial force that acts perpendicular to the vessel surface at every point along the inner surface of the sphere. Thus, the force due to pressure F_{out} is simply the product of pressure and the cross-sectional area of the sphere, or

$$F_{\text{out}} = P\pi r^2$$

Force F_{out} is countered by another force due to tension F_{down} acting downward along the wall of the sphere. Force F_{down} is the average tension force keeping the upper half of the elastic sphere attached to the lower half and is defined by

$$F_{\text{down}} = 2\pi r T$$

where T is the average tension. Equating these two forces and solving for T yield

$$P\pi r^2 = 2\pi r T$$
$$T = \frac{Pr}{2}$$

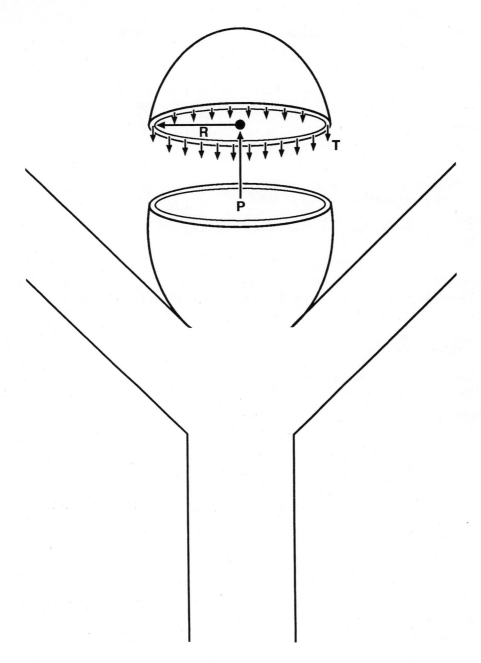

Fig. 11-3

Supplementary Problems

11.1. A rectangular steel bar of length 15 m has a cross-section 5 cm × 2.5 cm. When a 1500-kg load is suspended from one end of the vertical bar, it is elongated by 2 mm. Determine (1) the stress, (2) the strain, and (3) Young's modulus for the steel bar.

Solution

(1) $1.18 \times 10^7 \, \text{N m}^{-2}$; (2) 1.33×10^{-4}; (3) $8.9 \times 10^{10} \, \text{N m}^{-2}$

11.2. A steel column 5.0 m in length with a cross-sectional area of $A = 6.5 \times 10^{-4} \, \text{m}^2$ can support a load of $3.3 \times 10^4 \, \text{N}$. Given Young's modulus for steel $Y = 2 \times 10^{11} \, \text{N m}^{-2}$, determine the length by which the column will shorten.

Solution

1.27 mm

11.3. Determine the decrease in volume of $2.5 \times 10^{-4} \, \text{m}^3$ of mercury when subjected to a pressure of $20 \times 10^5 \, \text{N m}^{-2}$. The bulk modulus for mercury is $B = 2.8 \times 10^{10} \, \text{N m}^{-2}$.

Solution

$-1.78 \times 10^{-12} \, \text{m}^3$

11.4. A load of 2500 N is applied to the lower end of a bone 0.2 m in length and 1.5 cm^2 in cross-sectional area. Given that Young's modulus for bone is $2 \times 10^{10} \, \text{N m}^{-2}$, determine the change in length of bone.

Solution

0.16 mm

11.5. Determine the volumetric decrease of 1.8 m^3 of water when subjected to a pressure of $5 \times 10^6 \, \text{N m}^{-2}$. The bulk modulus for water is $B = 2.1 \times 10^{11} \, \text{N m}^{-2}$.

Solution

$4.3 \times 10^{-5} \, \text{m}^3$

11.6. Determine the pressure required to compress a given volume of water V_o to 99 percent of its volume. The bulk modulus for water is $2.2 \times 10^9 \, \text{N m}^{-2}$.

Solution

$2.2 \times 10^7 \, \text{N m}^{-2}$

11.7. Determine the maximum length of elongation that a steel wire 2 m in length and 6 mm in diameter can experience prior to exceeding its elastic limit. For the steel wire $Y = 2 \times 10^{11} \, \text{N m}^{-2}$, and the elastic limit is $2.5 \times 10^8 \, \text{N m}^{-2}$.

Solution

2.5 mm

11.8. A steel cube of side length $L = 0.5 \, \text{m}$ is subjected to a shearing force of 5000 N along the top surface. Given the shear modulus for steel $G = 8.3 \times 10^{10} \, \text{N m}^{-2}$, determine the shear angle.

Solution

$2.4 \times 10^{-7} \, \text{rad}$

11.9. Express Laplace's law for an elastic sphere in terms of its volume.

Solution

$0.31 P V^{1/3}$

11.10. Given an elastic sphere with an internal pressure of 150 mmHg and a radius $r = 0.4$ cm, what is the circumferential tension experienced by the wall of the elastic sphere?

Solution

5.0×10^5 dyn cm^{-1}

11.11. A large artery can be approximated as a tube of 2-cm diameter with a pressure of 100 mmHg. Determine the tension force exerted on the wall of the artery.

Solution

133 N m^{-1}

Chapter 12

Fluid Statics

The physical behavior of a solid particle can be represented and understood easily because it constitutes a single entity of sufficient size that we can visualize such behavior as well. Extension of the same observations becomes more complex when dealing with fluids since we are, in effect, dealing with an ensemble of "virtual" particles that cannot be visualized. The term *fluid* is used to describe an object or substance which must be in motion to resist externally applied forces or stresses. Note that although one tends to think of fluids primarily as liquids, fluids also describe the behavior of gases. This chapter presents the physical principles, concepts, and examples of a fluid at rest (fluid statics).

12.1 DEFINITIONS OF FLUID STATICS

Density ρ is a physical property of a fluid, given as mass per unit volume, or

$$\rho = \frac{\text{mass}}{\text{unit volume}} = \frac{m}{V}$$

Density represents the fluid equivalent of mass and is given in units of ML^{-3}.

Specific gravity (sp. gr.) of a given substance is the ratio of the density of the substance ρ_{sub} to the density of water ρ_w, or

$$\text{Sp. gr.} = \frac{\text{density of substance}}{\text{density of water}} = \frac{\rho_{sub}}{\rho_w}$$

where the density of water ρ_w is $1.0\,\text{g cm}^{-3}$ or $1 \times 10^3\,\text{kg m}^{-3}$. Assuming that equal volumes are chosen, the specific gravity can be expressed in terms of weight:

$$\text{Sp. gr.} = \frac{\text{weight of substance}}{\text{weight of water}} = \frac{w_{sub}}{w_w}$$

Specific gravity is a pure number that is dimensionless and unitless.

Pressure P is defined as a force F acting perpendicular to a surface area A and is given by

$$\text{Pressure} = \frac{\text{force}}{\text{area}} = \frac{F}{A}$$

Pressure is a scalar quantity and is expressed in dimensions of $ML^{-1}T^{-2}$. The SI units for pressure are N m^{-2}. Two specific types of pressure particularly applicable to fluids include atmospheric pressure and hydrostatic pressure.

Atmospheric pressure P_{atm} represents the average pressure exerted by the earth's atmosphere at sea level and is defined numerically as 1 atm but can also be expressed as $760\,\text{mmHg}$, $1.01 \times 10^5\,\text{Pa}$, or $14.7\,\text{lb in}^{-2}$.

Hydrostatic pressure P_{hyd} is the fluid pressure exerted at a depth h in a fluid of density ρ and is given by

$$P_{hyd} = \rho g h$$

If an external pressure P_{ext} is exerted on the contained fluid, then the total pressure P is the sum of the external pressure and the hydrostatic pressure:

$$P = P_{ext} + \rho g h$$

where atmospheric pressure, in most cases, is considered an external pressure.

Solved Problem 12.1.   Thirty milliliters of an anesthetic solution drawn into a 5-g syringe is found to have a combined mass of 80 g. Determine the density of the anesthetic solution.

> **Solution**
>
> The density of the anesthetic solution can be determined from
>
> $$\rho = \frac{m}{V}$$
>
> where ρ is density (g cm^{-3}), m is mass (g), and V is volume (cm^3 or mL). The mass of the anesthetic solution m_a is the total mass m_T minus that of the syringe m_s, or
>
> $$m_T = m_a + m_s$$
> $$m_a = m_T - m_s = 80 \text{ g} - 5 \text{ g} = 75 \text{ g}$$
>
> The volume V of anesthetic solution contained in the syringe is
>
> $$V = 30 \text{ mL} = 30 \text{ cm}^3$$
>
> Thus, the density of the anesthetic solution can be determined from
>
> $$\rho = \frac{75 \text{ g}}{30 \text{ mL}} = 2.5 \text{ g mL}^{-1} \text{ or } 2.5 \text{ g cm}^{-3}$$

Solved Problem 12.2.   The density of a radiopharmaceutical is 0.75 g cm^{-3}. Determine the mass of 2.0 L of this radiopharmaceutical. (*Note*: 1 L = 1000 cm^3.)

> **Solution**
>
> The mass of a liquid or fluid is related to the density and volume by
>
> $$\rho = \frac{m}{V}$$
>
> Rearranging and solving for the mass m yield
>
> $$m = \rho V = (0.75 \text{ g cm}^{-3})(2000 \text{ cm}^3) = 1500 \text{ g} = 1.5 \text{ kg}$$

Solved Problem 12.3.   Determine the appropriate size of container needed to hold 0.7 g of ether which has a density of 0.62 g cm^{-3}.

> **Solution**
>
> The volume of a liquid or fluid is related to mass and density by
>
> $$V = \frac{m}{\rho} = \frac{0.7 \text{ g}}{0.62 \text{ g cm}^{-3}} = 1.129 \text{ mL}$$

Solved Problem 12.4.   A solid aluminum cube has dimensions of length 6 in on each side. Given that the density of aluminum is 170 lb ft^{-3}, determine its mass.

> **Solution**
>
> By definition, $m = \rho V$. The volume of the cube is
>
> $$V = 0.5 \text{ ft} \times 0.5 \text{ ft} \times 0.5 \text{ ft} = 0.125 \text{ ft}^3$$
>
> Therefore, the mass is
>
> $$m = (170 \text{ lb ft}^{-3})(0.125 \text{ ft}^3) = 21.3 \text{ lb}$$

Solved Problem 12.5. Bone has a density of 1.06 g cm^{-3}. Determine the specific gravity of bone.

Solution

By definition, specific gravity is

$$\text{Sp. gr.} = \frac{\rho_{\text{bone}}}{\rho_w} = \frac{1.06 \text{ g cm}^{-3}}{1.00 \text{ g cm}^{-3}} = 1.06$$

Solved Problem 12.6. As calculated in the previous problem, the specific gravity of bone is 1.06. Determine the mass of 1 cm^3 of bone.

Solution

Mass is related to specific gravity by

$$\text{Sp. gr.} = 1.06 = \frac{\rho_{\text{bone}}}{\rho_w} = \frac{m_{\text{bone}}/V_{\text{bone}}}{1.00 \text{ g cm}^{-3}}$$

Solving for m_{bone} yields

$$m_{\text{bone}} = 1.06 \text{ g}$$

Solved Problem 12.7. In playing a record, a phonograph needle supports 3.5 g within a circular area of 0.30-mm radius. Determine the pressure exerted by the phonograph needle on the record.

Solution

The pressure exerted by the phonograph needle can be determined from the definition for pressure:

$$P = \frac{F}{A} = \frac{mg}{\pi r^2} = \frac{(3.5 \times 10^{-3} \text{ kg})(9.8 \text{ m s}^{-2})}{3.14(0.30 \times 10^{-3} \text{ m})^2} = 1.21 \times 10^5 \text{ N m}^{-2}$$

Solved Problem 12.8. Normal systolic blood pressure of human circulation is 120 mmHg. Determine the equivalent height of a column of water.

Solution

To determine the hydrostatic pressure exerted by a column of mercury of 120 mm:

$$P = \rho g h = (13.6 \text{ g cm}^{-3})(980 \text{ cm s}^{-2})(12 \text{ cm}) = 1.6 \times 10^5 \text{ dyn cm}^{-2}$$

We now want to determine the height of a column of water required to exert the same pressure as 120 mmHg:

$$1.6 \times 10^5 \text{ dyn cm}^{-2} = (1.0 \text{ g cm}^{-3})(980 \text{ cm s}^{-2})(h \text{ cmH}_2\text{O})$$

Solving for h yields

$$h = 163 \text{ cmH}_2\text{O}$$

Solved Problem 12.9. Assume that the area of a foot for an 80-kg person is $25 \text{ cm} \times 6 \text{ cm}$. Determine the pressure that the person exerts on the ground while standing.

Solution

Pressure is defined as force per unit area, where the force is the weight of the person W:

$$W = mg = (80 \text{ kg})(9.8 \text{ m s}^{-2}) = 784 \text{ N}$$

and the area is the total cross-sectional area over which this force is exerted:

$$A_{\text{foot}} = \pi(0.25 \text{ m} \times 0.06 \text{ m}) = 0.047 \text{ m}^2$$

Since the person typically stands on two feet, the total area is $2A_{foot} = 0.094\,\text{m}^2$. Thus, the presssure exerted by the person on the ground is

$$P = \frac{F}{A} = \frac{78\,\text{N}}{0.094\,\text{m}^2} = 8340\,\text{N}\,\text{m}^{-2}$$

Solved Problem 12.10.   Assuming the density of water is $1.0\,\text{g}\,\text{cm}^{-3}$, determine the pressure at the bottom of a swimming pool at depths of (1) 60 cm, (2) 120 cm, and (3) 180 cm.

Solution

There are two components of pressure acting on the bottom of the swimming pool: hydrostatic pressure due to the water and atmospheric pressure ($1.0\,\text{atm} = 1.013 \times 10^5\,\text{N}\,\text{m}^{-2}$), thus yielding the net or total pressure

$$P_{total} = P_{atm} + P_{hyd} = P_{atm} + \rho g h$$

(1)   At a depth of 60 cm, the total pressure exerted on the bottom of the pool is

$$P_{total} = P_{atm} + \rho g h = 1.013 \times 10^5\,\text{N}\,\text{m}^{-2} + (1000\,\text{kg}\,\text{m}^{-3})(9.8\,\text{m}\,\text{s}^{-2})(0.6\,\text{m}) = 1.07 \times 10^5\,\text{N}\,\text{m}^{-2}$$

(2)   At a depth of 120 cm, the total pressure exerted on the bottom of the pool is

$$P_{total} = P_{atm} + \rho g h = 1.013 \times 10^5\,\text{N}\,\text{m}^{-2} + (1000\,\text{kg}\,\text{m}^{-3})(9.8\,\text{m}\,\text{s}^{-2})(1.2\,\text{m}) = 1.13 \times 10^5\,\text{N}\,\text{m}^{-2}$$

(3)   At a depth of 180 cm, the total pressure exerted on the bottom of the pool is

$$P_{total} = P_{atm} + \rho g h = 1.013 \times 10^5\,\text{N}\,\text{m}^{-2} + (1000\,\text{kg}\,\text{m}^{-3})(9.8\,\text{m}\,\text{s}^{-2})(1.8\,\text{m}) = 1.19 \times 10^5\,\text{N}\,\text{m}^{-2}$$

ILLUSTRATIVE EXAMPLE 12.1.   Physical Properties of Human Blood

Human blood is a complex fluid with a physical density of $1.06\,\text{g}\,\text{cm}^{-3}$ and consists primarily of two components, a particulate component and an aqueous component. The particulate component consists of a suspension of red blood cells (or erythrocytes), white blood cells (or leukocytes), and platelets. Red blood cells are $8\,\mu\text{m}$ in diameter and are responsible for the transport of oxygen to tissues and the removal of CO_2. White blood cells are 10 to $20\,\mu\text{m}$ in diameter and protect the body from disease. Platelets are $2.5\,\mu\text{m}$ in diameter and play an integral role in blood clot formation or thrombosis. The aqueous component, or plasma, is a mixture consisting primarily of nutrients, salts, and blood proteins.

ILLUSTRATIVE EXAMPLE 12.2.   Pressure in the Eye and Glaucoma

In front of the human eye, clear fluid flows in and out of a small duct, called the anterior chamber. The purpose of this fluid is to cleanse and provide nutrients to adjacent tissues. The fluid is continuously produced in the eye and subsequently drained from the eye. Normal intraocular pressure, or pressure within the eye, is approximately 15 mmHg. If the drainage becomes obstructed, the fluid accumulates in the eye, corresponding to an increase of intraocular pressure to 25 to 50 mmHg and the clinical onset of glaucoma. The pressure increase in the eyeball increases wall stress in the eyeball and compresses the optic nerve, both combining to adversely affect vision and ultimately damage the eyeball. Treatment of glaucoma involves a surgical procedure to remove the drainage obstruction and reestablish drainage flow from the eyeball.

ILLUSTRATIVE EXAMPLE 12.3.   Pressure and Infection

Once a foreign, microbial agent has infiltrated the body's defense mechanisms, an infection evolves in the surrounding tissue. Once an infection occurs, the infected cells release a certain biochemical known as a *vasodilator* which acts to dilate or enlarge adjacent capillary vessels. Enlargement of the capillary vessels causes an increase in blood flow to the infected region accompanied by a concomitant increase in pressure. Pressure increases in the infected region represent the source of swelling and inflammation associated with infection. Although the increase in blood flow can function, in some instances, to accelerate the healing process, antibiotics are often required to eradicate the infected cells and restore the infected region to its normal state.

ILLUSTRATIVE EXAMPLE 12.4.   Intravenous Delivery

For intravenous (IV) delivery of nutrients, fluids, blood, and drugs, positive pressure is exerted from the substance in a container placed a distance h above a vein through a thin, flexible tube. The other end of the tube is inserted directly into the vein, which exhibits a venous pressure of 3 mmHg. The pressure due to the IV fluid is proportional to the height of the surface of the needle, that is, the vein, as shown in Figure 12-1. Consider, as an example, the pressure required for blood transfusion. Given the density of blood is $1.04\,\mathrm{g\,cm^{-3}}$, we want to determine the net pressure acting to transfer the blood into the vein.

$$P_{\mathrm{blood}} = \frac{\rho_{\mathrm{blood}}}{\rho_{\mathrm{Hg}}} \times 10^3\,\mathrm{mmHg} = \frac{1.04\,\mathrm{g\,cm^{-3}}}{13.6\,\mathrm{g\,cm^{-3}}} \times 10^3\,\mathrm{mmHg} = 76.5\,\mathrm{mmHg}$$

This is the pressure exerted by the blood on the vein. This pressure is counteracted by the venous pressure P_{vein}, which is $3\,\mathrm{mmHg}$. Thus, the net pressure or the difference between P_{blood} and P_{vein} is

$$P_{\mathrm{net}} = 76.5\,\mathrm{mmHg} - 3\,\mathrm{mmHg} = 73.5\,\mathrm{mmHg}$$

Delivery through the vein offers the least resistance to the transfer of blood or any other substance into the human body.

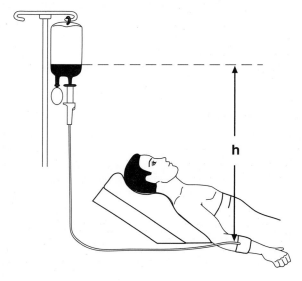

Fig. 12-1

ILLUSTRATIVE EXAMPLE 12.5.   Blood Pressure in Human Circulation

    Besides blood flow, the most important parameter in the circulation is the blood pressure. The blood pressure must be sufficient to drive the blood from the heart and into the blood vessels, but must also be low enough to create a pressure gradient and allow efficient draining of the blood back to the heart for additional cycles. As a consequence, the blood pressure varies significantly at various points in the circulation, allowing the different blood vessels to accommodate their assigned role and function.

    The left ventricle ejects blood from the heart into the arterial system under a pressure of approximately $120\,\mathrm{mmHg}$. This corresponds to the systolic phase of the cardiac cycle and is commonly referred to as *systolic blood pressure*. As blood penetrates further into the circulatory system, the pressures start to decline steadily. From the aorta, blood flows through the larger arteries at $110\,\mathrm{mmHg}$, through the medium arteries at $75\,\mathrm{mmHg}$, and through the smaller arteries or arterioles at $40\,\mathrm{mmHg}$ until it reaches the capillary bed. Blood enters the capillary bed under a pressure of $30\,\mathrm{mmHg}$ and exits under a pressure of $16\,\mathrm{mmHg}$. The low pressure is sufficient to generate movement yet slow enough for the blood to perform its physiological function of sustaining cellular metabolism. Blood drains from the capillary bed into the smallest veins or venules at $16\,\mathrm{mmHg}$, continuing into the medium-sized veins under a pressure of $12\,\mathrm{mmHg}$, and into the large veins at $4\,\mathrm{mmHg}$ before reentering the heart for another circulatory cycle. The pressures within the circulatory system range from 120 to $4\,\mathrm{mmHg}$ and maintain this range on a continual basis.

    Several specific pressures can be readily identified. The systolic blood pressure and diastolic blood pressure values for a normal person are 120 and 80 mmHg, respectively. During a physical examination, the systemic blood pressure is typically presented as

$$\text{Systemic blood pressure} = \frac{\text{systolic blood pressure}}{\text{diastolic blood pressure}} = \frac{120 \text{ mmHg}}{80 \text{ mmHg}}$$

The *pulse pressure* is the difference between systolic and diastolic pressures and is therefore normally about 50 mmHg. In most cases, it is more convenient to condense these two blood pressure readings into a single one which also represents the overall status of the blood pressure in a patient. This is accomplished through a *mean blood pressure* BP_{mean}, defined by

$$\text{BP}_{\text{mean}} = \text{systolic BP} + \frac{2 \times \text{diastolic BP}}{3}$$

The *mean pressure* is the average pressure throughout the cardiac cycle. As systole is shorter than diastole, the mean pressure is slightly less than the value halfway between the systolic and diastolic pressure. It can be determined only by integrating the area under a blood pressure-time curve; however, as an approximation, it is the diastolic pressure plus one-third of the pulse pressure.

ILLUSTRATIVE EXAMPLE 12.6.   The Sphygmomanometer and Blood Pressure Measurement

The most common method of blood pressure measurement involves an instrument called the *sphygmomanometer*. To obtain a measurement of blood pressure, a rubber cuff is wrapped around the upper part of the arm, directly over the artery to be compressed, that is, the brachial artery. The cuff is inflated with air by squeezing an attached bulb, compressing the artery until the pressure exerted by the cuff around the arm is greater than the expected pressure of the blood. At this point, the circulation is stopped. The pressure of the air in the cuff required to stop the circulation is transmitted to a mercury manometer where it is subsequently recorded. Pressure of the air is slowly released from the cuff until blood resumes flow through the artery. This pressure, indicated by a characteristic sound heard from a stethoscope placed over the brachial artery, is the systolic pressure. As the pressure in the cuff is released further, the sounds change until they ultimately disappear. At this point, the recorded pressure is the diastolic pressure.

12.2   SURFACE TENSION

Surface tension γ is the tension, or force per unit length, created by the cohesive forces of molecules on the surface of a liquid acting toward the interior. Surface tension is given as force per unit length and defined as the ratio of the surface force F to the length d along which the force acts, or

$$\gamma = \frac{F}{d}$$

Surface tension is given in units of mN m^{-1} or dyn cm^{-2}.

12.3   CAPILLARY ACTION

Capillary action refers to the rise or fall of a liquid in a narrow tube or capillary, as shown in Figure 12-2, causing the formation of a curved surface or the meniscus at the wall of the tube, with the height h given by

$$h = \frac{2\gamma \cos \theta}{r \rho g}$$

where γ is the surface tension, θ is the angle of the contact point between the capillary wall and the tangent to the liquid surface, and r is the radius of the capillary tube.

Solved Problem 12.11.   A capillary tube of inner radius $r = 0.30$ mm is partially submerged in water with a surface tension $\gamma = 72 \text{ dyn cm}^{-1}$ and an angle of contact $\theta = 0°$. Determine the height of the rise of water in the capillary tube.

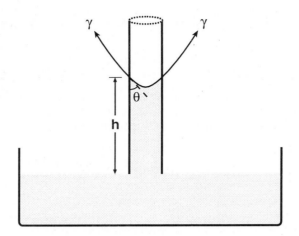

Fig. 12-2

Solution

The height of the rise of water through the capillary tube is

$$h = \frac{2\gamma \cos\theta}{\rho g r} = \frac{2 \times 72 \text{ dyn cm}^{-1} \times \cos 0°}{(1 \text{ g cm}^{-3})(980 \text{ cm s}^{-2})(0.03 \text{ cm})} = 4.89 \text{ cm}$$

ILLUSTRATIVE EXAMPLE 12.7.   Lung Surfactant and Respiratory Distress Syndrome

Respiratory distress syndrome (RDS) is characterized by difficulty of breathing due to underdeveloped lungs and is typically found in premature newborns. In normal respiratory physiology, air enters the trachea (windpipe), passes through bronchi, bronchioles, and eventually ends within alveoli, or tiny bubblelike sacs where the gaseous exchange of O_2 and CO_2 occurs. The alveoli possess a natural tendency to contract or become smaller in size due primarily to a layer of water mixed with a unique chemical called *lung surfactant* that lines the inner wall surface of the alveoli. Lung surfactant possesses a low surface tension, $\gamma = 25 \text{ mN m}^{-1}$, which is critically important in preventing collapse of the alveoli and maintaining blood flow to the capillary vessels. In premature newborns afflicted with RDS, a clear membrane is found within the alveoli, corresponding to reduced amounts of surfactant. If the surfactant is reduced or missing, the alveoli become lined with a higher proportion of water, which exhibits a much larger surface tension: $\gamma = 75 \text{ mN m}^{-1}$. The larger surface tension acts to contract the alveoli to a larger degree and, in most instances, to the point of collapse. Lungs with collapsed alveoli require increased work of respiration with a substantiated reduction in the volume of airflow exchanged.

12.4   PASCAL'S PRINCIPLE

Pascal's principle states, "An external pressure applied to a confined fluid will be transmitted equally to all points within the fluid."

Solved Problem 12.12.   An example of Pascal's principle is seen in the hydraulic jack, shown in Figure 12-3. If a force of 300 N is applied to a piston of 1-cm² cross-sectional area, determine the lifting force transmitted to a piston of cross-sectional area of 100 cm².

Solution

According to Pascal's principle,

$$P_{1 \text{ cm}^2} = P_{100 \text{ cm}^2}$$

$$\left(\frac{F}{A}\right)_{1 \text{ cm}^2} = \left(\frac{F}{A}\right)_{100 \text{ cm}^2}$$

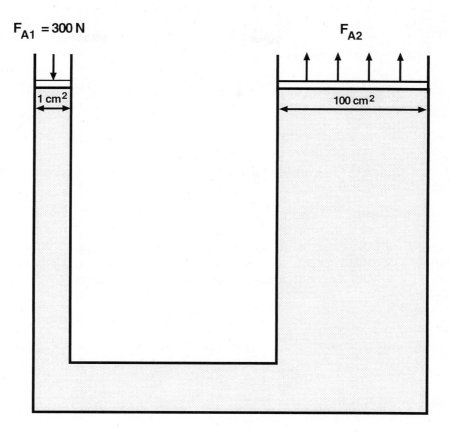

Fig. 12-3

Making the appropriate substitutions yields

$$\frac{300 \text{ N}}{1 \text{ cm}^2} = \frac{F}{100 \text{ cm}^2}$$

Solving for F yields

$$F = 30{,}000 \text{ N}$$

12.5 ARCHIMEDES' PRINCIPLE

Archimedes' principle states, "A body immersed wholly or partially in a fluid is subjected to a buoyant force that is equal in magnitude to the weight of the fluid displaced by the body," or

$$\text{Buoyant force} = \text{weight of displaced fluid}$$

If the buoyant force is equal to or greater than the weight of the displaced fluid, then the object remains afloat. However, if the buoyant force is less than the weight of the displaced fluid, then the object sinks.

Solved Problem 12.13.   A humpback whale weighs 5.4×10^5 N. Determine the buoyant force required to support the whale in its natural habitat, the ocean, when it is completely submerged. Assume that the density of seawater is 1030 kg m^{-3} and that the density of the whale ρ_{whale} is approximately equal to the density of water ($\rho_{\text{water}} = 1000 \text{ kg m}^{-3}$).

Solution

The volume of the whale can be determined from

$$m = \rho V = \frac{W}{g}$$

Solving for V yields

$$V_{\text{whale}} = \frac{W_{\text{whale}}}{\rho_{\text{whale}}g} = \frac{5.4 \times 10^5 \, \text{N}}{(1000 \, \text{kg m}^{-3})(9.8 \, \text{m s}^{-2})} = 55.1 \, \text{m}^3$$

The whale displaces $55.1 \, \text{m}^3$ of water when submerged. Therefore, the buoyant force (BF), which is equal to the weight of displaced water, is given by

$$\text{BF} = W_{\text{seawater}} = \rho_{\text{seawater}}gV_{\text{whale}} = (1030 \, \text{kg m}^{-3})(9.8 \, \text{m s}^{-2})(55.1 \, \text{m}^3) = 5.6 \times 10^5 \, \text{N}$$

Solved Problem 12.14. In a sideshow attraction at the county fair, one is asked to estimate the number of coins submerged in a large container of water. Suppose a 1-L container was filled with coins of mass $m = 3.0 \, \text{g}$, such that 0.1 L of water was displaced. Determine the number of coins in the container, assuming the density of the coin is $8.9 \, \text{g cm}^{-3}$.

Solution

Given the mass of the coin is $m = 3.0 \, \text{g}$ and the density is $\rho = 8.9 \, \text{g cm}^{-3}$, the volume of a single coin is

$$V_{\text{coin}} = \frac{m_{\text{coin}}}{\rho_{\text{coin}}} = \frac{3.0 \, \text{g}}{8.9 \, \text{g cm}^{-3}} = 0.337 \, \text{cm}^3 = 0.337 \, \text{mL}$$

The number of coins is thus given by

$$\text{No. of coins} = \frac{\text{volume displaced}}{V_{\text{coin}}} = \frac{0.1 \, \text{L}}{0.337 \times 10^{-3} \, \text{L}} = 297 \, \text{coins}$$

Supplementary Problems

12.1. Halothane, an anesthetic agent, has a density $\rho = 1.86 \, \text{g cm}^{-3}$ and is in a test tube of volume $V = 10 \, \text{mL}$. Determine the mass of halothane.

Answer

$18.6 \, \text{g}$

12.2. Given a 5-mL volume of ethyl alcohol ($\rho = 0.81 \, \text{g cm}^{-3}$), determine the volume of water ($\rho = 1.00 \, \text{g cm}^{-3}$) required for the mass to equal that of ethyl alcohol.

Answer

$4.05 \, \text{mL}$

12.3. Determine the force exerted by water on the bottom of an aquarium tank $0.9 \, \text{m} \times 0.5 \, \text{m}$ if the water level is at $0.85 \, \text{m}$.

Answer

$4.95 \times 10^4 \, \text{N}$

12.4. Determine the hydrostatic pressure due to a 50-cm column of (1) water ($\rho_{water} = 1.0 \text{ g cm}^{-3}$); and (2) mercury ($\rho_{mercury} = 13.6 \text{ g cm}^{-3}$).

Solution

(1) 4.9×10^4 dyn cm^{-2}; (2) 6.6×10^5 dyn cm^{-2}

12.5. Consider a mercury barometer or a U-shaped glass tube with one end or arm sealed and filled with mercury, as shown in Figure 12-4. Determine the height h of the mercury column if the pressure at the open-ended arm is (1) atmospheric pressure and (2) 100 mmHg.

Solution

(1) 760 mm; (2) 97 mm

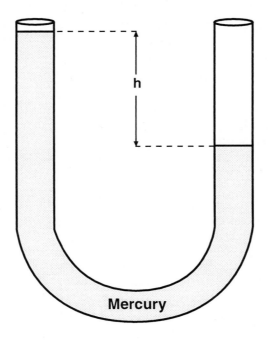

Fig. 12-4

12.6. Determine the hydrostatic pressure and total pressure at a depth of 4.0 m in a swimming pool.

Solution

3.92×10^4 N m^{-2}; 1.4×10^5 N m^{-2}

12.7. Consider the piston of cross-sectional area 10^{-2} m^2 exerting a force of 600 N on a water-filled container of height $h = 0.8$ m (see Figure 12-5). Given the density of water of $\rho = 1000$ kg m^{-3}, determine the pressure exerted on the bottom of the container of 10^{-1} m^2.

Solution

67,840 N m^{-2}

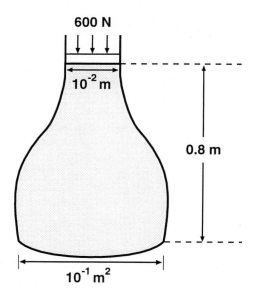

Fig. 12-5

12.8. Consider an inverted cone of height $h = 0.45$ m and radius $r = 0.15$ m filled with water. Determine the weight and force of the fluid acting downward at the base of the cone.

Solution

103.8 N; 311.6 N

12.9. Consider an aquarium tank with a base 3.5 ft by 2.0 ft filled with water to a depth of 4.0 ft. Determine the total pressure exerted on the bottom of the tank.

Solution

2367 lb ft^{-2}

12.10. Determine the pressure required to raise water to the top of a building 20 m high.

Solution

1.96×10^5 N m^{-2}

12.11. An aneurysm can be approximated as an elastic sphere with a small opening by which blood circulates and exerts pressure against the inner wall. Determine the force, in newtons, exerted by the blood on an aneurysm, given a blood pressure of 150 mmHg and a cross-sectional area of 25 cm^2.

Solution

49.8 N

12.12. A capillary tube of inner radius $r = 0.2$ mm is partially submerged in ethyl alcohol. The surface tension and density of ethyl alcohol are 22.3 dyn cm^{-1} and 8×10^2 kg m^{-3}, respectively. Given that the contact angle is 0°, determine the height of rise of ethyl alcohol in the capillary tube.

Solution

2.81 cm

12.13. Referring to the piston assembly of the hydraulic jack in Figure 12-3, determine the force required by a piston $A_1 = 1 \text{ cm}^2$ to lift an object of 1200 N over a surfacer area $A_2 = 100 \text{ cm}^2$.

Solution

12 N

12.14. A capillary tube of 1.2-mm inner diameter is partially submerged in water. Water rises in the capillary tube to a height of 18.5 mm. Given that the density and surface tension of water are 1.06 g cm^{-3} and 72.8 dyn cm^{-1}, respectively, determine (1) the surface force and (2) the weight of the water in the capillary tube.

Solution

(1) 13.7 dyn; (2) 13.7 dyn

12.15. For a solution in a test tube, determine the pressure energy (P), gravitational potential energy (GPE), and the total energy (TE) at (1) the surface level of the solution and (2) the bottom of the test tube.

Solution

(1) $P = P_{atm}$; GPE = 0; TE = P_{atm}; (2) $P = P_{atm} + \rho g h$; GPE = $-\rho g h$; TE = P_{atm}

12.16. For his trip around the world, Billy Bob will use a helium-filled balloon with a volume of 2300 m^3 and a total weight of 5500 N. Given that the density of helium is 0.178 kg m^{-3}, determine the maximum load that the balloon can lift.

Solution

$1.96 \times 10^4 \text{ N}$

Chapter 13

Fluid Dynamics

Although fluids differ from solids in terms of structure and composition, fluids possess inertia, as defined by their density, and are thus subject to the same physical interactions as solids. For example, if acted on by an external force, fluids will accelerate. Once in motion, fluids possess energy by which work can be done. All these dynamic interactions as they pertain to fluids in motion are discussed in this chapter.

13.1 DEFINITIONS OF FLUID DYNAMICS

Viscosity η of a fluid is an inherent property that exerts a resistance or frictional force against the fluid motion in response to shear stress. The SI unit for viscosity is $N\ s\ m^{-2}$ or $kg\ m^{-1}\ s^{-1}$. Viscosity is typically expressed in units of poise (P) where

$$1 \text{ poise (P)} = 0.1 \text{ kg m}^{-1}\text{ s}^{-1}$$

Fluid flow Q through a rigid, cylindrical tube of radius R and length L subjected to a constant external pressure gradient ΔP can be expressed as

$$Q = \frac{\pi}{8} \frac{\Delta P\, r^4}{L\eta}$$

where η is fluid viscosity. Fluid flow Q is the rate of volume flow through a pipe and is given in dimensions of $L^3 T^{-1}$ and units of $cm^3\ s^{-1}$ or $mL\ s^{-1}$. Fluid flow can also be expressed in terms of the velocity or speed of flow by

$$Q = Av$$

where A is the cross-sectional area of the pipe and v is flow velocity, expressed in units of $cm\ s^{-1}$.

Solved Problem 13.1. Determine the change in fluid flow for (1) a decrease in the pressure gradient by one-half, (2) an increase in viscosity by 2, (3) a decrease in vessel length by one-half, and (4) an increase in vessel radius by 2.

Solution

The effect of the various parameters on fluid flow can be determined by analysis of their qualitative dependence according to Poiseuille's law:

$$Q = \frac{\pi}{8} \frac{\Delta P\, r^4}{L\eta}$$

(1) Fluid flow Q is directly dependent on the pressure gradient ΔP. Thus, a decrease in pressure gradient by one-half implies

$$\Delta P = \frac{\Delta P}{2}$$

Substituting into Poiseuille's law gives

$$Q = \frac{\pi}{8} \frac{(\Delta P/2)\, r^4}{L\eta} = \frac{1}{2}\left(\frac{\pi}{8} \frac{\Delta P\, r^4}{L\eta}\right) = \frac{Q}{2}$$

Thus, a decrease in pressure gradient by one-half results in a decrease in fluid flow by one-half.

(2) Fluid flow Q is inversely dependent on the fluid viscosity η. Thus, an increase in fluid viscosity by 2 implies

$$\eta = 2\eta$$

Substituting into Poiseuille's law, we have

$$Q = \frac{\pi}{8} \frac{\Delta P \, r^4}{L(2\eta)} = \frac{1}{2}\left(\frac{\pi}{8} \frac{\Delta P \, r^4}{L\eta}\right) = \frac{Q}{2}$$

Thus, an increase in fluid viscosity by 2 results in a decrease in fluid flow by one-half.

(3) Fluid flow Q is inversely dependent on the vessel length L. Thus, a decrease in vessel length by one-half implies

$$L = \frac{L}{2}$$

Substituting into Poiseuille's law yields

$$Q = \frac{\pi}{8} \frac{\Delta P \, r^4}{(L/2)\,\eta} = 2\left(\frac{\pi}{8} \frac{\Delta P \, r^4}{L\eta}\right) = 2Q$$

Thus, a decrease in vessel length by one-half results in an increase in fluid flow by 2.

(4) Fluid flow Q is dependent on the vessel radius r to the fourth power. Thus, an increase in vessel radius by 2 implies

$$r = (2r)^4 = 16r^4$$

Substituting into Poiseuille's law, we find

$$Q = \frac{\pi}{8} \frac{\Delta P (2r)^4}{L\eta} = 16\left(\frac{\pi}{8} \frac{\Delta P \, r^4}{L\eta}\right) = 16Q$$

Thus, an increase in vessel radius by 2 results in an increase in fluid flow by 16.

Solved Problem 13.2. Gasoline flows from a pump 3.0 cm in diameter at an average velocity of 10 cm s^{-1}. Determine the flow rate of the gasoline.

Solution

Flow rate is related to velocity by

$$Q = Av = (\pi r^2)v = (3.14)(1.5 \text{ cm})^2(10 \text{ cm s}^{-1}) = 70.6 \text{ cm}^3 \text{ s}^{-1} = 70.6 \text{ mL s}^{-1}$$

Solved Problem 13.3. Oil is flowing through a circular tube of radius $r = 0.15$ m. If the volumetric flow rate is measured to be $0.50 \text{ m}^3 \text{ s}^{-1}$ at a certain point, determine: (1) the velocity of the oil at this point and (2) the volume of oil that passes this point over 1 min.

Solution

(1) The volumetric flow rate is related to the velocity by

$$Q = Av$$

Solving for v and making the appropriate substitutions, we have

$$v = \frac{Q}{A} = \frac{0.50 \text{ m}^3 \text{ s}^{-1}}{\pi(0.15 \text{ m})^2} = 7.1 \text{ m s}^{-1}$$

(2) The flow volume V is related to the volumetric flow rate Q by

$$V = Qt = (0.50 \text{ m}^3 \text{ s}^{-1})(60 \text{ s}) = 30 \text{ m}^3$$

Solved Problem 13.4.   Blood is pumped from the heart at a rate of 5 L min^{-1} into the aorta of radius 2.0 cm. Assuming that the viscosity and density of blood are 4×10^3 N s m^{-2} and 1×10^3 kg m^{-3}, respectively, determine the velocity of blood through the aorta.

Solution

The blood flow velocity is related to the volumetric flow rate by

$$\text{Velocity} = \frac{\text{flow rate}}{\text{cross-sectional area}}$$

where

$$\text{Flow rate} = \frac{5 \times 10^{-3}\,\text{m}^3}{\text{min}} \cdot \frac{1\,\text{min}}{60\,\text{s}} = 8.33 \times 10^{-5}\,\text{m}^3\,\text{s}^{-1}$$

$$\text{Cross-sectional area} = \pi r^2 = (3.14)(0.02\,\text{m})^2 = 1.26 \times 10^{-3}\,\text{m}^2$$

Therefore, the blood flow velocity is

$$v = 6.6 \times 10^{-2}\,\text{m s}^{-1} \text{ or } 6.6\,\text{cm s}^{-1}$$

Solved Problem 13.5.   The number of capillary vessels in the human circulation is approximately 1×10^9 with the diameter and length of each vessel being 8 μm and 1 mm, respectively. Assuming cardiac output is 5 L min^{-1}, determine (1) the average velocity of blood flow through the capillary vessels, (2) the time taken for blood to traverse a single capillary vessel, and (3) the time required for 1 mL of blood to flow through a single capillary vessel at a normal flow rate.

Solution

(1)   The blood flow velocity through the capillary vessels can easily be determined by

$$v = \frac{Q}{nA} = \frac{5 \times 10^3\,\text{mL min}^{-1}}{(1 \times 10^9\,\text{capillary vessels})(3.14)(4 \times 10^{-4}\,\text{cm})^2} = 9.9\,\text{cm min}^{-1} = 0.17\,\text{cm s}^{-1}$$

(2)   The time taken for the blood to traverse a single capillary vessel is given by

$$t = \frac{1\,\text{mL}}{(5 \times 10^3/10^9\,\text{capillary vessels})\,\text{mL min}^{-1}} = 2 \times 10^5\,\text{min} = 139\,\text{days}$$

(3)   The time required for 1 mL of blood to flow through a single capillary vessel at a normal flow rate is

$$t = \frac{1.0\,\text{mm}}{1.66\,\text{mm s}^{-1}} = 0.60\,\text{s}$$

ILLUSTRATIVE EXAMPLE 13.1.   Water Transport in Plants

The transport of water in plants occurs through extremely small capillary vessels known as *xylem*. Xylem vessels or elements are elongated tubules, typically of diameter 1 to 20 μm and length 100 to 500 μm, joined end to end. Since water transport through xylem tissue is similar to fluid flow through a vessel, it becomes useful to apply Poiseuille's law to determine the hydrostatic pressure gradient. To use Poiseuille's law as stated in the text, one must know the volume of fluid flowing through the stem during a given time interval, the viscosity of fluid (assumed to be water with a viscosity $\eta = 0.001$ Pa s at 20°C), and length and radius of the capillary vessels in the xylem tissue. Every parameter is readily available with the exception of flow volume. Flow volume is extremely difficult to determine since not all vessels measured within the cross section of a stem are functional in water transport. This limitation can be circumvented by expressing Poiseuille's law in terms of velocity, which can easily be determined directly from plants. Two expressions of volumetric flow rate are

$$Q = \frac{\pi}{8}\frac{\Delta P\,r^4}{L\eta} \qquad \text{and} \qquad Q = Av$$

Equating these two expressions and solving for v yields

$$v = \frac{\Delta P}{L} \frac{r^2}{8\eta}$$

or

$$\Delta P = \frac{8\eta v L}{r^2}$$

Assume for a given plant that a 10% sucrose solution, $\eta = 0.015$ dyn s cm^{-2}, is driven at a velocity of 325 cm h^{-1} through xylem vessels of radius 2.4 μm and length 5.0 μm. The hydrostatic pressure gradient can be calculated as

$$\Delta P = \frac{8\eta v L}{r^2} = \frac{8(0.015 \text{ dyn s cm}^{-2})(0.09 \text{ cm s}^{-1})(5.0 \times 10^{-4} \text{ cm})}{5.8 \times 10^{-8} \text{ cm}^2} = 93.1 \text{ dyn cm}^{-2}$$

13.2   EQUATION OF CONTINUITY

The *equation of continuity* states that (1) the total volume of an incompressible fluid, that is, fluid that maintains constant density regardless of changes in pressure and temperature, entering the tube will be equal to that exiting the tube and (2) flow measured at one point along the tube will be equal to the flow at another point along the tube, regardless of the cross-sectional area of the tube at each point. This can be expressed in equation form as

$$Q = A_1 v_1 = A_2 v_2 = \text{constant}$$

The equation of continuity is an illustration of the conservation of mass.

Solved Problem 13.6.   In one draining system, a pipe of 25-cm inner diameter drains into a connected pipe of 22-cm inner diameter. If the water velocity through the larger pipe is 5 cm s^{-1}, determine the average velocity in the smaller pipe.

   Solution

      Using the equation of continuity, we have

$$A_1 v_1 = A_2 v_2$$
$$\pi (12.5 \text{ cm})^2 (5 \text{ cm s}^{-1}) = \pi (11.0 \text{ cm})^2 (v_2)$$

   Solving for v_2 yields $v_2 = 6.46$ cm s^{-1}.

Solved Problem 13.7.   How would the results of Solved Problem 13.6 change if oil were used instead of water?

   Solution

      There would be no change in the results. The equation of continuity is applicable only to fluids that are incompressible, that is, of constant density. The density factors out of the relationship between the volumetric flow rate and flow velocity.

Solved Problem 13.8.   Assuming flow of an incompressible fluid, if the velocity measured at one point within a vessel is 40 cm s^{-1}, what is the velocity at a second point which is one-third the original radius?

   Solution

      This problem can be solved by using the continuity equation

$$\rho_1 A_1 v_1 = \rho_2 A_2 v_2$$

where ρ is the density of blood, A is the cross-sectional area of the vessel, v is velocity, and subscripts 1 and 2 refer to positional locations within the vessel. Since the blood flow is incompressible

$$\rho_1 = \rho_2$$

and $v_1 = 40 \text{ cm s}^{-1}$, $A_2 = A_1/3$ or $A_1/A_2 = 3$. Solving for v_2 gives

$$v_2 = \frac{A_1 v_1}{A_2} = 3v_1 = 3 \times 40 \frac{\text{cm}}{\text{s}} = 120 \frac{\text{cm}}{\text{s}}$$

ILLUSTRATIVE EXAMPLE 13.2.   Blood Flow in a Tapering Blood Vessel

One example of the continuity equation relates to blood flow in a vessel (Figure 13-1). Although the geometry of human blood vessels can be approximated by an elastic cylinder, anatomically it also tapers, or steadily decreases in size. The continuity equation guarantees that flow is equal at any point along the tapered tube, regardless of the degree of tapering. Thus, the relation between the flow at any two points within the tapered vessel is dependent on the cross-sectional area A of the vessel at the points of interest and their corresponding velocities v, which are given according to

$$A_1 v_1 = A_2 v_2$$

or

$$\frac{A_1}{A_2} = \frac{v_2}{v_1}$$

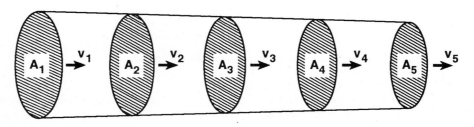

$$A_1 v_1 = A_2 v_2 = A_3 v_3 = A_4 v_4 = A_5 v_5$$

Fig. 13-1

ILLUSTRATIVE EXAMPLE 13.3.   Blood Flow in a Vessel Bifurcation

Another specific example of the continuity equation is blood flow at a vessel bifurcation. A vessel bifurcation in the circulatory system is a major junction or branching of a parent vessel into two other smaller, daughter vessels (Figure 13-2). Approximations of blood flow within branching arteries are possible through the equation of continuity:

Flow in parent artery = flow in branching or daughter arteries

$$A_p \mathbf{v}_p = A_{d1} \mathbf{v}_{d1} + A_{d2} \mathbf{v}_{d2}$$

where A = cross-sectional area of the vessel and $\mathbf{v}$ = blood flow velocity.

13.3   BERNOULLI'S PRINCIPLE

Bernoulli's principle, the fluid equivalent of conservation of energy, states that the energy of fluid flow through a rigid vessel by a pressure gradient is equal to the sum of the kinetic energy, gravitational potential energy, and the pressure energy, or

$$E_{\text{tot}} = P + \frac{1}{2}\rho v^2 + \rho gh = \text{constant}$$

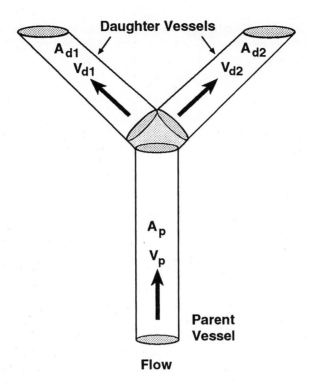

Fig. 13-2

An important application of Bernoulli's principle involves fluid flow through a vessel with a region of expansion or contraction. Bernoulli's principle describing fluid flow through a vessel with sudden changes in geometry can be expressed as

$$\left(P + \frac{1}{2}\rho v^2 + \rho gh\right)_1 = \left(P + \frac{1}{2}\rho v^2 + \rho gh\right)_2$$

where 1 describes the energy of fluid flow in the normal region of the vessel and 2 describes the energy of fluid flow in the obstructed or enlarged region.

Solved Problem 13.9.   Express Bernoulli's law, given as

$$P_1 - P_2 = \frac{\rho v_2^2}{2}\left[1 - \left(\frac{v_1}{v_2}\right)^2\right]$$

in terms of the cross-sectional area of the two points within the vessel assuming GPE = 0.

   Solution

      Based on the continuity equation,

$$A_1 v_1 = A_2 v_2$$

$$\left(\frac{v_1}{v_2}\right)^2 = \left(\frac{A_2}{A_1}\right)^2$$

Substituting into the expression for Bernoulli's principle gives

$$P_1 - P_2 = \frac{\rho v_2^2}{2}\left[1 - \left(\frac{A_2}{A_1}\right)^2\right]$$

Solved Problem 13.10. Bernoulli's principle can be applied to a syringe to describe the dynamics of an injection. Assuming that 1 refers to the position within the body of the syringe and 2 is the position within the throat of the syringe or region prior to entrance into the needle, derive an expression for the velocity of fluid exiting the throat and entering the needle.

Solution

This problem can be solved by starting with Bernoulli's principle

$$P_1 - P_2 = \frac{\rho v_2^2}{2}\left[1 - \left(\frac{A_2}{A_1}\right)^2\right]$$

and solving for v_2, yielding

$$v_2 = \sqrt{\frac{2(P_1 - P_2)}{\rho[1 - (A_2/A_1)^2]}}$$

Solved Problem 13.11. What is the kinetic energy per unit volume of blood that has a speed of 0.5 m/s? (*Note*: $\rho_{\text{blood}} = 1000 \text{ kg m}^{-3}$.)

Solution

The kinetic energy of the blood is given by

$$\text{KE} = \frac{1}{2}mv^2 = \frac{1}{2}(\rho V)v^2$$

Substituting the values given in the problem yields

$$\text{KE} = \frac{1}{2}\left(1000\,\frac{\text{kg}}{\text{m}^3}\,V\right)\left(0.5\,\frac{\text{m}}{\text{s}}\right)^2 = 500(0.25)\,\frac{\text{J}}{\text{m}^3} = 125\,\frac{\text{J}}{\text{m}^3}$$

ILLUSTRATIVE EXAMPLE 13.4.   Bernoulli's Principle and Vessel Disease

In the majority of cases of vessel disease, the arteries responsible for the transport of blood supply through the circulatory system suffer one of two fates: (1) expansion in the development of an aneurysm or ballooning of the vessel wall (Figure 13-3) or (2) constriction resulting in an obstructed or clogged vessel due primarily to atherosclerosis (Figure 13-4), commonly referred to as ''hardening of the arteries,'' which is the development of calcified, fibrotic growths that accumulate along the innermost layer of the vessel wall.

The premise behind applications of Bernoulli's principle to a vessel with an expansion or a constriction is that since conservation of mass holds, Bernoulli's principle can be applied to two points representing flow through different segments of a vessel. Therefore, Bernoulli's principle can be expressed for the two points according to the following: one point (1) represents flow through a normal region of the vessel, and another point (2) represents flow through either an obstructed or enlarged region at the maximum point of stenosis. Bernoulli's principle expressed for these two points can be equated, relating hemodynamic parameters characteristic of the two distinct regions of flow:

$$P_1 + \frac{\rho_1 v_1^2}{2} = P_2 + \frac{\rho_2 v_2^2}{2}$$

where the gravitational potential energy component for both regions is zero.

13.4   TORRICELLI'S THEOREM

Torricelli's theorem is a special case of Bernoulli's principle and describes the speed of a liquid flowing from an opening in a tank filled with liquid to a height h, as shown in Figure 13-5. The outward velocity v of a liquid from an opening a distance h from the surface level of the liquid is given by

$$v = \sqrt{2gh}$$

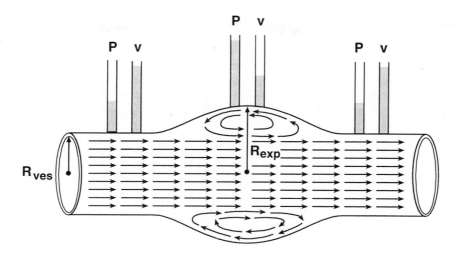

Fig. 13-3

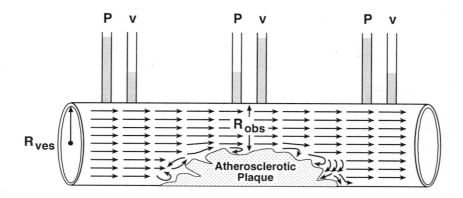

Fig. 13-4

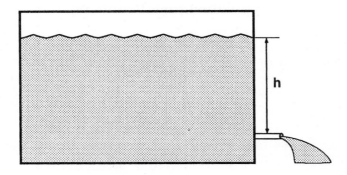

Fig. 13-5

Solved Problem 13.12.   Derive the basis for Torricelli's theorem, that is,

$$v = \sqrt{2gh}$$

Solution

Consider a water-filled tank with an opening a distance h from the surface of the water. The pressure from the water at its maximum height is great enough to produce an outflow exactly equal to the inflow. Application of Bernoulli's principle yields

$$\left(P + \frac{1}{2}\rho v^2 + \rho gh\right)_1 = \left(P + \frac{1}{2}\rho v^2 + \rho gh\right)_2$$

with point 1 at the surface of the water at a height h above the hole and point 2 at the hole itself.

$$(P_{atm} + \rho gh)_1 = \left(P_{atm} + \frac{1}{2}\rho v^2\right)_2$$

where v is the velocity of efflux from the hole. Hence, solving for v yields

$$v = \sqrt{2gh}$$

Solved Problem 13.13.   As part of his farm chores, Jake is filling up a trough with water at a rate of 10^{-4} m s^{-1}. Unbeknownst to Jake, there is a circular leak of a cross-sectional area of 1 cm^2 in the trough. Determine the maximum height of water rise in the trough.

Solution

This is an example of Torricelli's theorem. The velocity of the discharge from the leak is given by

$$v = \sqrt{2gh}$$

At equilibrium, v is the rate of influx divided by the area of the hole, or

$$v = \frac{10^{-4}\,\text{m}^3\,\text{s}^{-1}}{10^{-4}\,\text{m}^2} = 1.0\ \text{m s}^{-1}$$

Therefore, the maximum height of water in the tank is

$$h = \frac{v^2}{2g} = \frac{(1.0\ \text{m s}^{-1})^2}{2 \times 9.8\ \text{m s}^{-2}} = 5.1\ \text{cm}$$

Supplementary Problems

13.1.   The volume of fluid flowing through a tube of inner diameter $d = 2$ mm is 300 cm^3 after 2 min. Determine the average velocity of the fluid in the tube.

Solution

79 cm s^{-1}

13.2.   Oil is flowing through a circular pipe of radius 0.35 m. Determine the velocity of the flowing oil if the volumetric flow rate is measured at 0.70 m^3 s^{-1}.

Solution

1.82 m s^{-1}

13.3. Blood is flowing through an artery of radius 2 mm at a rate of 40 cm s^{-1}. Determine the (1) flow rate and (2) volume observed after 30 s.

Solution

(1) 5 cm^3 s^{-1}; (2) 150 cm^3

13.4. For a plastic syringe, determine the pressure gradient required to force a solution of viscosity of 0.015 dyn s cm^{-2} through a hypodermic needle of 20-mm length and diameter 0.50 mm at a flow rate 75 cm^3 s^{-1}.

Solution

1.47 × 10^7 dyn cm^{-2}

13.5. The velocity of water flowing through a square pipe of side $D = 0.20$ m is 4.5 m s^{-1}. Determine (1) the flow rate and (2) the volume of water measured after 1 min.

Solution

(1) 0.18 m^3 s^{-1}; (2) 10.8 m^3

13.6. Water is flowing through two connected pipes of different radii. Given that the radius of the first pipe is 0.45 m, determine the velocity of the water if the volumetric flow rate is 0.85 m^3 s^{-1}.

Solution

1.34 m s^{-1}

13.7. In Problem 13.6, if the velocity of water traveling through the second pipe is 2.0 m s^{-1}, determine the radius of the second pipe.

Solution

0.37 m

13.8. Determine the time needed for a fluid of velocity $v = 8$ m s^{-1} to traverse a vessel of length $L = 90$ cm.

Solution

0.11 s

13.9. The carotid artery of an individual was found to have atherosclerotic deposits along the inner wall. Assuming that the pressure gradient remains the same, determine the factor by which blood flow will be reduced if the atherosclerotic deposits reduce the diameter by (1) 25 percent, (2) 50 percent, and (3) 75 percent.

Solution

(1) 0.32; (2) 0.0625; (3) 0.0039

13.10. Water is flowing through a pipe of radius 1.0 cm with a speed of 0.25 m s^{-1}. Given that the viscosity of water is 1.005 cP, determine the pressure gradient over a 4-m section of pipe.

Solution

800 dyn

13.11. While chipping golf balls in his backyard, Dan strikes a golf ball that crashes through a patio window and into an aquarium tank 1.5 m below water level, leaving a circular hole of 1.0-cm diameter. Determine (1) the velocity of water exiting the aquarium and (2) the force that Dan must apply to the tank to prevent leakage.

Solution

(1) 5.4 m s^{-1}; (2) 2.31 N

13.12. Blood flow through the aorta of 1.2-cm radius is approximately $80 \text{ cm}^3 \text{ s}^{-1}$. Determine the average blood flow velocity through the aorta.

Solution

17.7 cm s^{-1}

13.13. In making a balloon animal at a birthday party, Sparky the clown had just blown up a balloon to a pressure of 10 dyn cm^{-2} when he inadvertently let go, causing the balloon to fly through the room. Given the density of air is $1.20 \times 10^{-3} \text{ g cm}^{-3}$, determine the velocity of the balloon following release.

Solution

1.29 m s^{-1}

Chapter 14

Thermal Physics

Temperature is qualitatively based on the sense of touch and whether a body feels hot or cold. Temperature is a property of all bodies and is a relative measure of the kinetic energy of the molecules and atoms of the body. In this chapter, we investigate the concept of temperature and the physics involved in dynamic interactions between bodies.

14.1 DEFINITIONS OF THERMAL PHYSICS

Temperature is a physical property of a body that reflects its warmth or coldness. Temperature is a scalar quantity and is expressed in units of degrees. Three common scales of temperature measurement are Fahrenheit (°F), Celsius (°C), and Kelvin (K), and they are defined according to three physical states of matter: absolute zero, freezing point of water, and boiling point of water. The actual temperature values for these conditions according to each temperature scale are displayed below:

Physical State	°F	°C	K
Absolute zero	−460	−273	0
Freezing point of water	32	0	273
Boiling point of water	212	100	373

Absolute zero is the temperature at which a substance no longer exerts pressure and has zero kinetic energy.

The temperature scales are related according to

$$T_F = \frac{9}{5} T_C + 32°$$
$$T_C = T_K - 273°$$

where T_C is degrees Celsius, T_F is degrees Fahrenheit, and T_K is kelvins.

Thermal expansion is a physical phenomenon in which increases in temperature can cause substances in the solid, liquid, and gaseous states to expand.

Linear Expansion of Solids

A solid subjected to an increase in temperature ΔT experiences an increase in length ΔL that is proportional to the original length L_o of the solid. The relation for the linear expansion of solids is

$$\Delta L = \alpha L_o \, \Delta T$$

where the proportionality constant α is the coefficient of linear expansion and is expressed in units of inverse temperature, Θ^{-1}.

Volumetric Expansion of Liquids

A liquid subjected to an increase in temperature ΔT experiences an increase in volume ΔV that is proportional to the original volume V_o of the liquid. The relation for the thermal expansion of liquids is

$$\Delta V = \beta V_o \, \Delta T$$

where the proportionality constant β is the coefficient of volumetric expansion and is equal to 3α. As in the case of α, the units of β are inverse temperature, Θ^{-1}.

Volumetric Expansion of Gases

The volumetric expansion of gas can be summarized by two important gas laws, Charles's law and Boyle's law, which further can be combined into the ideal gas law.

Charles's law states that at constant pressure, the volume V occupied by a given mass of gas is directly proportional to the absolute temperature T:

$$V = kT$$

or

$$\frac{V_1}{T_1} = \frac{V_2}{T_2} \qquad P = \text{constant}$$

where k is a proportionality constant.

Boyle's law states that at constant temperature, the volume V occupied by a given mass of gas is inversely proportional to the pressure P exerted on it:

$$PV = k$$

or

$$P_1 V_1 = P_2 V_2 \qquad T = \text{constant}$$

The *ideal gas law* combines the relations from Charles's law and Boyle's law to yield

$$PV = nRT$$

or

$$\frac{P_1 V_1}{T_1} = \frac{P_2 V_2}{T_2}$$

where R is the universal gas constant ($R = 0.082$ L atm mol K^{-1}) and n is the number of moles or gram molecular weight of the gas and represents the sum of all atomic weights of a molecule.

Solved Problem 14.1.   Derive a relation between the Fahrenheit and Kelvin temperature scales.

Solution

The Fahrenheit and Celsius temperature scales are related by the following:

$$T_F = \frac{9}{5} T_C + 32° \tag{1}$$

The Celsius and Kelvin temperature scales are related by

$$T_C = T_K - 273° \tag{2}$$

Substituting Eq. (2) into Eq. (1) gives

$$T_F = \frac{9}{5}(T_K - 273°) + 32° = \frac{9}{5} T_K - 460° \tag{3}$$

Solved Problem 14.2.   Normal body temperature is 98.6°F. Convert this temperature to degrees Celsius and kelvins.

Solution

(1)   To convert degrees Fahrenheit to degrees Celsius, we begin with the relation

$$T_F = \frac{9}{5} T_C + 32°$$

Solving for T_C gives

$$T_C = \frac{5}{9}(T_F - 32°) = \frac{5}{9}(98.6°F - 32°) = 37°C$$

(2)   To convert degrees Fahrenheit to kelvins, we can use Eq. (3) from Solved Problem 14.1:

$$T_F = \frac{9}{5}T_K - 460°$$

Algebraically solving for T_K, we find

$$T_K = \frac{5}{9}(T_F + 460°) = \frac{5}{9}(98.6°F + 460°) = 310\,K$$

Solved Problem 14.3.   Fever can result in an elevation of body temperature from 98.6 to 102°F. Determine the change in temperature in degrees Celsius and kelvins.

Solution

The change in temperature as a result of fever is

$$\Delta T_F = 102°F - 98.6°F = 3.4°F$$

(1)   To convert this temperature change from degrees Fahrenheit to degrees Celcius,

$$T_C = \frac{5}{9}(T_F - 32°) = \frac{5}{9}(3.4°F - 32°) = -15.9°C$$

(2)   To convert this temperature change from degrees Fahrenheit to kelvins,

$$T_K = \frac{5}{9}(3.4°F + 460°) = 257.4\,K$$

Solved Problem 14.4.   Heat stroke, caused by an extreme elevation in body temperature, affects the normal human when the body temperature reaches 42°C. What is this temperature on the Fahreinheit scale?

Solution

The Celsius and Fahrenheit temperature scales are related by

$$T_F = \frac{9}{5}T_C + 32° = \frac{9}{5}(42°C) + 32° = 107.6°F$$

Solved Problem 14.5.   At what temperature on the Celcius scale is the Celsius temperature equal to the (1) Fahrenheit temperature and (2) Kelvin temperature?

Solution

(1)   Defining the unknown temperature as T_x, we can rewrite the relation between temperature in degrees Fahrenheit and degrees Celcius as

$$T_x = \frac{9}{5}T_x + 32°$$

Solving for T_x, we get

$$-32° = \frac{9}{5}T_x - T_x = \frac{4}{5}T_x$$

$$T_x = \frac{-32° \times 5}{4} = -40°$$

(2)   We must first derive a relationship between degrees Fahrenheit and kelvins:

$$T_K = T_C + 273°$$

where

$$T_C = \frac{5}{9}(T_F - 32°)$$

Substituting T_C into T_K yields

$$T_K = \frac{5}{9}(T_F - 32°) + 273° = \frac{5}{9}T_F + 256°$$

Again, defining the unknown temperature as T_x, we can rewrite the relation between temperature in degrees Fahrenheit and kelvins as

$$T_x = \frac{5}{9}T_x + 256°$$

$$\frac{4}{9}T_x = 256°$$

$$T_x = 576°$$

Solved Problem 14.6.   A copper rod 1.0 m in length is maintained at 25°C. Assuming the coefficient of linear expansion for copper is 1.7×10^{-5} °C^{-1}, determine the increase in length when it is heated to 40°C.

Solution

The increase in length can be determined from

$$\Delta L = \alpha L_o \Delta T = (1.7 \times 10^{-5}\,°\text{C}^{-1})(1.0\,\text{m})(40°\text{C} - 25°\text{C}) = 2.55 \times 10^{-4}\,\text{m} = 0.25\,\text{mm}$$

Solved Problem 14.7.   A metal cylindrical column of length $L = 4.0$ m is heated from 25 to 255°C. Given a coefficient of linear expansion of 11×10^{-6} °C^{-1}, determine the change in length following expansion.

Solution

The relation to use is

$$\Delta L = \alpha L \Delta T = (11 \times 10^{-6}\,°\text{C}^{-1})(4.0\,\text{m})(225°\text{C} - 25°\text{C}) = 8.8\,\text{mm}$$

Solved Problem 14.8.   The volume of a brass sphere maintained at 110°C is 500 cm^3. Determine the change in volume of the sphere at −10°C, given $\beta = 57 \times 10^{-6}$ °C^{-1}.

Solution

The volumetric change due to temperature is

$$\Delta V = \beta V \Delta T = (57 \times 10^{-6}\,°\text{C}^{-1})(500\,\text{cm}^3)(-10°\text{C} - 110°\text{C}) = 3.42\,\text{cm}^3$$

Solved Problem 14.9.   Determine the volume occupied by 2 mol of oxygen maintained at a pressure of 3.5 atm and a temperature of 27°C.

Solution

From the ideal gas law,

$$PV = nRT$$

$$(3.5\,\text{atm})V = (2\,\text{mol})(0.082\,\text{L atm mol K}^{-1})(300\,\text{K})$$

$$V = 14.1\,\text{L}$$

ILLUSTRATIVE EXAMPLE 14.1.   Hypothermia: Human Response to Cold Temperatures

In hypothermia, the body temperature is lowered, due primarily to exposure to extremely cold weather (below 10°C). Since energy radiates from heat to cold, the body begins to lose heat to the cold environment. As the body loses too much heat, circulatory mechanisms are activated that cause a reduction in blood flow to the skin. In addition, the physiological response to cold—shivering—results in an increase in heat production. This source of internal heating continues until carbohydrate reserves are depleted. Carbohydrates serve as a biochemical supply of readily available stored energy. Once the reserves are depleted, the body temperature begins to drop. If the body temperature drops below 33°C, external heat must be supplied to the person to prevent potential harm and severe thermal injuries.

14.2   HEAT

Heat Q is a form of energy that can be converted to work and other forms of energy. When heat is transferred between a system and its surroundings as a result of differences in temperature,

$$Q < 0 \qquad \text{when heat is transferred from surroundings to system}$$
$$Q > 0 \qquad \text{when heat is transferred from system to surroundings}$$

Heat is measured in units of calories. One calorie is the amount of heat required to raise the temperature of 1 g of water by 1°C or 1 K. Since a temperature change of 1°C is the same as a change of 1 K, the temperature dependence of heat can be interchanged between degrees Celsius and kelvins. Mechanical energy that is equivalent to a given amount of heat energy includes

$$4.186 \text{ J} = 1 \text{ cal}$$
$$1 \text{ Btu} = 1055 \text{ J} = 777.9 \text{ ft lb}$$
$$3413 \text{ Btu} = 1 \text{ kWh} = 3.6 \times 10^6 \text{ J}$$

Specific heat relates the change in temperature ΔT experienced by a body to the heat gained or lost Q by

$$Q = mc \, \Delta T$$

where m is the mass of the body and c is a proportionality constant termed the *specific heat capacity*.

Specific heat capacity is the quantity of heat required to raise the temperature of a unit quantity of substance by 1° and is dependent only on the material of the object. Units of specific heat capacity are kcal kg^{-1}°C^{-1} or Btu lb^{-1} ft^{-1}. The specific heat capacity of water is 1.0 kcal kg^{-1}°C^{-1}.

Heat of transformation is similar to the specific heat but accounts for changes in the phase of the body. The specific heat of a body assumes that no change in phase occurs during a temperature change. In order for a substance to change states of matter, that is, from solid to liquid or from liquid to gas, heat energy must be added to the substance. The amount of heat required to change the phase of 1 kg of a substance is the heat of transformation L. Thus, the total amount of heat Q gained or lost by a substance of mass m during a change between phases is

$$Q = mL$$

where L is the heat of transformation unique to the substance. The heat of transformation exists in two forms, according to the particular phase transformation:

Heat of fusion L_f is the amount of heat energy required to change 1 g of solid matter to liquid.

Heat of vaporization L_v is the amount of heat energy required to change 1 g of liquid matter to gas.

Phase Change	Temperature	Heat of Transformation
Liquid → solid		
Solid → liquid	Melting point	Heat of fusion L_f
Liquid → gas		
Gas → liquid	Boiling point	Heat of vaporization L_v

Solved Problem 14.10. A 75-kg woman suffering from the flu experiences a fever resulting in a temperature of 103°F. Assuming that the human body is composed primarily of water, determine the amount of heat required to raise the body temperature from normal temperature at 98.6°F to 102°F, in joules and in calories. The specific heat capacity of water is $4190 \, \text{J kg}^{-1} \, °\text{C}^{-1}$.

Solution

We must first convert the temperature values from Fahrenheit to Celsius.

$$T_C = \frac{5}{9} T_F - 32° = \frac{5}{9}(98°\text{F}) - 32° = 22.4°\text{C}$$

$$T_C = \frac{5}{9} T_F - 32° = \frac{5}{9}(102°\text{F}) - 32° = 24.6°\text{C}$$

Therefore, the temperature difference ΔT is $24.6°\text{C} - 22.4°\text{C} = 2.2°\text{C}$. The amount required to raise the body temperature is given by

$$Q = mc \, \Delta T = (75 \, \text{kg})(4190 \, \text{J kg}^{-1} \, °\text{C}^{-1})(2.2°\text{C}) = 6.9 \times 10^5 \, \text{J}$$

To convert from joules to calories,

$$(6.9 \times 10^5 \, \text{J}) \frac{1 \, \text{cal}}{4.186 \, \text{J}} = 165 \, \text{kcal}$$

Solved Problem 14.11. Consider a certain heater in which a tank of water of volume $V = 1 \, \text{m}^3$ and mass $m = 120 \, \text{kg}$ is maintained at 100°C overnight and cools to 20°C during the following day. Given that the specific heat capacity is $4186 \, \text{J kg}^{-1} \, \text{C}^{-1}$, determine the heat energy generated during the day.

Solution

$$Q = mc \, \Delta T = (120 \, \text{kg})(4186 \, \text{J kg}^{-1} \, °\text{C}^{-1})(100°\text{C} - 20°\text{C}) = 4.02 \times 10^7 \, \text{J}$$

Solved Problem 14.12. Juanita Valdez, an avid tea drinker, heated 0.5 kg of water to 90°C and then poured it into a 300-g pot ($c_{\text{pot}} = 0.25 \, \text{kcal kg}^{-1} \, °\text{C}^{-1}$) at 20°C for brewing. The specific heat capacity of water is $1 \, \text{kcal kg}^{-1} \, °\text{C}^{-1}$. Determine the temperature of the water in the pot.

Solution

$$\text{Heat gained by pot} = \text{heat lost by water}$$
$$m_{\text{pot}} c_{\text{pot}} \Delta T_{\text{pot}} = m_{\text{water}} c_{\text{water}} \Delta T_{\text{water}}$$
$$(0.3 \, \text{kg})(0.25 \, \text{kcal kg}^{-1} \, °\text{C}^{-1})(T - 20°\text{C}) = (0.5 \, \text{kg})(1 \, \text{kcal kg}^{-1} \, °\text{C}^{-1})(90°\text{C} - T)$$
$$T \times 0.075 \, \text{kcal} \, °\text{C}^{-1} - 1.5 \, \text{kcal} = 45 \, \text{kcal} - T \times 0.5 \, \text{kcal} \, °\text{C}^{-1}$$
$$T = 80.9°\text{C}$$

Solved Problem 14.13. Determine the amount of heat energy required to completely melt a 0.05-kg coin, initially maintained at a temperature of 20°C. The melting point of silver is 1000°C, the heat of fusion of silver is $1 \times 10^5 \, \text{J kg}^{-1}$, and the specific heat capacity of silver is $250 \, \text{J kg}^{-1} \, °\text{C}^{-1}$.

Solution

This problem must be solved in two steps. The first step is to determine the amount of heat energy required to reach the melting-point temperature, which can be calculated by

$$Q = mc \, \Delta T = (0.05 \, \text{kg})(250 \, \text{J kg}^{-1} \, °\text{C}^{-1})(980°\text{C}) = 12.3 \, \text{kJ}$$

This is the amount of heat required for the coin to reach its melting point. However, the coin is still in the solid phase. Additional heat energy must be supplied to transform the coin from solid to liquid:

$$Q = mL_f = (0.05 \text{ kg})(1 \times 10^5 \text{ J kg}^{-1}) = 5 \text{ kJ}$$

Thus, the total heat energy required to completely melt the silver coin is

$$Q = 12.3 \text{ kJ} + 5 \text{ kJ} = 17.3 \text{ kJ}$$

14.3   MECHANISMS OF HEAT TRANSFER

There are three mechanisms of heat transfer, all of which are important in maintaining normal human body temperature.

Conduction

Heat energy is transferred by molecular collisions between the rapidly moving molecules of the hot region and the slower-moving molecules of the cooler region. A portion of the kinetic energy generated by the rapidly moving molecules is transferred to the slower-moving molecules, causing an increase in heat energy at the cooler end and a subsequent increase in the flow of heat.

Consider a block of material of thickness h and a surface area of each side A. If one side is heated to a temperature T_2 such that a difference in temperature of the object exists between the other side, originally held at T_2, then the rate of heat energy transfer, or heat energy transfer per unit time t, is

$$\text{Rate of heat energy conduction} = \frac{Q}{t} = \frac{kA \, \Delta T}{h}$$

where k is a proportionality constant known as the *thermal conductivity* of the material. Here k reflects the ability of a material to conduct heat and is typically expressed in units of $\text{J s}^{-1} \text{m}^{-1} \, ^\circ\text{C}^{-1}$. Conduction is the primary source of heat transfer in solids.

Convection

Convection represents the transfer of heat energy due to the physical motion or flow of the heated substance. In contrast to conduction, convection is the primary mechanism of heat transfer in fluids.

Radiation

Radiation represents the transfer of heat energy by electromagnetic waves which are emitted by rapidly vibrating electrically charged particles. The electromagnetic waves propagate from the heated body or source at the speed of light ($3 \times 10^8 \text{ m s}^{-1}$). Consider an object placed at room temperature. If it is subjected to electromagnetic waves, the charged particles or electrons of the object gain kinetic energy. Thus, the electromagnetic radiation is transformed to internal energy. Two examples of heat transfer by radiation are (1) the heating of food by using a microwave oven and (2) the heat energy received by the sun.

Consider an object maintained at a temperature T. The rate of emission of heat energy by radiation of the body is

$$H = Ae\sigma T^4$$

where A is the surface of the object emitting radiation energy (cm^2), e is the emissivity of the object surface [e is unitless and varies between 0 (poor absorber of radiation) and 1 (excellent absorber of radiation or blackbody)], σ is the Stefan-Boltzmann constant ($\sigma = 5.67 \times 10^{-8} \text{ W m}^{-2} \text{ K}^{-4}$ or $5.67 \times 10^{-5} \text{ erg cm}^{-2} \text{ K}^{-4} \text{ s}^{-1}$), and T is the absolute temperature, in kelvins. If the object initially held at a temperature T is subjected to environmental surroundings at temperature T_s, then the net flow of heat from the body to its surroundings is

$$H = Ae\sigma(T^4 - T_s^4)$$

Solved Problem 14.14.   Wilfred the witch is preparing a potion in a steel cauldron placed on an open fire. The bottom of the cauldron is 5 mm thick with an effective surface area of 200 cm². The bottom surface of the cauldron is at a temperature of 375°C while the inner surface is at 125°C. Given that the thermal conductivity of steel is 0.11 cal cm^{-1} s^{-1} °C^{-1}, determine the rate of heat conduction per second through the bottom of the cauldron.

Solution

$$Q = \frac{ktA\,\Delta T}{h} = \frac{(0.11 \text{ cal cm}^{-1}\text{ s}^{-1}\text{ °C}^{-1})(1 \text{ s})(200 \text{ cm}^2)(375°\text{C} - 125°\text{C})}{0.5 \text{ cm}} = 1.1 \times 10^4 \text{ cal}$$

Since this calculation was based on 1 s, the rate of heat energy transfer due to conduction is 1.1×10^4 cal s^{-1}.

Solved Problem 14.15.   The human body of surface area 1.5 m² with a surface temperature of 23°C (296 K) radiates heat in the form of energy. If the body is surrounded by an environment of 20°C, determine the (1) total rate of radiation of heat energy and (2) net rate of heat loss from the body. Assume the emissivity e is equal to 1.

Solution

(1)   The rate of radiation of heat energy is given by

$$H = Ae\sigma T^4 = (1.5 \text{ m}^2)(1)(5.67 \times 10^{-8} \text{ W m}^{-2}\text{ K}^{-4})(296 \text{ K})^4 = 654.9 \text{ W}$$

(2)   The net rate of heat energy loss due to radiation is given by

$$H = Ae\sigma(T^4 - T_s^4) = (1.5 \text{ m}^2)(1)(5.67 \times 10^{-8} \text{ W m}^{-2}\text{ K}^{-4})(296 \text{ K}^4 - 293 \text{ K}^4) = 25.5 \text{ W}$$

ILLUSTRATIVE EXAMPLE 14.2.   Heat Stroke

Normal physiological function can be achieved only if the body temperature is maintained at constant levels about 98.6°F. If subjected to hot temperatures for an extended period, the body becomes overheated, and the body temperature begins to rise. Heat stroke is a clinical condition caused by a severe and prolonged elevation of body temperature above 40°C (or 104°F). The effects of heat stroke are especially important in the thermal regulation of physiological processes as they relate to metabolic activity. The internal mechanisms of metabolism involve the conversion of the energy supplied by food intake to heat. At rest, the normal human metabolizes at a rate of 75 W. Under prolonged periods of strenuous exercise or labor, the rate of heat production can approach 800 W. If it is performed outdoors under intense sunlight, this rate increases to 950 W.

To maintain constant body temperature, the body must possess the ability to efficiently dissipate the generated heat via numerous mechanisms including radiation, convection, conduction, respiration, and evaporation. Evaporation through sweating is the body's main method of heat transfer to the environment. The maximum rate of energy dissipation through sweating is 650 W. Consequences of an elevated body temperature and heat stroke include cessation of cell growth at 40.6°C and irreversible chemical damage to vital organs such as the brain and kidneys at 42°C.

14.4   THERMODYNAMICS

First Law of Thermodynamics

The total energy of a body—the sum of potential energy and kinetic energy—is represented by its internal energy U. The internal energy of a body can be altered either by performing work on it or by adding heat to it. Thus the change in internal energy is related to the heat energy transferred to the body Q and the work done by the system W according to

$$\Delta U = Q - W$$

It represents the principle of conservation of energy. The proper sign conventions for each of the quantities represented in the above equation can be summarized below:

W: $+$ for work done *by* system

         $-$ for work done *on* system

Q: $+$ for heat flow *into* system

         $-$ for heat flow *out of* system

Second Law of Thermodynamics

A system subjected to a spontaneous change will respond such that its disorder or entropy will increase or at least will remain the same. For example, heat energy from a hot object will always flow toward the cold object, but not vice versa. For a given system in equilibrium, the change in entropy ΔS is related to the added heat ΔQ by

$$\Delta S = \frac{\Delta Q}{T}$$

Units of entropy are expressed in J K^{-1}.

Solved Problem 14.16. A gas maintained under a constant pressure of 4.5×10^5 Pa is subjected to 800 kJ of heat, inducing a volumetric expansion from 0.5 to $2.0\,\text{m}^3$. Determine the (1) work done by the gas on the system and (2) change in the internal energy of the gas.

Solution

(1)   The work done by the gas can be calculated from

$$W = P\Delta V = (4.5 \times 10^5\,\text{Pa})(2.0\,\text{m}^3 - 0.5\,\text{m}^3) = 6.75 \times 10^5\,\text{J}$$

(2)   The change in internal energy of the gas can be determined from the first law of thermodynamics:

$$\Delta U = Q - W = 8.0 \times 10^5\,\text{J} - 6.75 \times 10^5\,\text{J} = 1.25 \times 10^5\,\text{J}$$

Solved Problem 14.17. Thirty-five grams of ice initially at 0°C begins to melt. Determine the entropy of the melting process, given the heat of fusion for water is $80\,\text{cal}\,\text{g}^{-1}$.

Solution

The heat needed to melt the ice is

$$\Delta Q = mL_f = (35\,\text{g})(80\,\text{cal}\,\text{g}^{-1}) = 2800\,\text{cal}$$

The entropy can now be calculated as

$$\Delta S = \frac{\Delta Q}{T} = \frac{280\,\text{cal}}{273\,\text{K}} = 10.25\,\text{cal}\,\text{K}^{-1}$$

Converting to joules gives

$$10.25\,\text{cal}\,\text{K}^{-1} \times 4.184\,\text{J}\,\text{cal}^{-1} = 42.9\,\text{J}\,\text{K}^{-1}$$

Supplementary Problems

14.1.   Convert $-100°\text{F}$ to degrees Celcius.

   Answer

   $-73.3°\text{C}$

14.2.   Convert 30°C to degrees Fahrenheit.

Solution

86°F

14.3.   Convert 80°C to kelvins.

Solution

353 K

14.4.   Convert 102°F to kelvins.

Solution

312.2 K

14.5.   The melting point of gold is 1338 K. Convert this temperature to degrees Celsius and degrees Fahrenheit.

Solution

1065°C, 1949°F

14.6.   The length of a steel rod is 25 m at a temperature of 20°C. If it is placed in an environment with a temperature of 35°C, determine (1) the change in length of the rod and (2) the new length of the rod. The coefficient of linear expansion for steel is $1.2 \times 10^{-5}\,\text{K}^{-1}$.

Solution

(1) 4.5 mm; (2) 25.0045 m

14.7.   An aluminum bar and a steel bar, both of length 6.0 m and at an initial temperature of 20°C, are placed outside at a temperature of 32°C. Which bar will expand more? The coefficients of linear expansion for aluminum and steel are 2.4×10^{-5} and $1.2 \times 10^{-5}\,\text{K}^{-1}$, respectively.

Solution

Aluminum bar

14.8.   A steel rod of length 5.0 m at an initial temperature of 20°C is used as structural support for a building. The steel rod is placed such that 1.5 mm separates it from an adjacent rod. If the rod could be subjected to temperatures as high as 40°C, determine (1) if this is an appropriate separation distance for the steel rods and (2) the change in length from the distance of separation.

Solution

(1) Yes; (2) 0.3 mm

14.9.   A thermometer with 1 mL of mercury initially at room temperature 20°C is inserted into the mouth of a patient experiencing a 25°C fever. Given that the coefficient of volume expansion of mercury is $1.8 \times 10^{-4}\,\text{K}^{-1}$, determine the volume expansion of mercury in the thermometer.

Solution

9×10^{-4} mL

14.10. Determine the temperature required for 1.5 mol of NH_3 to occupy 20 L under a pressure of 1100 mmHg.

Solution

234.9 K

14.11. Given 2.0 mol of gas X maintained under a pressure of 1700 mmHg and at a temperature of 27°C, determine the volume it will occupy.

Solution

17.3 L

14.12. In sterilization procedures, the temperature of a scalpel can reach 150°C. Given a 50-g steel scalpel, determine the amount of heat that must be removed to reduce its temperature from 150 to 20°C, given that $c_{steel} = 0.11$ cal g^{-1} °C^{-1}

Solution

715 cal

14.13. Suppose that the tank described in Solved Problem 14.11 were filled with mercury so that the mass of the heater now becomes 1360 kg and $c = 138$ J kg^{-1} °C^{-1}. Determine the heat energy generated during the day.

Solution

1.5×10^7 J

14.14. Ten grams of ice at 10°C is placed into 30 g of water in a cup at a temperature of 20°C. Ignoring the effects of the container, determine the number of grams of ice that melt given that $c_{water} = 1$ cal g^{-1} °C^{-1}, for ice $L_f = 80$ cal g^{-1}, and $c_{ice} = 0.5$ cal g^{-1} °C^{-1}.

Solution

6.9 g

14.15. Elevated body temperature in a patient can be rapidly reduced with an alcohol rub. The heat energy lost by the person is transformed to the evaporation of alcohol, which takes place rapidly. Determine the number of grams of alcohol that must be evaporated from the surface of a 75-kg person to reduce body temperature by 2.0°C. The heat of vaporization for alcohol is 204 cal g^{-1}. The specific heat capacity for the human body is 0.83 cal g^{-1} °C^{-1}.

Solution

457 g

14.16. An aluminum ice cube tray filled with 300 g of water is held initially at 0°C. If heat is being removed from the water at a rate of 20 cal s^{-1}, determine the time required for the water to freeze. The heat of fusion for water is 80 cal g^{-1}.

Solution

20 min

14.17. Assuming that body core temperature is 37°C and skin surface temperature is 33°C, determine the rate of heat conduction of the human body. The surface area of the body is 1.5 m^2, and the thickness of human tissue h averages 1.0 cm. Thermal conductivity for tissue is 0.2 J s^{-1} m^{-1} °C^{-1}.

Solution

$120 \, \text{J s}^{-1}$

14.18. Consider an unclothed person in a dark room with a temperature of 20°C. Given that the skin surface temperature is 33°C and the exposed area of the skin is 1.5 m^2, determine the heat transfer by radiation. Assume that emissivity e is 0.95.

Solution

111.2 W

14.19. Determine the change in internal energy of 5 kg of hydrogen heated from 15 to 50°C while maintained at constant volume. The specific heat capacity of hydrogen is 144 J kg^{-1} °C^{-1}.

Solution

$2.52 \times 10^4 \, \text{J}$

14.20. A 20-g ice cube is added to 150 g of water in a cup. The water and ice are in equilibrium at 0°C. Over a 10-min period, heat is allowed to enter the system, causing 10 g of the ice cube to melt. Determine the change in entropy of the system, given the heat of fusion for water is 80 cal g^{-1}.

Solution

$12.3 \, \text{J K}^{-1}$

Chapter 15

Waves and Sound

In Chapter 10, the concept of oscillatory motion and its periodic or cyclical nature was introduced and characterized in mathematical form as a sinusoidal function. Although the net displacement of a particle subjected to oscillatory motion varies between zero and the amplitude, the distance traveled by the particle is no greater than $2\pi A$. If one were to draw an analogy to a person running around a track, the maximum distance traversed by the runner would be the circumference of the circular track. Sound is the propagation of a disturbance in an elastic medium and can be represented by a wave. A wave is a physical entity that also travels in a periodic nature, similar to an oscillation, but carries energy with a velocity in a given direction. This chapter addresses the physical concepts and interactions of waves and sound.

15.1 DEFINITIONS OF WAVES AND SOUND

Waves are physical oscillations that occur as a direct result of a disturbance in an elastic medium and can be represented by a sinusoidal function (see Figure 15-1). The physical characteristics of a wave include the following:

Amplitude A of a wave represents the maximum displacement of particles within the elastic medium from its point of equilibrium. Amplitude is a measure of displacement and is therefore expressed in units of length.

Frequency f of a wave represents the number of complete wave cycles that pass a defined point per second. Frequency is expressed in units of hertz (Hz), or cycles per second.

Period T is the time required to observe one complete wave cycle. The period is related to the frequency by

$$T = \frac{1}{f}$$

Period is expressed in units of time, that is, seconds.

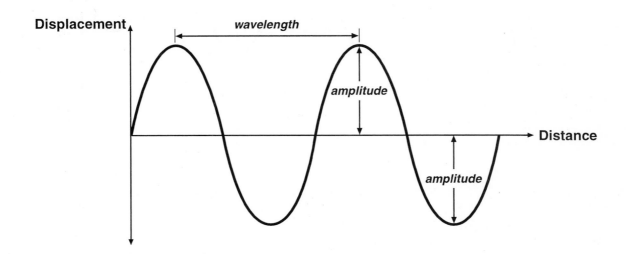

Fig. 15-1

Wavelength λ of a wave is the distance between adjacent peaks. As the wavelength decreases, more complete cycles are observed per unit time and therefore the frequency of the wave increases. The relationship between wavelength and frequency is given by

$$\lambda = \frac{v}{f}$$

Wavelength is expressed in units of length or L.

Speed v is the speed of the propagating wave and is defined by

$$v = \lambda f$$

and it is expressed in units of length per unit time, or LT^{-1}.

Phase ϕ represents the angular difference in position for two points on a vibrating wave. Let's assume, for the moment, that particles are positioned at all points along the wave displayed in Figure 15-1. Two particles at the crest, or peak, of each successive cycle are separated in terms of a wavelength but are in phase because they move simultaneously as the wave propagates. A particle on the crest of a wave and another particle in the trough, or valley, of the cycle move opposite to each other and are, thus, 180° or one-half cycle, out of phase.

Two types of waves are transverse and longitudinal waves.

Transverse waves are waves generated by a transverse force, that is, force acting up and down with respect to the string, causing the motion of particles vibrating in a direction up and down at right angles (perpendicular) to the direction of wave propagation. Transverse waves are illustrated in Figure 15-2. Examples of transverse waves include light waves and waves in a string (guitar string).

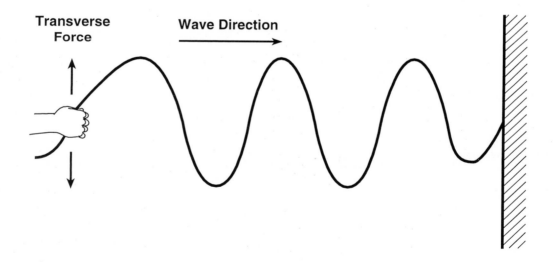

Fig. 15-2

Speed of a Transverse Wave

The speed of a transverse wave v is related to the magnitude of the transverse force or tension in the string F applied to the string and the linear mass density of the string μ, and it is defined as

$$v = \sqrt{\frac{\text{transverse force}}{\text{linear mass density}}} = \sqrt{\frac{F}{\mu}}$$

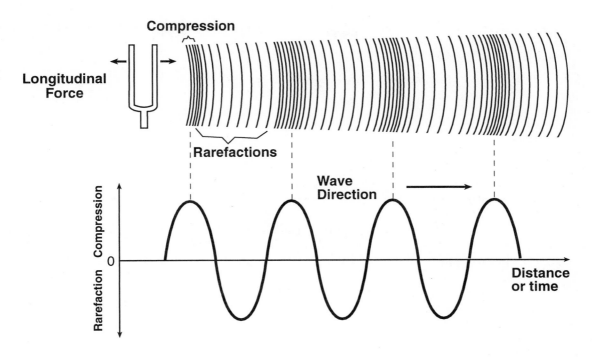

Fig. 15-3

Longitudinal waves are waves generated by the motion of molecules or particles of a medium vibrating back and forth in (parallel to) the direction of wave propagation. Longitudinal waves are illustrated in Figure 15-3. Examples of longitudinal waves include sound waves and emissions from a tuning fork.

Speed of a Longitudinal Wave

Longitudinal waves as well as transverse waves can propagate through a stretched spring. However, fluids (liquids and gases) can accommodate only longitudinal waves. Accordingly, the speed of a longitudinal wave is dependent on parameters unique to a solid, liquid, and gas and is expressed below for all three cases:

$$v = \sqrt{\frac{\text{Young's modulus}}{\text{density of solid}}} = \sqrt{\frac{Y}{\rho}} \qquad \text{solids}$$

$$v = \sqrt{\frac{\text{bulk modulus}}{\text{density of liquid}}} = \sqrt{\frac{B}{\rho}} \qquad \text{liquids}$$

$$v = \sqrt{\frac{\gamma \times \text{pressure of gas}}{\text{density of gas}}} = \sqrt{\frac{\gamma P}{\rho}} \qquad \text{gases}$$

where γ is a constant ($\gamma = 1.40$ for air). The speed of longitudinal waves in a gas can also be expressed as

$$v = \sqrt{\frac{\gamma R T}{M}}$$

where R is the ideal gas constant ($= 8.315 \text{ J mol}^{-1} \text{ K}^{-1}$), T is temperature in kelvins, and M is the mean molecular mass of the gas in kg mol^{-1}.

Solved Problem 15.1.   Determine the frequency of a wave with a period of (1) 1 ms, (2) 10 ms, (3) 1 μs, and (4) 10 μs.

   Solution

   By definition, the frequency f of a wave is related to its period T by

$$f = \frac{1}{T}$$

(1)   For a period of 1 ms,

$$f = \frac{1}{1 \times 10^{-3}\,\text{s}} = 1000\,\frac{\text{cycles}}{\text{s}} = 1\,\text{kHz}$$

(2)   For a period of 10 ms,

$$f = \frac{1}{10 \times 10^{-3}\,\text{s}} = 100\,\frac{\text{cycles}}{\text{s}} = 100\,\text{Hz}$$

(3)   For a period of 1 μs,

$$f = \frac{1}{1 \times 10^{-6}\,\text{s}} = 1 \times 10^{6}\,\frac{\text{cycles}}{\text{s}} = 1\,\text{MHz}$$

(4)   For a period of 10 μs,

$$f = \frac{1}{10 \times 10^{-6}\,\text{s}} = 100{,}000\,\frac{\text{cycles}}{\text{s}} = 100\,\text{kHz}$$

Solved Problem 15.2.   Radio station KGKL broadcasts at a frequency of 960 kHz. What is the period of the radio wave transmissions?

   Solution

   The frequency is related to the period by

$$T = \frac{1}{f} = \frac{1}{960{,}000\,\text{Hz}} = 1 \times 10^{-6}\,\text{s} = 1\,\mu\text{s}$$

Solved Problem 15.3.   Middle C of the musical scale has a frequency of 256 Hz. Determine its period.

   Solution

   The period is related to the frequency by

$$T = \frac{1}{f} = \frac{1}{256\,\text{Hz}} = 3.9 \times 10^{-3}\,\text{s} = 3.9\,\text{ms}$$

Solved Problem 15.4.   Determine the wavelength for waves of frequency (1) 30 Hz, (2) 500 Hz, (3) 8 kHz, and (4) 60 kHz. Assume the wave velocity is 344 m s^{-1}.

   Solution

   By definition, velocity = wavelength × frequency. Therefore,

(1) $\lambda = \dfrac{v}{f} = \dfrac{344\,\text{m s}^{-1}}{30\,\text{Hz}} = 11.5\,\text{m}$

(2) $\lambda = \dfrac{v}{f} = \dfrac{344\,\text{m s}^{-1}}{500\,\text{Hz}} = 0.688\,\text{m} = 68.8\,\text{cm}$

(3) $\lambda = \dfrac{v}{f} = \dfrac{344 \text{ m s}^{-1}}{8000 \text{ Hz}} = 0.043 \text{ m} = 4.3 \text{ cm}$

(4) $\lambda = \dfrac{v}{f} = \dfrac{344 \text{ m s}^{-1}}{60{,}000 \text{ Hz}} = 0.0057 \text{ m} = 5.7 \text{ mm}$

Solved Problem 15.5.   Hans and Franz are playing on a mountain in the Alps when they hear their mother from an adjacent mountain yodeling—a sign that dinner is ready. If the mother hears an echo of her yodeling in 10 s, how far away is the mountain where Hans and Franz are?

Solution

The speed of sound in air is approximately 760 mi h^{-1}. Converting to mi s^{-1}, we have

$$760\,\frac{\text{mi}}{\text{h}} \cdot \frac{1 \text{ h}}{3600 \text{ s}} = 0.21 \text{ mi s}^{-1}$$

The distance that the sound waves originally traveled is

$$x = vt = (0.21 \text{ mi s}^{-1})(10 \text{ s}) = 2 \text{ mi}$$

Since the sound had to go to the mountain and then return, 2 mi is the round-trip distance of the echoes. Therefore, the distance to the mountain is one-half this distance, or $x = 1$ mi.

Solved Problem 15.6.   When plucked, a guitar string, with a linear mass density of 0.0012 kg m^{-1} and tightened to a tension of 3.5 N, generates a 250-Hz wave with a wavelength of 0.45 m. Determine the (1) velocity and (2) period of the transverse wave.

Solution

(1)   The velocity of a transverse wave through a tight string can be determined by

$$v = \sqrt{\frac{F}{\mu}} = \sqrt{\frac{3.5 \text{ N}}{0.0012 \text{ kg m}^{-1}}} = 54 \text{ m s}^{-1}$$

(2)   The period of the wave is related to the frequency by

$$T = \frac{1}{f} = \frac{1}{250 \text{ Hz}} = 4 \times 10^{-3} \text{ s} = 4 \text{ ms}$$

Solved Problem 15.7.   Determine the velocity of longitudinal mechanical waves through a copper rod, given that Young's modulus and the density for copper are 1.1×10^{11} Pa and 8.9×10^{3} kg m^{-3}, respectively.

Solution

The wave velocity through the copper rod is

$$v = \sqrt{\frac{Y}{\rho}} = \sqrt{\frac{1.1 \times 10^{11} \text{ Pa}}{8.9 \times 10^{3} \text{ kg m}^{-3}}} = 3.52 \times 10^{3} \text{ m s}^{-1}$$

Solved Problem 15.8.   Given that the bulk modulus of water is $B = 2.2 \times 10^{9}$ N m^{-2}, determine the speed of longitudinal waves in water.

Solution

The speed of longitudinal waves in water can be determined by

$$v = \sqrt{\frac{B}{\rho}} = \sqrt{\frac{2.2 \times 10^{9} \text{ N m}^{-2}}{1 \times 10^{3} \text{ kg m}^{-3}}} = 1483 \text{ m s}^{-1}$$

Solved Problem 15.9. Given that the mean molecular mass for air is $28.8 \times 10^{-3}\,\mathrm{kg\,mol^{-1}}$ and $\gamma = 1.40$, determine the speed of longitudinal waves in air at 20°C.

    Solution

    The speed of longitudinal waves through a gas can be determined by

$$v = \sqrt{\frac{\gamma RT}{M}} = \sqrt{\frac{(1.40)(8.315\,\mathrm{J\,mol^{-1}\,K^{-1}})(293\,\mathrm{K})}{28.8 \times 10^{-3}\,\mathrm{kg\,mol^{-1}}}} = 344\,\mathrm{m\,s^{-1}}$$

15.2 STANDING WAVES

    Consider a wave generated on a stretched string with one side connected to a rigid surface. As the wave propagates toward an end, it reflects, inverts, and continues back and forth across the string, causing the string to vibrate. In such instances, two waves (original and reflected waves) of equal frequencies and amplitudes are moving in opposite directions along the string, creating a standing wave, as shown in Figure 15-4. The points along the horizontal axis where the two waves intersect are termed *displacement nodes*, and the points that are maximally displaced from the horizontal axis, that is, points at the peaks and valleys of each wave, are termed *displacement antinodes*. The distance between two adjacent displacement nodes is one-half wavelength. The frequency of a standing wave is given by

$$f = \frac{v}{\lambda}$$

where v is the velocity of the transverse wave.

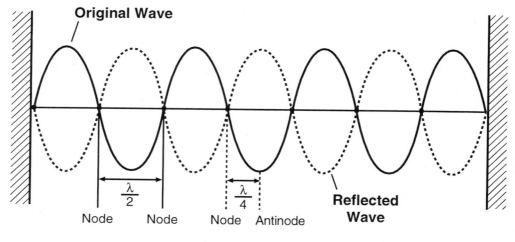

Fig. 15-4

15.3 RESONANCE

    The presence of a standing wave implies that the vibrating system is in resonance. In the stretched string, as is the case in a vibrating system, there exist one or more natural frequencies at which the system will vibrate. Two fundamental vibrating systems are the stretched string and a pipe.

Resonance Frequencies in a Stretched String

    A stretched string of length L with both ends attached to a rigid surface possesses displacement nodes at both ends. Since the distance between two adjacent nodes is $\lambda/2$, for this case, $\lambda/2 = L$, or $\lambda = 2L$. Substituting this value into the expression for the frequency yields

$$f = \frac{v}{\lambda} = \frac{v}{2L}$$

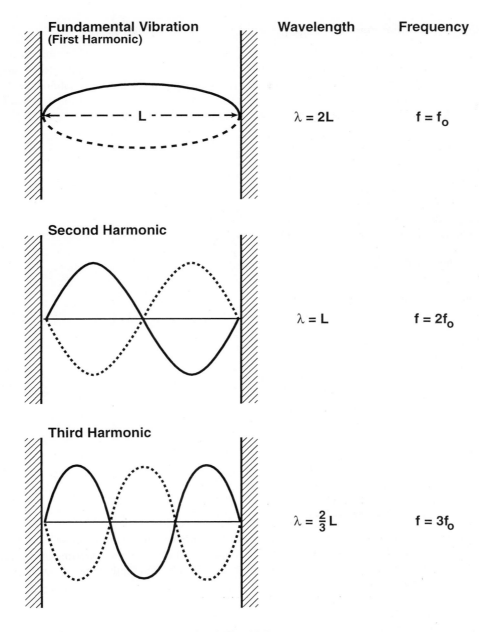

Fig. 15-5

This is the lowest frequency that the string can accommodate and is termed the *fundamental frequency*, or *first harmonic*. Harmonic frequencies, second and greater, are integer multiples of the first harmonic and can be determined from the generalized relation

$$f = \frac{nv}{2L} \qquad n = 1, 2, 3, \ldots$$

The first, second, and third harmonics are illustrated in Figure 15-5.

Resonance Frequencies in a Pipe

The resonance frequencies in a pipe can be derived in a fashion similar to that for the stretched string. However, the pipe offers more conditions to consider, such as both ends open, both ends closed, and one end open. A displacement node can be found at the end of a closed pipe while a displacement antinode will be present toward the open end of the pipe. Given that the distance from node to node is $\lambda/2$ and the distance from node to antinode is $\lambda/4$, the resonant frequencies for a pipe of length L are as follows:

Pipe with both ends open:
$$L = 2\frac{\lambda}{4} = \frac{\lambda}{2}$$

Fundamental frequency
$$f_1 = \frac{v}{2L}$$

Harmonics
$$f_n = \frac{nv}{2L} \qquad (n = 1, 2, 3, \ldots)$$

Pipe with both ends closed:
$$L = \frac{\lambda}{2}$$

Fundamental frequency
$$f_1 = \frac{v}{2L}$$

Harmonics
$$f_n = \frac{nv}{2L} \qquad (n = 1, 2, 3, \ldots)$$

Pipe with one end open:
$$L = \frac{\lambda}{4}$$

Fundamental frequency
$$f_1 = \frac{v}{4L}$$

Harmonics
$$f_n = \frac{nv}{4L} \qquad (n = 1, 2, 3, \ldots)$$

15.4 PRINCIPLE OF SUPERPOSITION

Principle of superposition states that the resultant wave disturbance of a point in space from two or more waves is the algebraic sum of the individual waves. The frequency of the resultant superimposed wave, known as the *beat frequency* f_{beat}, is equal to the difference in the frequencies of the individual waves f_1 and f_2, or

$$f_{\text{beat}} = f_1 - f_2$$

Solved Problem 15.10. A plucked guitar string vibrates such that seven nodes are generated, including one node at each end of the string. Determine the resonant frequency of the guitar string, and identify its harmonic.

Solution

By definition, the harmonic frequencies of a stretched string (including a guitar string) are

$$f_n = \frac{nv}{2L}$$

We now need to find the value of the integer n. Referring to the case for the fundamental frequency, we see that two nodes imply $n = 1$. Following the mathematical progression, three nodes imply $n = 2$, and four nodes imply $n = 3$, leading to the expression

$$n = \text{number of nodes} - 1$$

Therefore, since we have seven nodes, $n = 7 - 1 = 6$, and the frequency is

$$f_n = \frac{6v}{2L} = \frac{3v}{L}$$

Since this frequency is six times the fundamental frequency, it is the sixth harmonic.

Solved Problem 15.11.   Consider a cello string with length $L = 1.8$ m and a mass $m = 7$ g. Determine the tension required to produce a fundamental frequency of 460 Hz.

Solution

By definition, the fundamental frequency of a standing wave through a string is

$$f = \frac{v}{2L}$$

where v is the velocity of a transverse wave. Solving for v gives

$$v = f(2L) = \sqrt{\frac{F}{\mu}}$$

where μ is the linear mass density of the cello string, given by

$$\mu = \frac{m}{L} = \frac{7 \times 10^{-3}\,\text{kg}}{1.8\,\text{m}} = 3.88 \times 10^{-3}\,\text{kg m}^{-1}$$

Solving for the force in terms of the string length, string mass, and frequency, we have

$$F = \mu f^2 (2L)^2 = (3.88 \times 10^{-3}\,\text{kg m}^{-1})(460\,\text{s}^{-1})^2 (2 \times 1.8\,\text{m})^2 = 2.955 \times 10^3\,\text{N}$$

Solved Problem 15.12.   Determine the velocity of waves in an open pipe of 1.3-m length if the fundamental frequency is 225 Hz.

Solution

For an open pipe, the fundamental frequency is given by

$$f = \frac{v}{2L}$$

Solving for v gives

$$v = f(2L) = (225\,\text{Hz})(2)(1.3\,\text{m}) = 585\,\text{m s}^{-1}$$

Solved Problem 15.13.   The frequency of the original wave generated in a stretched string is 275 Hz, and the frequency of the reflected wave is 255 Hz. Determine the beat frequency.

Solution

The beat frequency is the difference of the frequencies of the original and reflected waves, or

$$f_{\text{beat}} = f_1 - f_2 = 275\,\text{Hz} - 255\,\text{Hz} = 20\,\text{Hz}$$

15.5   SOUND

Sound waves are longitudinal waves that can propagate through all forms of matter—solids, liquids, and gases. The speed of sound through air is approximately 344 m s^{-1} (760 mi h^{-1}). In contrast, the speed of sound in water is 1480 m s^{-1}. The speed of sound depends on the spacing of particles within the propagating medium. Therefore, the speed of sound is the fastest in solids and faster in liquids than in gases.

The range of frequencies of sound audible or detectable by the human ear is 20 Hz to 20 kHz. Frequencies above 20 kHz are termed *ultrasonic waves*.

Sound intensity represents the rate of energy transported by the wave per unit area perpendicular to its direction of motion. The level of sound intensity β, expressed in units of decibels (dB), is defined in terms of intensity I as

$$\beta = 10 \log \frac{I}{I_o}$$

where I_o is the threshold for human hearing ($= 1 \times 10^{-12}$ W m^{-2}). And I_o represents the intensity of the weakest sound detectable by the human ear. Listed below are representative values of sound intensity I and their corresponding sound level β.

Intensity (W m^{-2})	Sound Level (dB)	Sound Description
1×10^{-12}	0	Threshold of hearing at 1000 Hz
1×10^{-10}	20	Distant whisper
1×10^{-8}	40	Average home
1×10^{-6}	60	Normal conversation
1×10^{-4}	80	Loud radio
1×10^{-2}	100	Siren at 30 m
1	120	Loud indoor rock concert Threshold of pain
1×10^{2}	140	Jet airplane at 30 m
1×10^{4}	160	Bursting of eardrums

15.6 DOPPLER EFFECT

Doppler effect refers to the shift in frequency of a transmitted sound caused by a change in path length between source and observer. Consider a source emitting a sound with a frequency f_s detected by an observer with a frequency f_o. If the source is moving toward the observer, the observer will perceive an increase in the sound frequency. Conversely, if the source is moving away from the observer, the observer will perceive a decrease in frequency. The altered frequency detected by the observer is known as the *Doppler-shifted frequency*. These qualitative relationships can be expressed by

$$f_o = f_s \left(\frac{v \pm v_o}{v \mp v_s} \right)$$

where v is the speed of sound in the given medium. These are several specific cases of the Doppler effect:

(1) Moving sound source toward a fixed observer

$$f_o = f_s \left(\frac{v}{v - v_s} \right)$$

(2) Moving sound source away from a fixed observer

$$f_o = f_s \left(\frac{v}{v + v_s} \right)$$

(3)   Fixed sound source with observer moving toward source

$$f_o = f_s \left(\frac{v + v_o}{v} \right)$$

(4)   Fixed sound source with observer moving away from source

$$f_o = f_s \left(\frac{v - v_o}{v} \right)$$

Solved Problem 15.14.   At a rock concert, the Rolling Pebbles played their music at a sound level of 5 W m^{-2}. Determine the sound level of the music in decibels.

Solution

By definition, the sound level is related to the sound intensity by

$$\beta = 10 \log \frac{I}{I_o} = 10 \log \frac{5 \text{ W m}^{-2}}{1 \times 10^{-12} \text{ W m}^{-2}} = 127 \text{ dB}$$

Solved Problem 15.15.   A particular sound was measured at 75 dB. Determine its sound intensity.

Solution

Using the expression for the sound level intensity, we have

$$\beta = 10 \log \frac{I}{I_o}$$

$$75 \text{ dB} = 10 \log \frac{I}{1 \times 10^{-12} \text{ W m}^{-2}}$$

Solving for I yields

$$I = 1 \times 10^{-12} \text{ W m}^{-2} \times \text{antilog} \frac{75 \text{ dB}}{10} = 3.16 \times 10^{-5} \text{ W m}^{-2}$$

Solved Problem 15.16.   A police car, in pursuit of a driver suspected of speeding, is traveling at 30 m s^{-1} when its siren is turned on, operating at a frequency of 1.2 kHz. Given that the speed of sound in air is 344 m s^{-1}, determine the frequency heard by a stationary witness as (1) the police car approaches the observer and (2) the police car passes the observer.

Solution

(1)   For the case of the moving source approaching a stationary observer, the Doppler-shifted frequency detected by the observer is given by

$$f_o = f_s \left(\frac{v}{v - v_s} \right) = 1200 \text{ Hz} \left(\frac{344 \text{ m s}^{-1}}{344 \text{ m s}^{-1} - 30 \text{ m s}^{-1}} \right) = 1315 \text{ Hz}$$

(2)   For the case of the moving source passing a stationary observer, the Doppler-shifted frequency detected by the observer is given by

$$f_o = f_s \left(\frac{v}{v + v_s} \right) = 1200 \text{ Hz} \left(\frac{344 \text{ m s}^{-1}}{344 \text{ m s}^{-1} + 30 \text{ m s}^{-1}} \right) = 1104 \text{ Hz}$$

Solved Problem 15.17.   Ultrasound is commonly used to assess the development of a fetus within the womb of an expectant mother. In such an examination, an 8-MHz ultrasonic beam is focused on the abdomen, emitting ultrasonic waves that propagate through the abdomen and reflect from the wall of the fetal heart, which is moving toward the ultrasonic receiver as the heart beats. The reflected sound wave is then mixed with the transmitted sound wave, resulting in a Doppler-shifted frequency by 10 kHz. The speed of sound in body tissue is 1500 m s^{-1}. Determine the speed of the fetal heart.

Solution

This problem requires the use of Doppler's equation for a fixed source and observer moving toward the source, or

$$f_o = f_s \left(\frac{v + v_o}{v} \right)$$

where $f_s = 8$ MHz $(8 \times 10^6$ Hz), $f_o = 8$ MHz $+ 10$ kHz $= 8.01$ MHz $(8.01 \times 10^6$ Hz), and $v = 1500$ m s^{-1}. The Doppler shift is added to the original frequency since, for the case of an observer moving toward a stationary source, the observed frequency is higher than the original frequency. Substituting into the Doppler equation above, we have

$$v_o = \frac{f_o}{f_s} v - v = \left(\frac{8.01 \text{ MHz}}{8.00 \text{ MHz}} \right) 1500 \frac{\text{m}}{\text{s}} - 1500 \frac{\text{m}}{\text{s}} = 1501.875 \frac{\text{m}}{\text{s}} - 1500 \frac{\text{m}}{\text{s}} = 1.875 \frac{\text{m}}{\text{s}}$$

ILLUSTRATIVE EXAMPLE 15.1.   Dolphins and Echolocation

Echolocation is a physical process used by dolphins and other animals such as bats to navigate through their surroundings and search for prey. In using echolocation, dolphins emit high-frequency, ultrasonic waves ($\approx 10^6$ Hz) in the form of whistles or clicks, sending as many as 2000 clicks per second. The sounds originate in air sacs in the nasal cavity, then move forward through a structure in the dolphin's forehead called the *melon*. The echoes from the surrounding object or prey bounce back to the dolphin and travel through the lower jaw to the inner ear, where they are registered. Echolocation is particularly efficient for dolphins since sound travels almost 5 times faster in water than in air.

ILLUSTRATIVE EXAMPLE 15.2.   Ultrasound and Assessment of Stroke Risk

Stroke is a general clinical term describing the severe reduction and/or cessation of blood flow to the brain. A major cause of stroke is the obstruction of the major arteries in the neck (carotid arteries) by atherosclerotic calcified deposits along the inner wall of the vessels which directly supply the brain with blood. Knowledge of blood flow through pertinent vessels reveals critical information regarding the presence and severity of abnormal obstructions, possibly leading to stroke.

Ultrasound is used to directly obtain blood flow measurements of the major vessels of the circulatory system, particularly the carotid arteries of the neck. Doppler ultrasound utilizes sound waves in the ultrasonic range (20 kHz to 20 MHz) to obtain blood flow measurements based on the Doppler effect, as illustrated in Figure 15-6. In a Doppler ultrasound procedure, an ultrasonic transducer is placed directly over the region of interest, that is, the carotid arteries. The transducer generates sounds waves which propagate from the transducer through the skin of the patient. Ultrasonic waves of an original frequency f_o and corresponding speed v are directed toward the surface of a blood vessel, interact with the red blood cells contained in the flowing blood, and are reflected back toward the transducer at an altered, Doppler-shifted frequency f_D with a velocity v_D. The Doppler-shifted frequency is related to the aforementioned parameters by the relation

$$f_o = \frac{f_s v}{(v \pm v_s)(2 \cos \theta)}$$

where v is the speed of sound in blood (1540 m s^{-1}), θ is the angle between the axis of sound propagation and the flowing blood, and 2 is a factor to correct for the transit time both to and from the source. The reflected sound propagates back toward the transducer in the form of "echoes" and is subsequently detected, registered, and used to calculate the speed of blood flow through the vessel of interest.

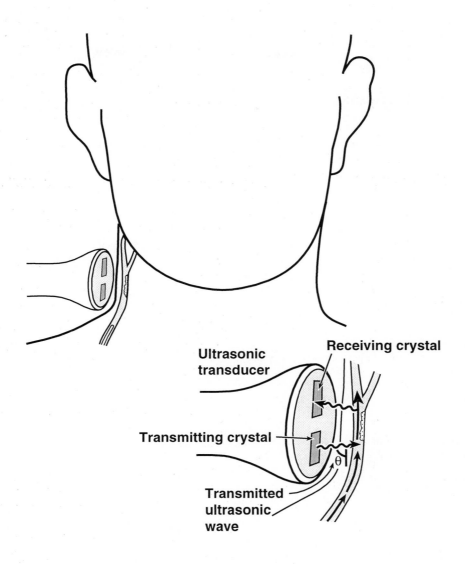

Fig. 15-6

Supplementary Problems

15.1. Determine the frequency of a wave with a period of (1) 1 s, (2) 100 ms, and (3) 100 μs.

Solution

(1) 1 Hz; (2) 10 Hz; (3) 10 kHz

15.2. Radio station WXYZ broadcasts at a frequency of 1060 kHz. If the speed of the waves is 3×10^8 m s^{-1}, determine the (1) wavelength and (2) period of the radio waves.

Solution

(1) 283 m; (2) 0.9 μs

15.3. Given the wave velocity is 344 m s^{-1}, determine the frequency for waves of wavelength (1) 1 m, (2) 1 cm, and (3) 1 mm.

Solution

(1) 344 Hz; (2) 34.4 kHz; (3) 344 kHz

15.4. In confirming reports of an approaching storm, Mrs. Raines, the meteorologist, notices that 10 s passes between observing lightning and hearing thunder. How far away is the storm?

Solution

2 mi

15.5. The operational wavelength of a radar detection system is 4.8 cm. Determine the frequency of the radar waves.

Solution

6.25×10^9 Hz

15.6. Consider a steel piano wire of length 80 cm and mass 0.5 g that corresponds to middle C on the musical scale. The frequency of middle C is 256 Hz. If the velocity of a typical longitudinal wave is 320 m s^{-1}, determine the (1) required tension of the piano string, (2) wavelength, and (3) fundamental frequency of the transverse wave.

Solution

(1) 64.0 N; (2) 1.25 m; (3) 200 Hz

15.7. Determine the speed of a longitudinal wave in an aluminum rod whose density is $\rho = 2.7 \times 10^3$ kg m^{-3} and whose Young's modulus is $Y = 7.0 \times 10^{10}$ N m^{-2}.

Solution

5091 m s^{-1}

15.8. Given an aluminum rod and a copper rod, both of equal size and length, through which rod will a longitudinal wave travel faster? For aluminum, $Y = 7.0 \times 10^{10}$ N m^{-2} and $\rho = 2700$ kg m^3; for copper, $Y = 12.5 \times 10^{10}$ N m^{-2} and $\rho = 8920$ kg m^{-3}.

Solution

Aluminum rod

15.9. A metal wire is 2 ft in length and weighs 0.035 lb. If a 3.5-lb weight is hung from one end of the wire, determine the speed of the transverse waves along the wire.

Solution

80 ft s^{-1}

15.10. Determine the speed of compressional waves in water and copper. The bulk moduli for water and copper are 2.2×10^9 N m^{-2} and 1.25×10^{11} N m^{-2}, respectively. The densities for water and copper are 1×10^3 kg m^{-3} and 8.9×10^3 kg m^{-3}, respectively.

Solution

Water: 1483 m s^{-1}; copper: 3747 m s^{-1}

15.11. Determine the speed of compressional waves through helium gas held in a closed container at 700°C and a pressure of 2.5 atm. For helium gas, $M = 4\,\mathrm{kg\,mol^{-1}}$ and $\gamma = 1.66$.

Solution

$1832\,\mathrm{m\,s^{-1}}$

15.12. When Tiny Tim plays "Tiptoe through the Tulips," he strums a ukelele whose strings of length $L = 0.40\,\mathrm{m}$ and linear mass density $5 \times 10^{-4}\,\mathrm{kg\,m^{-1}}$ are under a tension of 250 N. Determine the (1) wavelength, (2) wave velocity, and (3) frequencies of the first three harmonics of the ukelele strings.

Solution

(1) $\lambda_1 = 0.80\,\mathrm{m}$, $\lambda_2 = 0.40\,\mathrm{m}$, $\lambda_3 = 0.27\,\mathrm{m}$; (2) $707\,\mathrm{m\,s^{-1}}$; (3) $f_1 = 883.8\,\mathrm{Hz}$, $f_2 = 1767.5\,\mathrm{Hz}$, $f_3 = 2651.3\,\mathrm{Hz}$

15.13. In failing to get the attention of her friend with a whisper, Daphne raises her voice by 15 dB. Determine the change in sound intensity of Daphne's voice.

Solution

31.6

15.14. For a sound of intensity of $10^{-6}\,\mathrm{W\,m^{-2}}$ corresponding to sound characteristic of normal conversation, calculate the sound level in decibels. (Assume that the threshold of hearing is $1 \times 10^{-12}\,\mathrm{W\,m^{-2}}$.)

Solution

60 dB

15.15. The sound level of 120 dB corresponds to the threshold of pain while that at 160 dB is sufficient to rupture eardrums. For both sound levels, calculate the intensity in $\mathrm{W\,m^{-2}}$.

Solution

(1) $1\,\mathrm{W\,m^{-2}}$; (2) $1 \times 10^4\,\mathrm{W\,m^{-2}}$

15.16. While in pursuit, a police officer turns on the siren that operates at 1100 Hz. If the police car is approaching a stationary witness at $120\,\mathrm{ft\,s^{-1}}$, determine the frequency of the siren heard by the observing witness.

Solution

1236.6 Hz

15.17. In trailing a car whose velocity is $50\,\mathrm{ft\,s^{-1}}$ for a suspected traffic violation, a police car traveling at a velocity of $100\,\mathrm{ft\,s^{-1}}$ operates its siren whose frequency is 1350 Hz. Determine the frequency of the siren that the driver of the car hears.

Solution

1418 Hz

15.18. A guitar string of length 0.75 m, fixed at both ends, exhibits two successive frequencies of 250 and 280 Hz. Determine the (1) beat frequency, (2) fundamental frequency, and (3) speed of the wave.

Solution

(1) 30 Hz; (2) 30 Hz; (3) $45\,\mathrm{m\,s^{-1}}$

15.19. From the general form of the Doppler equation, derive an expression for the Doppler-shifted frequency, given (1) a moving observer following a moving source, (2) a moving source following a moving observer, (3) an observer and source moving away from each other in opposite directions, and (4) an observer and source moving toward each other.

Solution

(1) $f_o = f_s \left(\dfrac{v + v_o}{v + v_s} \right)$

(2) $f_o = f_s \left(\dfrac{v - v_o}{v - v_s} \right)$

(3) $f_o = f_s \left(\dfrac{v - v_o}{v + v_s} \right)$

(4) $f_o = f_s \left(\dfrac{v + v_o}{v - v_s} \right)$

Chapter 16

Electricity

Electricity represents a form of energy due to the presence and/or flow of electrically charged particles. Electricity can be produced in matter by six individual physical mechanisms: friction, heat, light, pressure, chemical interactions, and magnetism. As was the case for macroscopic objects, the physics of the electrically charged particle can be presented with respect to statics and dynamics. The statics of charged particles, or electrostatics, is presented in this chapter.

16.1 DEFINITIONS OF ELECTRICITY

Electric charge q is a physical property of elementary particles of the atom, a fundamental building block of all matter. The SI unit of charge is the coulomb, abbreviated C. Although charge can be either positive or negative, the magnitude of charge is $e = 1.6 \times 10^{-19}$ C. For two of the most common particles of the atom, the charge of the positively charged proton is $+1.6 \times 10^{-19}$ C, and the charge of the negatively charged electron is -1.6×10^{-19} C. Two like charges (either two positive charges or two negative charges) repel each other. A positive and negative charge attract each other.

Conservation of charge means that the net charge of a system remains constant. In other words, the amount of positive charge equals the amount of negative charge regardless of its physical state.

Coulomb's law describes the electrostatic force F between two charged particles q_1 and q_2 separated by a distance r:

$$F = \frac{1}{4\pi\varepsilon_o} \frac{q_1 q_2}{r^2} = k \frac{q_1 q_2}{r^2}$$

where ε_o is the permittivity constant, defined as $\varepsilon_o = 8.854 \times 10^{-12}\,\text{C}^2\,\text{N}^{-1}\,\text{m}^{-2}$. Values for k are

$$k = \frac{1}{4\pi\varepsilon_o} = 9.0 \times 10^9\,\text{N}\,\text{m}^2\,\text{C}^{-2}$$

Electrostatic force, as is the case for all types of force, is a vector quantity and is expressed in units of newtons. The direction of the electrostatic force is based on the charges involved. Opposite charges generate an attractive (negative) force, and the direction is toward the other charge; like charges generate a repulsive (positive) force, and the direction is away from the other charge.

Electric field E defines the electric force exerted on a positive test charge positioned within any given space. A positive test charge, q_o, is similar in most respects to a true charge except that it does not exert an electrostatic force on any adjacent or nearby charges. Thus, the electric field of a positive test charge provides an idealized distribution of electrostatic force generated by the test charge:

$$E = \frac{F}{q_o}$$

Since E is a vector quantity, the direction is dependent on the identity of the charge. Since the test charge is positive, if the other charge is negative, an attractive force is generated and the direction of E is toward the test charge. Likewise, if the other charge is positive, a repulsive force is generated and the direction of E is toward the test charge. Electric field E is expressed in units of force charge^{-1} (N C^{-1}). If E is known, it is possible to determine the electrostatic force exerted on any charge q placed at the same position from

$$F = Eq$$

An electric field can be produced by one or more electric charges. The electric field of a point charge always points away from a positive charge but toward a negative charge.

Electric field for a point charge can be derived by substituting the expression for Coulomb's law into that for the electric field, resulting in

$$E = \frac{1}{4\pi\varepsilon_o}\frac{q}{r^2} = \frac{kq}{r^2}$$

where E is the electric field, q is a positive point charge, and r is the distance from the test charge q_o to the point charge. For more than one charge in a defined region of space, the total electric field E_{tot}, since it is a vector quantity itself, is the vector sum of the electric field generated by each charge E_q in the distribution, or

$$E_{tot} = E_{q1} + E_{q2} + E_{q3} + E_{q4} + \cdots$$

Electric field lines of force make a visual display of the electric field that uses imaginary lines to represent the magnitude and direction of the electric field or the distribution of the electrostatic force over a region in space. The lines of force from the positive test charge are directed away from the charge while the lines of force of a negative charge are directed toward the charge. The magnitude of the force is greater in the region closer to the charge and becomes weaker farther from the charge. This is illustrated in Figure 16-1.

Repulsive Electrostatic Force of Like Charges

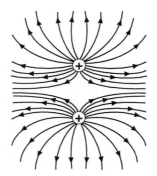

Attractive Electrostatic Force of Unlike Charges

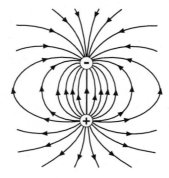

Fig. 16-1

Electric potential difference V between two points within an electric field represents the work W required to move an electric test charge q_o of 1 C from one point to the other, or

$$\text{Potential difference} = \frac{\text{work}}{\text{charge}}$$

$$V = \frac{W}{q_o} = k\frac{q_o}{r}$$

where the unit of potential difference is expressed in volts (V).

Given two points within a uniform electric field E, the potential difference V between the two points is the product of E and the distance d between the two points in a parallel direction to E or

$$V = Ed$$

Electrical work W is the energy required to move a charge q through a potential difference V, and it is given by the inverse of potential difference, that is,

$$W = qV$$

Work is typically expressed in units of joules. However, the work done in moving a charge $+e$ through a potential of 1 V is defined as 1 electronvolt (1 eV). Units of electronvolts and joules are related by

$$1 \text{ eV} = (1.6 \times 10^{-19} \text{ C})(1 \text{ V}) = 1.6 \times 10^{-19} \text{ J}$$

Electric power P is the energy expended by an energy source in transporting a charge q through a potential difference V within a time t:

$$\text{Power} = \frac{\text{work}}{\text{time}} = \frac{qV}{t}$$

Electric potential energy U is defined as the amount of work W required to move a charge q from an infinite distance to its final position within an electric field E, and it is given by

$$\text{Potential energy} = \text{work} = (\text{charge})(\text{potential difference})$$
$$U = W = qV$$

where V represents the electric potential difference generated by the charges, q, comprising the electric field. Units of electric potential energy are joules.

If only one charge is responsible for the potential difference of the electric field, then the above expression can be rewritten as

$$U = qV = qk\frac{q_o}{r}$$

Solved Problem 16.1. Two electric charges $q_1 = +9.5 \times 10^{-3} \text{ C}$ and $q_2 = -3.6 \times 10^{-4} \text{ C}$ are separated by a distance $r = 10$ cm. Determine the magnitude and direction of the electrostatic force F that each point charge exerts on the other.

Solution

This problem can be solved by using Coulomb's law:

$$F = k\frac{q_1 q_2}{r^2} = \left(9 \times 10^9 \frac{\text{N m}^2}{\text{C}^2}\right)\frac{(9.5 \times 10^{-3} \text{ C})(-3.6 \times 10^{-4} \text{ C})}{(0.1 \text{ m})^2} = -3.1 \times 10^6 \text{ N}$$

Since the two charges are oppositely charged, the direction of the electrostatic force is toward the opposing charge.

Solved Problem 16.2.   Two electrostatically charged spheres ($q_1 = 4e$ and $q_2 = -2e$, $e = 1.6 \times 10^{-19}$ C) are separated by a distance $r = 4$ nm. Determine the electrostatic force for (1) the given separation distance d, (2) one-half the separation distance, or $r = r/2$; (3) two times the separation distance, or $r = 2r$.

Solution

This problem can be solved by using Coulomb's law, or

$$F = k\frac{q_1 q_2}{r^2}$$

(1)   The electrostatic force for the separation distance $r = 4$ nm is

$$F = \left(9 \times 10^9 \, \frac{\text{N m}^2}{\text{C}^2}\right) \frac{(6.4 \times 10^{-19}\,\text{C})(-3.2 \times 10^{-19}\,\text{C})}{(4 \times 10^{-9}\,\text{m})^2} = -1.2 \times 10^{-8}\,\text{N}$$

where F is an attractive force and is thus directed toward the opposing charge.

(2)   The electrostatic force for one-half the separation distance, or $r = 2$ nm, is

$$F = \left(9 \times 10^9 \, \frac{\text{N m}^2}{\text{C}^2}\right) \frac{(6.4 \times 10^{-19}\,\text{C})(-3.2 \times 10^{-19}\,\text{C})}{(2 \times 10^{-9}\,\text{m})^2} = -4.6 \times 10^{-8}\,\text{N}$$

where F is an attractive force and is thus directed toward the opposing charge.

(3)   The electrostatic force for twice the separation distance or $r = 8$ nm, is

$$F = \left(9 \times 10^9 \, \frac{\text{N m}^2}{\text{C}^2}\right) \frac{(6.4 \times 10^{-19}\,\text{C})(-3.2 \times 10^{-19}\,\text{C})}{(8 \times 10^{-9}\,\text{m})^2} = -2.9 \times 10^{-9}\,\text{N}$$

where F is an attractive force and is thus directed toward the opposing charge.

Solved Problem 16.3.   Determine the electrostatic force between two α particles of charge $+2e$ (3.2×10^{-19} C) separated by 10^{-13} m.

Solution

Using Coulomb's law, we have

$$F = \left(9 \times 10^9 \, \frac{\text{N m}^2}{\text{C}^2}\right) \frac{(3.2 \times 10^{-19}\,\text{C})(3.2 \times 10^{-19}\,\text{C})}{(1 \times 10^{-13}\,\text{m})^2} = 9.22 \times 10^{-2}\,\text{N}$$

where F is a repulsive force and is thus directed away from the opposing charge.

Solved Problem 16.4.   Three charges $-q_1$, $+q_2$, and $+q_3$ are spaced equidistant from each other by a distance d. Determine the relative magnitudes of the three charges needed for them to remain in equilibrium.

Solution

Using Coulomb's law, we need to determine the electrostatic forces exerted between the three charges:

$$F_{12} = k\frac{(-q_1)(q_2)}{d^2} = k\frac{-q_1 q_2}{d^2}$$

$$F_{23} = k\frac{(q_2)(q_3)}{d^2} = k\frac{q_2 q_3}{d^2}$$

$$F_{13} = k\frac{(-q_1)(q_3)}{(2d)^2} = k\frac{-q_1 q_3}{(2d)^2}$$

Setting $F_{23} = F_{13}$, we have

$$k\frac{q_2 q_3}{d^2} = k\frac{-q_1 q_3}{4d^2}$$

This reduces to

$$q_1 = -4q_2$$

Setting $F_{12} = F_{23}$ gives

$$k\frac{-q_1q_2}{d^2} = k\frac{q_2q_3}{d^2}$$

This reduces to

$$q_3 = -q_1 = -(-4q_2) = 4q_2$$

Therefore, in terms of q_2, $q_1 = -4q_2$ and $q_3 = 4q_2$.

Solved Problem 16.5.   Four charges $q_1 = +2e$, $q_2 = +e$, $q_3 = -e$, and $q_4 = -2e$ are placed clockwise from the bottom left-hand corner at the four corners of a square with the length of one side equal to 5.0 cm. Determine the electrostatic force acting on q_1.

Solution

Using Coulomb's law, we must determine the electrostatic force exerted by q_2, q_3, and q_4 on q_1.

$$F_{12} = k\frac{q_1q_2}{r^2} = \left(9 \times 10^9 \frac{\text{N m}^2}{\text{C}^2}\right)\frac{(3.2 \times 10^{-19}\,\text{C})(1.6 \times 10^{-19}\,\text{C})}{(0.05\,\text{m})^2} = 1.84 \times 10^{-25}\,\text{N}$$

where F_{12} is a repulsive force due to like charges.

To determine the electrostatic force exerted by q_3 on q_1 requires calculation of the distance based on the pythagorean theorem:

$$r = \sqrt{(0.05\,\text{m})^2 + (0.05\,\text{m})^2} = 0.0707\,\text{m}$$

$$F_{13} = k\frac{q_1q_2}{r^2} = \left(9 \times 10^9 \frac{\text{N m}^2}{\text{C}^2}\right)\frac{(3.2 \times 10^{-19}\,\text{C})(-1.6 \times 10^{-19}\,\text{C})}{(0.0707\,\text{m})^2} = 9.4 \times 10^{-26}\,\text{N}$$

Force F_{13} is an attractive force due to like charges.

$$F_{14} = k\frac{q_1q_2}{r^2} = \left(9 \times 10^9 \frac{\text{N m}^2}{\text{C}^2}\right)\frac{(3.2 \times 10^{-19}\,\text{C})(-3.2 \times 10^{-19}\,\text{C})}{(0.05\,\text{m})^2} = 3.7 \times 10^{-25}\,\text{N}$$

Force F_{14} is an attractive force due to like charges.

We now resolve the force vectors into their x and y components to determine the resultant vector.

x component: $1.84 \times 10^{-25}\,\text{N} + (9.4 \times 10^{-26}\,\text{N})(\cos 45°) = 25.1 \times 10^{-26}\,\text{N}$

y component: $3.7 \times 10^{-25}\,\text{N} + (9.4 \times 10^{-26}\,\text{N})(\sin 45°) = 43.7 \times 10^{-26}\,\text{N}$

$$R = \sqrt{X^2 + Y^2} = \sqrt{(25.1 \times 10^{-26}\,\text{N})^2 + (43.7 \times 10^{-26}\,\text{N})^2} = 5.0 \times 10^{-25}\,\text{N}$$

$$\theta = \tan^{-1}\frac{Y}{X} = \tan^{-1}\frac{43.7 \times 10^{-26}\,\text{N}}{25.1 \times 10^{-26}\,\text{N}} = 60.1°$$

Solved Problem 16.6.   For an electron of charge $-e$ and a proton of charge $+e$ separated a distance 0.53 Å, compare the electrostatic and gravitational forces between the two particles. The masses of the electron and proton are $m_e = 9.1 \times 10^{-31}$ kg and $m_p = 1.7 \times 10^{-27}$ kg.

Solution

Computing the electrostatic force F_e and gravitational force F_g yields

$$F_e = k\frac{q_e q_p}{r^2} = \left(9 \times 10^9 \frac{\text{N m}^2}{\text{C}^2}\right)\frac{(-1.6 \times 10^{-19}\,\text{C})(1.6 \times 10^{-19}\,\text{C})}{(0.53 \times 10^{-10}\,\text{m})^2} = 8.2 \times 10^{-8}\,\text{N}$$

$$F_g = G\frac{m_e m_p}{r^2} = \left(6.67 \times 10^{-11}\frac{\text{m}^3}{\text{s}^2\,\text{kg}}\right)\frac{(9.1 \times 10^{-31}\,\text{kg})(1.7 \times 10^{-27}\,\text{kg})}{(0.53 \times 10^{-10})^2} = 3.67 \times 10^{-47}\,\text{N}$$

Taking the ratio of the two forces, we can see that the electrostatic force is 2.23×10^{39} times greater than the gravitational force, which can be considered negligible.

Solved Problem 16.7. The Bohr model of the hydrogen atom describes electron motion in a circular orbit of radius 0.53 Å about the nuclear proton. Determine the (1) electric field and (2) potential of the orbiting electron.

Solution

(1) The electric field E experienced by the orbiting electron is

$$E = \frac{kq}{r^2}$$

where $q = +e(1.6 \times 10^{-19}\,\text{C})$ and r is the distance between the electron and proton. Substituting values yields

$$E = \left(9 \times 10^9 \frac{\text{N m}^2}{\text{C}^2}\right)\frac{1.6 \times 10^{-19}\,\text{C}}{(0.53 \times 10^{-10}\,\text{m})^2} = 5.14 \times 10^{11}\,\text{N C}^{-1}$$

(2) The electric potential V of the electron is

$$V = \frac{kq}{r}$$

Substituting known values gives

$$V = \left(9 \times 10^9 \frac{\text{N m}^2}{\text{C}^2}\right)\frac{1.6 \times 10^{-19}\,\text{C}}{0.53 \times 10^{-10}\,\text{m}} = 27.2\,\text{N m C}^{-1}$$

Solved Problem 16.8. From Solved Problem 16.7 concerning the Bohr model of the hydrogen atom, determine (1) the electrostatic force between the proton and electron and (2) the velocity of the orbiting electron. The mass of the electron $m_e = 9.1 \times 10^{-31}$ kg.

Solution

(1) The electrostatic force can be determined from Coulomb's law:

$$F = \left(9 \times 10^9 \frac{\text{N m}^2}{\text{C}^2}\right)\frac{(1.6 \times 10^{-19}\,\text{C})(-1.6 \times 10^{-19}\,\text{C})}{(0.53 \times 10^{-10}\,\text{m})^2} = -8.21 \times 10^{-8}\,\text{N}$$

(2) The magnitude of the electrostatic force is the centripetal force of the orbiting electron. Thus,

$$F = \frac{mv^2}{r} \Rightarrow V = \sqrt{\frac{Fr}{m}} = \sqrt{\frac{(8.21 \times 10^{-8}\,\text{N})(0.53 \times 10^{-10}\,\text{m})}{9.1 \times 10^{-31}\,\text{kg}}} = 2.2 \times 10^6\,\text{m s}^{-1}$$

Solved Problem 16.9. (1) Determine the magnitude of the electric field at a point 2.5 Å from a proton. (2) Determine the potential at this point.

Solution

(1) The electric field can be determined by

$$E = \frac{kq}{r^2}$$

$$= \left(9 \times 10^9 \frac{\text{N m}^2}{\text{C}^2}\right)\frac{1.6 \times 10^{-19}\,\text{C}}{(2.5 \times 10^{-10}\,\text{m})^2} = 2.3 \times 10^{10}\,\text{N C}^{-1}$$

(2)   The potential at this point is

$$V = \frac{kq}{r}$$

$$= \left(9 \times 10^9 \, \frac{\text{N m}^2}{\text{C}^2}\right) \frac{1.6 \times 10^{-19} \, \text{C}}{2.5 \times 10^{-10} \, \text{m}} = 5.76 \, \text{N m C}^{-1}$$

Solved Problem 16.10.   Determine the number of electrons that must be added to an electrostatically charged sphere of 5-cm radius to generate an electric field of $5 \times 10^{-4} \, \text{N C}^{-1}$ just above the sphere's surface.

Solution

We first must use the definition of the electric field, given as

$$E = \frac{kq}{r^2}$$

Solving for q yields

$$q = \frac{r^2 E}{k} = \frac{(5 \times 10^{-2} \, \text{m})^2 (5 \times 10^{-4} \, \text{N C}^{-1})}{9 \times 10^9 \, \text{N m}^2 \, \text{C}^2} = 1.38 \times 10^{-16} \, \text{C}$$

The number of electrons n can be determined by

$$n = \frac{1.38 \times 10^{-16} \, \text{C}}{1.6 \times 10^{-19} \, \text{C/electron}} = 860 \text{ electrons}$$

Solved Problem 16.11.   Starting from rest, a proton drifts through a potential difference of 10^5 V. Determine the final kinetic energy and final velocity of the proton.

Solution

As for any physical object, the proton possesses only potential energy and no kinetic energy at rest. At its final position, the proton possesses only kinetic energy with no potential energy. Thus, by the law of conservation of energy

$$\text{Energy before motion} = \text{energy after motion}$$
$$\text{PE} + 0 = 0 + \text{KE}$$

By definition, the potential V is

$$V = \frac{E}{q}$$

where E is the potential energy. Therefore,

$$\text{PE} = qV$$
$$10^5 \, \text{eV} = \frac{1}{2} m_p v^2 = \text{KE}_{\text{final}}$$

Thus, the final kinetic energy is

$$\text{KE}_{\text{final}} = 10^5 \, \text{eV} = (10^5 \, \text{eV}) \left(\frac{1.6 \times 10^{-12} \, \text{erg}}{1 \, \text{eV}}\right) = 1.6 \times 10^{-7} \, \text{erg}$$

To compute the final velocity, we use the expression for the final kinetic energy:

$$\text{KE}_{\text{final}} = 1.6 \times 10^{-7} \, \text{erg} = \frac{1}{2} m_p v^2$$

Solving for v gives

$$v = \sqrt{\frac{2(1.6 \times 10^{-7}\,\text{erg})}{1.67 \times 10^{-24}\,\text{g}}} = 4.38 \times 10^8\,\text{cm s}^{-1}$$

Solved Problem 16.12.   Three electric charges $q_1 = 1e$, $q_2 = 2e$, and $q_3 = 1e$ are spaced 0.10 m equidistant along the base of an isosceles triangle, as shown in Figure 16-2. Calculate the electrostatic force acting on point A of charge $+1e$.

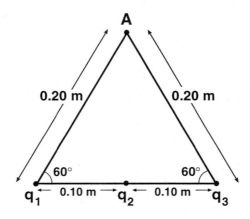

Fig. 16-2

Solution

The contribution by each charge to the electrostatic force acting on A can be determined from Coulomb's law:

$$F_{q_1 A} = \left(9 \times 10^9\,\frac{\text{N m}^2}{\text{C}^2}\right)\frac{(1.6 \times 10^{-19}\,\text{C})(1.6 \times 10^{-19}\,\text{C})}{(0.20\,\text{m})^2} = 5.76 \times 10^{-27}\,\text{N}$$

The distance from q_2 to point A can be calculated from elementary trigonometric relations:

$$q_2 A = (0.20\,\text{m})(\sin 60°) = 0.17\,\text{m}$$

$$F_{q_2 A} = \left(9 \times 10^9\,\frac{\text{N m}^2}{\text{C}^2}\right)\frac{(3.2 \times 10^{-19}\,\text{C})(1.6 \times 10^{-19}\,\text{C})}{(0.17\,\text{m})^2} = 1.59 \times 10^{-26}\,\text{N}$$

$$F_{q_3 A} = \left(9 \times 10^9\,\frac{\text{N m}^2}{\text{C}^2}\right)\frac{(1.6 \times 10^{-19}\,\text{C})(1.6 \times 10^{-19}\,\text{C})}{(0.20\,\text{m})^2} = 5.76 \times 10^{-27}\,\text{N}$$

The resultant force F_{res} on A can be determined by first resolving the force vectors into their x and y components:

x component: $(5.76 \times 10^{-27}\,\text{N})(\cos 60°) + 0 + (-5.76 \times 10^{-27}\,\text{N})(\cos 60°) = 0$

y component: $(5.76 \times 10^{-27}\,\text{N})(\sin 60°) + 1.59 \times 10^{-26}\,\text{N} + (-5.76 \times 10^{-27}\,\text{N})(\sin 60°) = 1.59 \times 10^{-26}\,\text{N}$

Force $F_{\text{res}} = 1.59 \times 10^{-26}\,\text{N}$ is directed upward along the y axis.

Solved Problem 16.13.   Determine the potential at point A as a result of the distribution of three charges at the base of the triangle in Figure 16-2.

Solution

$$V_{q_1} = \frac{kq_1}{r} = \frac{(9 \times 10^9 \, \text{N m}^2 \, \text{C}^{-2})(1.6 \times 10^{-19} \, \text{C})}{0.20 \, \text{m}} = 7.2 \times 10^{-9} \, \text{N m C}^{-1}$$

$$V_{q_2} = \frac{kq_2}{r} = \frac{(9 \times 10^9 \, \text{N m}^2 \, \text{C}^{-2})(1.6 \times 10^{-19} \, \text{C})}{0.1733 \, \text{m}} = 8.3 \times 10^{-9} \, \text{N m C}^{-1}$$

$$V_{q_3} = \frac{kq_3}{r} = \frac{(9 \times 10^9 \, \text{N m}^2 \, \text{C}^{-2})(1.6 \times 10^{-19} \, \text{C})}{0.20 \, \text{m}} = 7.2 \times 10^{-9} \, \text{N m C}^{-1}$$

The net potential V_{total} is simply the sum of the potentials from the three charges:

$$V_{\text{total}} = V_{q_1} + V_{q_2} + V_{q_3} = 7.2 \times 10^9 \, \text{N m C}^{-1} + 8.3 \times 10^{-9} \, \text{N m C}^{-1} + 7.2 \times 10^{-9} \, \text{N m C}^{-1}$$
$$= 2.3 \times 10^{-8} \, \text{N m C}^{-1}$$

Solved Problem 16.14. A proton $(+e)$ and an alpha particle $(+2e)$ are separated by 10 cm. Determine the point along the axis of separation where the electric field is zero.

Solution

To locate the point where the electric field is zero, the electric field vectors for each point charge must be equated. Because they are positive charges, the associated electric field vectors point in opposite directions. Since the charges are unequal in magnitude, the distance between the proton and the desired point along the line of separation is defined as x, and the corresponding distance for the alpha particle is $10 \, \text{cm} - x$. Therefore,

$$E_{\text{proton}} = E_{\text{alpha}}$$

$$\frac{ke}{x^2} = \frac{k(2e)}{(10 \, \text{cm} - x)^2}$$

$$\frac{1}{x^2} = \frac{2}{(10 \, \text{cm} - x)^2}$$

$$2x^2 = (10 \, \text{cm} - x)^2$$

Taking the square root of both sides gives

$$\sqrt{2} \, x = 10 \, \text{cm} - x$$

$$\sqrt{2} \, x + x = 10 \, \text{cm}$$

$$x(\sqrt{2} + 1) = 10 \, \text{cm}$$

$$x = \frac{10 \, \text{cm}}{\sqrt{2} + 1} = 4.15 \, \text{cm}$$

This point is positioned closer to the proton since the alpha particle has a greater electric field.

ILLUSTRATIVE EXAMPLE 16.1.   Electric Signal Transmission through Nerves

The human body is able to receive, interpret, and respond to sensory stimuli by transmission of electric signals through the nervous system along nerve fibers or networks of individual cells, known as *neurons*. The structure of a neuron, shown in Figure 16-3, consists of a cell body with a nucleus connected to dendrites, serving as an input for the signal. Also attached to the neuron is an axon, or a long conducting tail which propagates the electric signal or impulse away from the neuron. The axon consists of a series of connected segments approximately 2.0 mm in length and 20 μm in diameter. In addition, the segments of the axon are encased in a myelin sheath, which acts as a good insulator. Segments of the axon tail are separated by small uninsulated gaps known as the *nodes of Ranvier*. The end of the axon branches into nerve endings that transmit the signals across small gaps to other neurons. Electric signals travel along the axon followed by communication with other neighboring neurons, eventually making it through the nervous system along networks of neurons until they are registered in the brain. The propagation velocity of electric signals through the myelinated fibers of the axon is approximately 100 m s^{-1}. Each nerve can accommodate simultaneously many signals, analogous to a number of phone calls transmitted along a telephone cable.

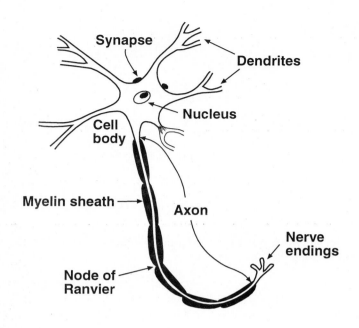

Fig. 16-3

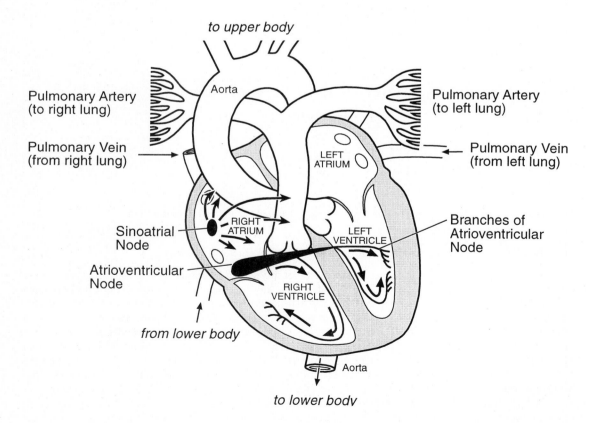

Fig. 16-4

ILLUSTRATIVE EXAMPLE 16.2.   Electrical Origin of the Heartbeat

The process of the human heartbeat is synchronized and regulated by electrical impulses generated by unique muscle cells in the right atrium of the heart. These muscle cells constitute the sinoatrial node or pacemaker of the heart. This impulse is transferred over the atria to the atrioventricular node in the base of the right ventricle, as shown in Figure 16-4, and then relayed to the muscular ventricles by bundles of specialized conducting fibers, known as *Purkinje fibers*. Assuming normal conditions, the heart beats or contracts almost as a single entity or mass.

ILLUSTRATIVE EXAMPLE 16.3.   Electrical Potential of Cellular Membranes

Many types of cells, encapsulated by a cellular membrane, contain primarily potassium chloride in their interior and primarily sodium chloride in their exterior. In the normal resting state, the cellular membrane is much more permeable to potassium ions than to sodium ions, resulting in an outward diffusion of potassium ions. The outward diffusion of potassium ions generates a negative charge in the interior of the cell which continues until the voltage inside the cell is -85 mV. This voltage is known as the *resting potential* of the cell.

If the cell is subjected to a stimulus of a mechanical, chemical, or electrical nature, the permeability of the cellular membrane changes such that sodium ions diffuse more readily into the cell. Following the inward diffusion of a small amount of sodium ions, the interior voltage of the cell increases to $+60$ mV, known as the *action potential* of the cell. Once the cell has achieved its action potential, the membrane again changes permeability, such that potassium ions readily diffuse outward, so the cell returns to its resting potential. Thus, depending on the state of the cell, the interior voltage of the cell varies from its resting potential (-85 mV) to its action potential ($+60$ mV), resulting in a net voltage change of 145 mV in the cell interior.

Supplementary Problems

16.1.   Two electric charges $q_1 = +2.1 \times 10^{-3}$ C and $q_2 = +4.2 \times 10^{-4}$ C are separated by a distance $d = 0.10$ cm. Determine the electrostatic force F that each point charge exerts on the other.

Solution

7.94×10^9 N; repulsive force

16.2.   Given two charges $q_1 = +4.2 \times 10^{-19}$ C and $q_2 = -5.0 \times 10^{-19}$ C separated by 1 nm, determine the (1) electrostatic force and (2) potential energy between the two charges.

Solution

(1) -1.89×10^{-9} N, attractive force; (2) -1.89×10^{-18} J

16.3.   Determine the magnitude of a charge if 1.6×10^{-15} J is required to move it through a potential difference of 1×10^5 V.

Solution

1.6×10^{-10} C

16.4.   Given a point charge of 0.2 C, determine the electric field intensity at a point 50 cm from the charge.

Solution

7.2×10^9 N C^{-1}

16.5.   Determine the potential difference if 15 J of work is performed in moving (1) $+1$ C and (2) -1 C across a distance d.

Solution

(1) 15 V; (2) -15 V

16.6.   Operation of a 60-W lightbulb requires $0.5 \, \text{C s}^{-1}$ from a 120-V electric outlet. Determine the work done per second by this electrical source.

Solution

60 J

16.7.   Determine the magnitude of the electric field at a distance 1 Å from (1) a proton and (2) an electron. $(1 \, \text{Å} = 10^{-10} \, \text{m.})$

Solution

(1) $1.44 \times 10^{11} \, \text{N C}^{-1}$; (2) $-1.44 \times 10^{11} \, \text{N C}^{-1}$

16.8.   For the proton and electron described in Supplementary Problem 16.7, determine the potential difference at these points.

Solution

(1) 14.1 V; (2) -14.4 V

16.9.   From Solved Problem 16.8 concerning the hydrogen atom, the electrostatic force exerted on the electron by the proton is $-8.21 \times 10^{-8} \, \text{N}$. Determine the electric field that acts on the electron.

Solution

$-5.13 \times 10^{11} \, \text{N C}^{-1}$

16.10.   Determine the resultant charge if 10^{12} electrons are added to an originally neutral metal sphere.

Solution

$1.6 \times 10^{-7} \, \text{C}$

16.11.   Determine the number of electrons required to produce a net charge of $3.2 \times 10^{-12} \, \text{C}$.

Solution

2.0×10^{7} electrons

16.12.   Two electrons exert a repulsive electrostatic force of $2 \times 10^{-7} \, \text{N}$ on each other. Determine the distance of separation.

Solution

$1.07 \times 10^{-11} \, \text{m}$

16.13.   Determine the force exerted on an electron by an electric field of $100 \, \text{V m}^{-1}$.

Solution

$1.6 \times 10^{-17} \, \text{N}$

16.14.   From Supplementary Problem 16.13, determine the acceleration of the electron.

Solution

$1.76 \times 10^{13} \, \text{m s}^{-2}$

16.15. Determine the kinetic energy of a proton accelerated by a potential difference of 25,000 V.

Solution

4.0×10^{-15} J

16.16. An alpha particle $(+2e)$ moves from rest in an electric field of 2000 V m^{-1}. (1) Determine the force exerted on the alpha particle by the electric field. (2) Determine the kinetic energy of the alpha particle after it moves 5 cm in this electric field.

Solution

(1) 6.4×10^{-16} N; (2) 3.2×10^{-17} J

16.17. Given that the work required to move a charge through a potential difference of 5×10^8 V is 1.6×10^{-10} J, determine the magnitude of charge.

Solution

3.2×10^{-19} C

16.18. Two point charges $q_1 = +e$ and $q_2 = -e$ are separated by 0.5 m along the x axis. Determine the magnitude and direction of the electric field at a point A, that is, 0.5 m along the y axis.

Solution

8.17×10^{-9} N C^{-1}; 75.1°

16.19. In a typical cathode-ray tube, electrons are accelerated through a potential difference of 3.0×10^4 V. Given that the mass of the electron is 9.1×10^{-31} kg, determine the speed of the electron.

Solution

1.02×10^8 m s^{-1}

Chapter 17

Direct-Current Circuits

In a direct-current (dc) circuit, electricity or electric charge flows continuously through connected circuit elements in a direction from the positive terminal to the negative terminal of a voltage source. This chapter addresses the physics involved in the flow of electrically charged particles, or electrodynamics in a dc circuit.

17.1 DEFINITIONS OF DC CIRCUITS

In a *dc circuit*, electricity generated by a voltage source or battery flows as current through an arrangement of circuit elements or components connected by a conductor or a material that provides minimal resistance to the flow of electrons. In contrast to conductors, insulators are materials that permit minimal current flow. These circuit elements may be connected in series or parallel. In a series circuit, two or more circuit elements are connected in a direct line or sequence. In a parallel circuit, two or more circuit elements are connected in a branching arrangement. Regardless of the type of circuit, these elements interact with and can directly influence the flow of charge through the circuit and are described below.

Current i is the rate or motion of electricity or, more specifically, electric charge. Current is expressed in units of charge per unit time or $C\,s^{-1}$. This unit is also known as an *ampere* (A).

Current in Series

Current that flows through one circuit element connected in series must also flow through the remaining elements connected in the series circuit. Therefore,

$$i = i_1 = i_2 = i_3 = \cdots = i_n$$

where n refers to the nth element in the series circuit.

Current in Parallel

In a parallel circuit, current flow begins at point A and continues through circuit elements connected in parallel before entering point B. The net current leaving point A must equal the net current entering point B, implying that

$$i = i_1 + i_2 + i_3 + \cdots + i_n$$

where n refers to the nth element in the series circuit.

Resistance R is an inherent property of a conductor that resists flow of electric current and represents a measure of the potential difference V that must be supplied to a circuit to drive current i through the circuit. Resistance is

$$\text{Resistance} = \frac{\text{potential difference}}{\text{current}}$$

$$R = \frac{V}{i}$$

As current flows through a resistor, electric energy or power P is dissipated into the resistor according to the following equivalent expressions:

$$P = iV = i^2 R = \frac{V^2}{R}$$

Resistance is expressed in units of volts per ampere, also known as the *ohm* (Ω).

The resistance of a conductor depends on the resistivity ρ unique to a material, its length L, and cross-sectional area A, or

$$R = \rho \frac{L}{A}$$

Resistivity is given in units of ohm $\cdot$ meter (Ω m)

Resistors in Series

In a circuit consisting of three resistors connected in series, as in Figure 17-1, the current must flow through the path presented by the resistors in series. To simplify circuit calculations, the equivalent resistance R_{eq} can be calculated in terms of the resistance of the individual components:

$$R_{eq} = R_1 + R_2 + R_3 + \cdots + R_n$$

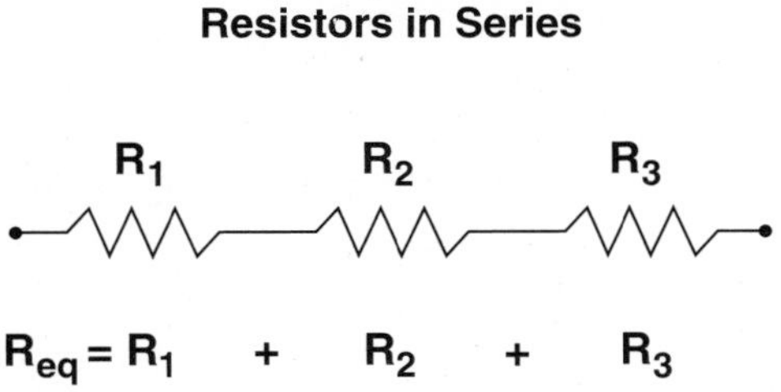

Resistors in Series

$$R_{eq} = R_1 \quad + \quad R_2 \quad + \quad R_3$$

Fig. 17-1

Resistors in Parallel

In a circuit consisting of three resistors connected in parallel, as in Figure 17-2, the current must flow through the path presented by the resistors in parallel. To simplify circuit calculations, the equivalent resistance R_{eq} can be calculated in terms of the resistance of the individual components:

$$\frac{1}{R_{eq}} = \frac{1}{R_1} + \frac{1}{R_2} + \frac{1}{R_3} + \cdots + \frac{1}{R_n}$$

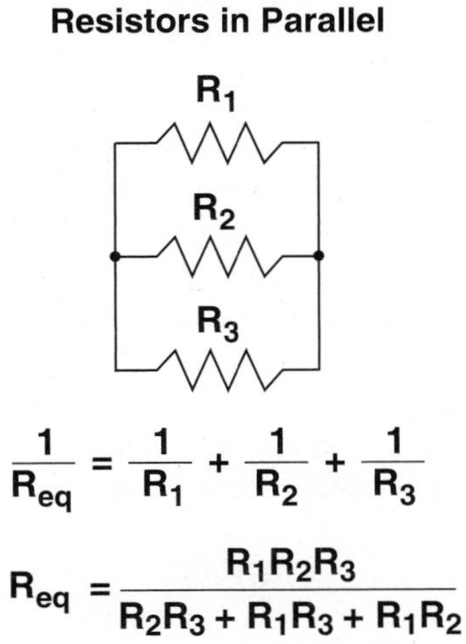

Resistors in Parallel

$$\frac{1}{R_{eq}} = \frac{1}{R_1} + \frac{1}{R_2} + \frac{1}{R_3}$$

$$R_{eq} = \frac{R_1 R_2 R_3}{R_2 R_3 + R_1 R_3 + R_1 R_2}$$

Fig. 17-2

where the equivalent resistance R_{eq} is *always less than* the smallest value of resistance of the individual components.

Electromotive force E is the voltage or potential difference generated between a positive terminal a and a negative terminal b of a battery or power source when no current is flowing. When current is flowing through a conductor, the conductor generates an internal resistance, thereby yielding a voltage drop of iR unique to the conductor. Therefore, the voltage V_{ab} generated by an emf is

$$V_{ab} = E - iR$$

Electromotive force is expressed in units of volts.

Voltage in Series

The net voltage of electromotive forces connected in series is the sum of the individual voltages of each electromotive force, or

$$V_{net} = V_1 + V_2 + V_3 + \cdots + V_n$$

Voltage in Parallel

The net voltage of electromotive forces connected in parallel is identical to that of the voltage from an individual electromotive force, or

$$V_{net} = V_1 = V_2 = V_3 = \cdots = V_n$$

Ohm's law states that the voltage V across a resistor R is proportional to the current i through it and can be written in equation form as

$$V = iR$$

An increase in voltage will drive more electrons through the wire (conductor) at a greater rate, thus the current will increase. If the voltage remains the same and the wire is a poor conductor, the resistance against the flow of electrons increases, thereby decreasing the current. It is possible to change i by manipulating V and R, but it is not possible to change V by manipulating i because flow is due to the difference in voltage, not vice versa. The current is a completely dependent variable.

Capacitors are circuit elements that store charge and consist typically of two conductors of arbitrary shape carrying equal and opposite charges separated by an insulator. Capacitance C depends on the shape and position of the capacitors and is defined as

$$C = \frac{q}{V}$$

where q is the magnitude of charge on either of the two conductors and V is the magnitude of potential difference between the two conductors. The SI unit of capacitance is coulomb/volt collectively known as the farad (F). Since capacitors store positive and negative charge, work is done in separating the two types of charge, which is stored as electric potential energy W in the capacitor, given by

$$W = \frac{1}{2}qV = \frac{1}{2}CV^2 = \frac{1}{2}\frac{q^2}{C}$$

where V is the potential difference and q is charge.

The most common type of capacitor is the parallel-plate capacitor consisting of two large conducting plates of area A and separated by a distance d. The capacitance of a parallel-plate capacitor is

$$C = K\varepsilon_o \frac{A}{d}$$

where K is a dielectric constant (dimensionless) and $\varepsilon_o = 8.85 \times 10^{-12}\,\mathrm{C^2\,N^{-1}\,m^{-2}} = 8.85 \times 10^{-12}\,\mathrm{F\,m^{-1}}$. For a vacuum, $K = 1$.

Capacitors in Series

The effective capacitance C_{eff} of capacitors connected in series, as shown in Figure 17-3, is given by

$$\frac{1}{C_{\text{eff}}} = \frac{1}{C_1} + \frac{1}{C_2} + \frac{1}{C_3} + \cdots + \frac{1}{C_n}$$

where C_n is the nth capacitor connected in series.

Capacitors in Series

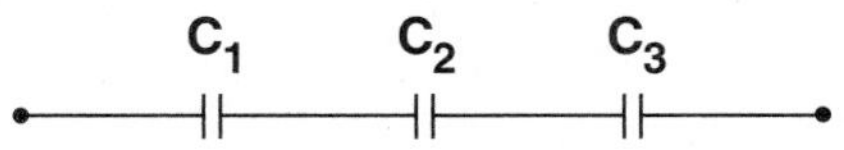

$$\frac{1}{C_{\text{eff}}} = \frac{1}{C_1} + \frac{1}{C_2} + \frac{1}{C_3}$$

$$C_{\text{eff}} = \frac{C_1 C_2 C_3}{C_2 C_3 + C_1 C_3 + C_1 C_2}$$

Fig. 17-3

Capacitors in Parallel

The effective capacitance C_{eff} of capacitors connected in parallel, as shown in Figure 17-4, is given by

$$C_{\text{eff}} = C_1 + C_2 + C_3 + \cdots + C_n$$

where C_n is the nth capacitor connected in parallel.

Solved Problem 17.1. Determine the number of electrons that pass a certain point each second in a wire that carries (1) 1 A and (2) 25 A.

Capacitors in Parallel

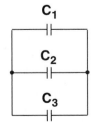

$$C_{\text{eff}} = C_1 + C_2 + C_3$$

Fig. 17-4

Solution

Current is defined as

$$1\,\mathrm{A} = 1\,\frac{\mathrm{C}}{\mathrm{s}}$$

(1)   No. of electrons $= \dfrac{1\,\mathrm{C\,s^{-1}}}{1.6 \times 10^{-19}\,\mathrm{C\,electron^{-1}}} = 6.25 \times 10^{18}\,\text{electrons}\,\mathrm{s^{-1}}$

(2)   No. of electrons $= \dfrac{25\,\mathrm{C\,s^{-1}}}{1.6 \times 10^{-19}\,\mathrm{C\,electron^{-1}}} = 1.56 \times 10^{20}\,\text{electrons}\,\mathrm{s^{-1}}$

Solved Problem 17.2.   Two resistances $R_1 = 8\,\Omega$ and $R_2 = 6\,\Omega$ are connected in series. Determine the equivalent resistance.

Solution

$$R_{eq} = R_1 + R_2 = 8\,\Omega + 6\,\Omega = 14\,\Omega$$

Solved Problem 17.3.   Three resistances $R_1 = 15\,\Omega$, $R_2 = 7\,\Omega$, and $R_3 = 4\,\Omega$ are connected in parallel. Determine the equivalent resistance.

Solution

$$\frac{1}{R_{eq}} = \frac{1}{R_1} + \frac{1}{R_2} + \frac{1}{R_3} = \frac{1}{15\,\Omega} + \frac{1}{7\,\Omega} + \frac{1}{4\,\Omega}$$
$$R_{eq} = 2.17\,\Omega$$

Solved Problem 17.4.   A circuit with four resistances connected in parallel yields an equivalent resistance of $1\,\Omega$. If $R_1 = 5\,\Omega$, $R_2 = 5\,\Omega$, and $R_3 = 10\,\Omega$, determine R_4.

Solution

Four resistances connected in parallel are related to the equivalent resistance by

$$\frac{1}{R_{eq}} = \frac{1}{R_1} + \frac{1}{R_2} + \frac{1}{R_3} + \frac{1}{R_4}$$
$$\frac{1}{1\,\Omega} = \frac{1}{5\,\Omega} + \frac{1}{5\,\Omega} + \frac{1}{10\,\Omega} + \frac{1}{R_4}$$

Solving for R_4 yields

$$R_4 = 2\,\Omega$$

Solved Problem 17.5.   Given three identical resistances in parallel, compare the equivalent value of the resistance to that of a single one.

Solution

The equivalent resistance of three resistors in parallel is

$$\frac{1}{R_{eq}} = \frac{1}{R_1} + \frac{1}{R_2} + \frac{1}{R_3} = \frac{3}{R}$$
$$R_{eq} = \frac{R}{3}$$

Thus, the equivalent resistance is one-third that of an individual resistor.

Solved Problem 17.6.   For the arrangement of resistances in Figure 17-5, determine (1) a general expression and (2) the numerical value for the equivalent resistance.

Solution

(1)   The arrangement of resistances can be simplified by first determining the equivalent resistance of the parallel arrangement of R_1, R_2, and R_3. Since they are connected in parallel,

$$\frac{1}{R_{\text{eq},123}} = \frac{1}{R_1} + \frac{1}{R_2} + \frac{1}{R_3}$$

$$R_{\text{eq},123} = \frac{R_1 R_2 R_3}{R_2 R_3 + R_1 R_3 + R_1 R_2}$$

We now have a parallel circuit where one side is a series connection of $R_{\text{eq},123}$ and R_4 and the other side is a series connection of R_5 and R_6. For a series connection, the equivalent resistance is

$$R_{\text{eq},1234} = R_{\text{eq},123} + R_4$$

$$R_{\text{eq},56} = R_5 + R_6$$

Therefore, the equivalent resistance of the entire circuit is

$$\frac{1}{R_{\text{eq},123456}} = \frac{1}{R_{\text{eq},1234}} + \frac{1}{R_{\text{eq},56}}$$

$$R_{\text{eq},123456} = \frac{R_{\text{eq},1234} R_{\text{eq},56}}{R_{\text{eq},1234} + R_{\text{eq},56}}$$

Substituting from the above expressions, we find

$$R_{\text{eq},123456} = \frac{(R_{\text{eq},123} + R_4)(R_5 + R_6)}{(R_{\text{eq},123} + R_4) + (R_5 + R_6)} = \frac{\left(\dfrac{R_1 R_2 R_3}{R_2 R_3 + R_1 R_3 + R_1 R_2} + R_4\right)(R_5 + R_6)}{\left(\dfrac{R_1 R_2 R_3}{R_2 R_3 + R_1 R_3 + R_1 R_2} + R_4\right) + (R_5 + R_6)}$$

(2)   This part of the problem can be solved by simply substituting the numerical values of all resistances in Figure 17-5 into the expression derived in part (1) of this problem.

$$R_{\text{eq},123456} = \frac{550}{171}\,\Omega = 3.2\ \Omega$$

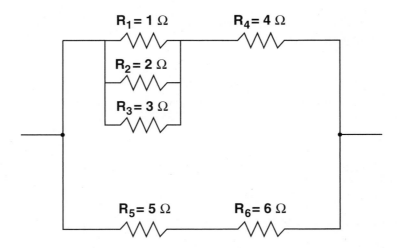

Fig. 17-5

Solved Problem 17.7.   The current through a cardiac defibrillator is 15 A. Determine the number of electrons that flow during 1 h of operation.

Solution

The charge is related to current by

$$q = it = (15 \text{ A})(1 \text{ h}) = (15 \text{ C s}^{-1})(3600 \text{ s}) = 54,000 \text{ C}$$

$$\text{No. electrons} = 54,000 \text{ C} \frac{1 \text{ electron}}{1.6 \times 10^{-19} \text{ C}} = 3.37 \times 10^{23} \text{ electrons}$$

Solved Problem 17.8.   Given that the resistivity of copper is $1.7 \times 10^{-8} \ \Omega \text{ m}$, determine the resistance of a copper wire 0.30 mm in diameter and 5 m in length.

Solution

The resistance of the copper wire is related to its diameter and length by

$$R = \frac{\rho L}{A} = \frac{\rho L}{\pi r^2}$$

$$= \frac{(1.7 \times 10^{-8} \ \Omega \text{ m})(5.0 \text{ m})}{3.14(0.15 \times 10^{-3} \text{ m})^2} = 1.2 \ \Omega$$

Solved Problem 17.9.   Determine the diameter of an aluminum wire 30 cm in length if the resistance is 0.5 Ω. The resistivity of aluminum is $2.6 \times 10^{-8} \ \Omega \text{ m}$.

Solution

$$R = \frac{\rho L}{A} = \frac{\rho L}{\pi r^2}$$

Solving for r yields

$$r = \sqrt{\frac{\rho L}{\pi R}} = \sqrt{\frac{(2.6 \times 10^{-8} \ \Omega \text{ m})(30 \times 10^{-2} \text{ m})}{3.14 \times 0.5 \ \Omega}} = 2.2 \times 10^{-5} \text{ m}$$

The diameter d of the wire is

$$d = 2r = 2(2.2 \times 10^{-5} \text{ m}) = 4.4 \times 10^{-5} \text{ m} = 0.04 \text{ mm}$$

Solved Problem 17.10.   The filament of a lightbulb has a resistance of 250 Ω. Determine the current through the filament if 120 V is applied to the lamp.

Solution

The current is related to the voltage by Ohm's law:

$$i = \frac{V}{R} = \frac{120 \text{ V}}{250 \ \Omega} = 0.48 \text{ A}$$

Solved Problem 17.11.   Determine the resistance of an electric can opener if 4 A is generated when it is connected to a 120-V circuit.

Solution

Using Ohm's law, we have

$$R = \frac{V}{i} = \frac{120 \text{ V}}{4 \text{ A}} = 30 \ \Omega$$

Solved Problem 17.12.   A circuit consists of three capacitors $C_1 = 5\ \mu\text{F}$, $C_2 = 8\ \mu\text{F}$, and $C_3 = 15\ \mu\text{F}$ and a voltage source of 25 V. Determine the equivalent capacitance, charge, and potential difference if the capacitors are connected (a) in series and (b) in parallel.

Solution

(a) $\dfrac{1}{C_{\text{eq}}} = \dfrac{1}{C_1} + \dfrac{1}{C_2} + \dfrac{1}{C_3} = \dfrac{1}{5\ \mu\text{F}} + \dfrac{1}{8\ \mu\text{F}} + \dfrac{1}{15\ \mu\text{F}} = 0.392\ \mu\text{F}^{-1}$

$C_{\text{eq}} = 2.5\ \mu\text{F}$

$q = C_{\text{eq}} V = (2.5\ \mu\text{F})(25\ \text{V}) = 6.25 \times 10^{-5}\ \text{C}$

$V_1 = \dfrac{q}{C_1} = \dfrac{6.25 \times 10^{-5}\ \text{C}}{5\ \mu\text{F}} = 12.5\ \text{V}$

$V_2 = \dfrac{q}{C_2} = \dfrac{6.25 \times 10^{-5}\ \text{C}}{8\ \mu\text{F}} = 7.8\ \text{V}$

$V_3 = \dfrac{q}{C_3} = \dfrac{6.25 \times 10^{-5}\ \text{C}}{15\ \mu\text{F}} = 4.2\ \text{V}$

(b) $C_{\text{eq}} = C_1 + C_2 + C_3 = 5\ \mu\text{F} + 8\ \mu\text{F} + 15\ \mu\text{F} = 28\ \mu\text{F}$

$q_1 = C_1 V = (5\ \mu\text{F})(25\ \text{V}) = 1.25 \times 10^{-4}\ \text{C}$

$q_2 = C_2 V = (8\ \mu\text{F})(25\ \text{V}) = 2.00 \times 10^{-4}\ \text{C}$

$q_3 = C_3 V = (15\ \mu\text{F})(25\ \text{V}) = 3.75 \times 10^{-4}\ \text{C}$

$V = 25\ \text{V}$

Solved Problem 17.13.   A voltage of 10 V is applied across a 5-μF capacitor. If a material of dielectric constant $K = 3$ is positioned between the plates, determine (1) the new charge, (2) the new capacitance, and (3) the new voltage across the capacitor.

Solution

(1) $q = CV = (5 \times 10^{-6}\ \text{F})(10\ \text{V}) = 50 \times 10^{-6}\ \text{C} = 50\ \mu\text{C}$

(2) $C' = KC = (3)(5\ \mu\text{F}) = 15\ \mu\text{F}$

(3) $V' = \dfrac{q}{C'} = \dfrac{50 \times 10^{-6}\ \text{C}}{15 \times 10^{-6}\ \text{F}} = 3.3\ \text{V}$

Solved Problem 17.14.   To restore cardiac function to a heart attack victim, a cardiac defibrillator is applied to the chest in an attempt to stimulate electrical activity of the heart and initiate the heartbeat. A cardiac defibrillator consists of a capacitor charged to approximately 7.5×10^3 V with stored energy of 500 W s. Determine the charge on the capacitor in the cardiac defibrillator.

Solution

The energy stored in a capacitor is

$$E = \frac{1}{2} CV^2$$

and the charge on the capacitor is

$$q = CV$$

Solving for C gives

$$C = \frac{q}{V}$$

Substituting into the expression for E, we have

$$E = \frac{1}{2}\left(\frac{q}{V}\right)V^2 = \frac{1}{2}qV$$

Solving for q leaves

$$q = \frac{2E}{V} = \frac{2 \times 500 \text{ W s}}{7.5 \times 10^3 \text{ V}} = 0.13 \text{ C}$$

ILLUSTRATIVE EXAMPLE 17.1.   Electrical Analogs and Blood Flow

Fluid flow through a rigid tube is driven by a difference of pressures between the two ends of the tube, or a pressure gradient ΔP. The fluid is subjected to a resistive force R dictated by the tube radius, tube length, and viscosity of the fluid. In terms of the pressure gradient and resistance, the volumetric rate of fluid flow Q is given by

$$Q = \frac{\Delta P}{R}$$

This is known as *Poiseuille's law*. For an isolated rigid vessel, Poiseuille's law explains fluid flow in a logical, straightforward manner.

Fluid flow through a rigid tube is analogous to current flow through a wire, and thus an electrical analogy can be drawn. Using electrical analogies, current is analogous to fluid flow rate, voltage is analogous to the pressure gradient, resistance remains the same concept in both cases, and conductance is analogous to vascular compliance or elasticity of the blood vessel. Table 17-1 summarizes the fluid flow parameters and corresponding electrical entities. As the resistance increases in a wire, the current will be reduced just as a corresponding increase in a tube will decrease the fluid flow rate. Thus, by employing common electrical principles, it becomes a simple task to investigate flow through an isolated tube or vessel.

More importantly, electrical principles are often applied to the calculation of current (flow), not primarily through single wires (vessels), but through networks of wires (vessels) connected as circuits. The multiple arrangement of tubes representing the vessels in a vascular system, in effect, resembles a complex circuit, such as the vessels presented in Figure 17-6. A voltage source V (pressure gradient ΔP) is driving current (flow) through a single wire (vessel) which branches into three vessels connected in parallel with respect to the circuit.

The primary goal in solving this network is to determine the total flow Q_T and the individual flow (Q_1, Q_2, Q_3) through each vessel. It is known that the total flow entering the arrangement of vessels is equal to the flow exiting the system. Because the vessels are in parallel

$$Q_T = Q_1 + Q_2 + Q_3$$

The total pressure gradient is identical to the pressure gradient through each vessel connected in parallel:

$$\Delta P_T = \Delta P_1 = \Delta P_2 = \Delta P_3$$

The effective resistance R_{eff} of three vessels connected in parallel is

$$\frac{1}{R_{\text{eff}}} = \frac{1}{R_1} + \frac{1}{R_2} + \frac{1}{R_3}$$

Table 17-1.   Electrical Analogs of Hemodynamic Parameters

Hemodynamic Parameter	Electrical Analog
Volumetric blood flow rate Q, cm^3 s^{-1}	Current i, A
Pressure gradient ΔP, dyn cm^{-2}	Voltage V, V
Vascular resistance R, dyn s cm^{-5}	Resistance R, Ω
Compliance C, cm^5 dyn^{-1} s^{-1}	Capacitance C, F

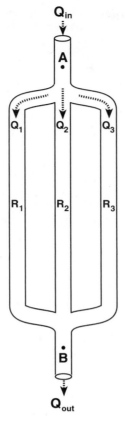

Fig. 17-6

which can be rearranged to yield

$$R_{\text{eff}} = \frac{R_1 R_2 R_3}{R_2 R_3 + R_1 R_3 + R_1 R_2}$$

These equations can be used to approximate blood flow through any type of vessel arrangement in the human circulation and are limited only by the complexity of the desired arrangement.

ILLUSTRATIVE EXAMPLE 17.2.   Electric Shock and the Human Body

The devastating effects of electric shock on the human body can be seen by elementary calculations with Ohm's law. The human body is a volume conductor whose resistance is approximately $1500\,\Omega$ when dry. However, when the body becomes wet, water is an excellent conductor, reducing the resistance of the human body to $500\,\Omega$. Assuming that the voltage source from the electric outlet is $120\,\text{V}$, the current through the dry body is

$$i = \frac{V}{R} = \frac{120\,\text{V}}{1500\,\Omega} = 0.08\,\text{A} = 80\,\text{mA}$$

while the current through the wet body is

$$i = \frac{V}{R} = \frac{120\,\text{V}}{500\,\Omega} = 0.24\,\text{A} = 240\,\text{mA}$$

The harmful effects of exposure to electricity, particularly if the electrical contact is wet, are evident from the given calculations. Ventricular fibrillation is the development of an irregular heartbeat and can occur at currents of $100\,\text{mA}$.

Currents greater than this can be potentially lethal by either of two mechanisms: (1) ventricular fibrillation or electrical overload of the nerves responsible for regulating heart function and (2) respiratory arrest due to paralysis of the nerves regulating the breathing process.

Summary of DC Circuits

Circuit Element	Series	Parallel
Voltage	$V = V_1 + V_2 + V_3 + \cdots + V_n$	$V = V_1 = V_2 = V_3 = \cdots = V_n$
Current	$i = i_1 = i_2 = i_3 = \cdots = i_n$	$i = i_1 + i_2 + i_3 + \cdots + i_n$
Resistance	$R_{\mathrm{eq}} = R_1 + R_2 + R_3 + \cdots + R_n$	$\dfrac{1}{R_{\mathrm{eq}}} = \dfrac{1}{R_1} + \dfrac{1}{R_2} + \dfrac{1}{R_3} + \cdots + \dfrac{1}{R_n}$
Capacitance	$\dfrac{1}{C_{\mathrm{eff}}} = \dfrac{1}{C_1} + \dfrac{1}{C_2} + \dfrac{1}{C_3} + \cdots + \dfrac{1}{C_n}$	$C_{\mathrm{eff}} = C_1 + C_2 + C_3 + \cdots + C_n$

17.2 KIRCHHOFF'S RULES OF CIRCUIT ANALYSIS

In a simple circuit, e.g., a voltage source connected to a single resistance, one can easily determine the current through the single resistance by direct observation and a single elementary calculation. However, as the elements in a given circuit increase in number and complexity, it becomes a much more difficult task to calculate current and voltage through all circuit elements. Analysis of complex electric circuits is typically performed according to Kirchhoff's two rules:

Node or Junction Rule

The algebraic sum of all currents entering a junction or node must be equal to the sum of all currents exiting the node.

Loop or Circuit Rule

The algebraic sum of potential drops encountered while traversing the circuit in either a counterclockwise or clockwise direction must be equal to zero.

Within a current loop, if a resistor is traversed in the direction of current, the change in voltage is $-iR$; in the opposite direction of current, the change in voltage is $+iR$. If a seat of voltage (emf) is traversed in the direction of voltage, the change in $+iR$ is $+V$; in the opposite direction of voltage, it is $-V$.

Consider, for example, the circuit displayed in Figure 17-7. The first step is to identify all nodes and loops within the circuit. Of the four nodes in the circuit, the current entering three of the nodes (A, C, and D) is identical to the current exiting the node. For node B, nodal analysis reveals the following equation:

$$i_1 + i_2 - i_3 = 0$$

Applying the loop theorem to the circuit yields

Loop 1: $$E_1 - i_1 R_1 + i_2 R_2 = 0$$

Loop 2: $$-E_2 - i_2 R_2 - i_3 R_3 = 0$$

We now have three equations which can be used to solve for the three unknowns: i_1, i_2, i_3.

Solved Problem 17.15. For the electric circuit in Figure 17-7, determine i_1, i_2, and i_3 given $R_1 = 5\ \Omega$, $R_2 = 8\ \Omega$, $R_3 = 11\ \Omega$, $E_1 = 1$ V, and $E_2 = 5$ V.

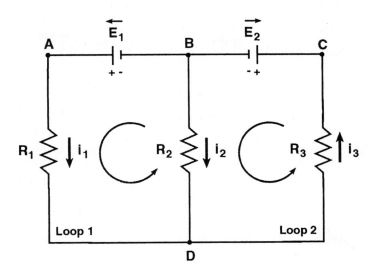

Fig. 17-7

Solution

Nodal and loop analysis of the electric circuit yields the following equations:

$$i_1 + i_2 - i_3 = 0$$

Loop 1: $$E_1 - i_1 R_1 + i_2 R_2 = 0$$
Loop 2: $$-E_2 - i_2 R_2 - i_3 R_3 = 0$$

Making the appropriate substitutions gives

$$i_1 + i_2 - i_3 = 0 \qquad\qquad\qquad (1)$$

$$1\,\text{V} - 5\,\Omega i_1 + 8\,\Omega i_2 = 0 \qquad\qquad\qquad (2)$$

$$-5\,\text{V} - 8\,\Omega i_2 - 11\,\Omega i_3 = 0 \qquad\qquad\qquad (3)$$

From Eq. (1), we can write

$$i_1 = -i_2 + i_3$$

Substituting into Eq. (2) and solving for i_2 yields

$$i_2 = 0.38 i_3 - 0.077 \qquad\qquad\qquad (4)$$

Substituting into Eq. (3), we obtain the value for i_3:

$$i_3 = 0.39\,\text{A}$$

Substituting into Eq. (4) yields

$$i_2 = 0.073\,\text{A}$$

Substituting values for i_2 and i_3 into Eq. (1) yields i_1:

$$i_1 = 0.32\,\text{A}$$

Solved Problem 17.16.   For the electric circuit in Figure 17-7, apply Kirchhoff's laws to derive the nodal and loop equations if the direction of current i_2 is reversed.

Solution

The nodal and loop equations of the electric circuit are

$$i_1 - i_2 - i_3 = 0$$

Loop 1: $$E_1 - i_1 R_1 - i_2 R_2 = 0$$

Loop 2: $$-E_2 + i_2 R_2 - i_3 R_3 = 0$$

Solved Problem 17.17.   One wire A, connected to a voltage source on one end, is connected at the other end to a branching arrangement of three wires B, C, and D. If the voltage source sends 15 A of current through A, and 5 A and 4 A are measured flowing from B and C, respectively, determine the magnitude and direction of current flowing through D.

Solution

A node is defined at the point of branching between A and B, C, D. Total current flowing into the node must equal total current flowing away from the node. Thus,

$$i_A = i_B + i_C + i_D$$
$$15\,\text{A} = 5\,\text{A} + 4\,\text{A} + i_D$$
$$i_D = 15\,\text{A} - 9\,\text{A} = 6\,\text{A} \qquad \text{flowing away from node}$$

Solved Problem 17.18.   Assume that nodal and loop analysis of a particular electrical network reveals the following equations:

Nodal analysis: $$i_1 = i_2 + i_3$$

Loop analysis: $$V_1 - i_1 R_1 + V_2 - i_2 R_2 = 0$$
$$V_2 - i_2 R_2 - V_3 + i_3 R_3 = 0$$

where $V_1 = 4$ V, $V_2 = 5$ V, $V_3 = 2$ V, $R_1 = 3\ \Omega$, $R_2 = 4\ \Omega$, and $R_3 = 7\ \Omega$. Determine the currents i_1, i_2, and i_3.

Solution

This problem involves elementary algebraic manipulations where we have three equations and three unknowns. The three equations can be written as

$$i_1 = i_2 + i_3 \qquad\qquad (1)$$
$$9\,\text{V} - 3\,\Omega i_1 - 4\,\Omega i_2 = 0 \qquad\qquad (2)$$
$$3\,\text{V} - 4\,\Omega i_2 + 7\,\Omega i_3 = 0 \qquad\qquad (3)$$

Substituting Eq. (1) into Eq. (2) yields an expression for i_3:

$$i_3 = -3\,\text{A} + \frac{7\,\Omega}{3\,\Omega} i_2 \qquad\qquad (4)$$

Substituting Eq. (4) into Eq. (3) yields i_2:

$$i_2 = 1.46\,\text{A}$$

Substituting the value for i_2 into Eq. (4) yields the value for i_3:

$$i_3 = 0.41\,\text{A}$$

From Eq. (1), we get the value for i_1:

$$i_1 = i_2 + i_3 = 1.46\,\text{A} + 0.41\,\text{A} = 1.87\,\text{A}$$

ILLUSTRATIVE EXAMPLE 17.3.   Arteriovenous Fistula: Short Circuit of the Human Circulation

In the human circulation, blood ejected from the heart flows according to a circuit of high peripheral resistance and high arterial pressure. Peripheral resistance represents the cumulative effect of resistance from all vessels in the human circulation. An arteriovenous fistula is a direct connection between artery and vein, resulting in the presence of a second circuit of low peripheral resistance and low venous pressure. A volume of blood from the normal circulation is forced continously through the fistula by the high arterial pressure generated by the arterial circulation.

The presence of a fistula translated to a serious overloading of the heart and a decrease of blood flow to adjacent tissue and organs, resulting in lack of oxygen (or ischemia) and ultimately tissue death (or infarct) if left untreated. Treatment of a fistula involves surgical removal of the fistula and redirection of blood flow through the normal circulation.

Supplementary Problems

17.1. For a wire carrying 50 A of current, determine the number of electrons that pass a given point in (1) 30 s and (2) 1 min.

Answer

(1) 9.38×10^{21} electrons; (2) 1.88×10^{22} electrons

17.2. An appliance that requires 120 V to operate has a resistance of 10 Ω. Determine the current drawn when it is in use.

Answer

12 A

17.3. A circuit consists of four identical resistances of 10 Ω connected in series. Determine the equivalent resistance.

Answer

40 Ω

17.4. In a given circuit, six identical resistances of 25 Ω are connected in parallel. Determine the equivalent resistance.

Answer

4.17 Ω

17.5. A circuit with three resistances connected in parallel yields an equivalent resistance of 2 Ω. If $R_1 = 15 \, \Omega$ and $R_2 = 15 \, \Omega$, determine R_3.

Answer

2.7 Ω

17.6. Given three identical resistances in series, compare the equivalent value of the resistance to that of a single one.

Answer

$R_{eq} = 3R$

17.7. The current through a cardiac pacemaker is 0.010 A. Determine the number of electrons that flow during (1) 1 hour and (2) 1 day of operation.

Answer

(1) 2.25×10^{20} electrons; (2) 5.4×10^{21} electrons

17.8. The operational current of a cardiac pacemaker is 0.010 A. Given that the operational current of an ECG monitor is 2.0 A, determine the time needed for the pacemaker to acquire the same charge as that acquired by the ECG monitor in 1 s.

Answer

3.33 min

17.9. Measurements of voltage potentials of the cell are typically performed by using micropipette electrodes, the tips of which consist of a glass capillary tube of length 1.0 mm and inner diameter of 4.0 μm. Given that the capillary tube is filled with NaCl solution of resistivity 0.04 Ωm, determine the resistance of the column of electrolyte.

Answer

3.18 MΩ

17.10. Given the resistivity of an object $\rho = 1 \times 10^{-4}\,\Omega$ cm, determine the resistance of a wire 2.0 m in length with a cross-sectional area of 0.02 cm^2.

Answer

1 Ω

17.11. Given that the resistivity of tungsten is $5.51 \times 10^{-8}\,\Omega$ m, determine the resistance of a tungsten wire 0.50 mm in diameter and 5 m in length.

Answer

1.41 Ω

17.12. Given a copper wire of length 10.0 m and diameter 0.2 mm, determine the diameter of aluminum wire of the same length that yields the same resistance. The resistivity of copper is $1.72 \times 10^{-8}\,\Omega$ m, and the resistivity of aluminum is $2.63 \times 10^{-8}\,\Omega$ m.

Answer

0.248 mm

17.13. Determine the diameter of a steel wire 50 cm in length if the resistance is 0.5 Ω. The resistvity of steel is $20 \times 10^{-8}\,\Omega$ m.

Answer

0.50 mm

17.14. For the arrangement of resistances displayed in Figure 17.8, determine the equivalent resistance.

Answer

21.75 Ω

17.15. Determine the voltage required to send 2 A of current through a circuit given its resistance of 5 Ω.

Answer

10 V

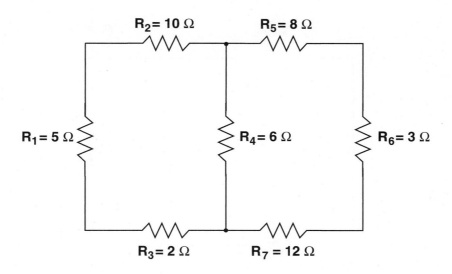

$R_2 = 10\,\Omega$ $R_5 = 8\,\Omega$

$R_1 = 5\,\Omega$ $R_4 = 6\,\Omega$ $R_6 = 3\,\Omega$

$R_3 = 2\,\Omega$ $R_7 = 12\,\Omega$

Fig. 17-8

17.16. A conducting wire with a resistance of $5\,\Omega$ is connected to a voltage source of $10\,\mathrm{V}$ which drives $2\,\mathrm{A}$ of current through the wire. Determine the power dissipated by the resistor.

Answer

20 W

17.17. Given three identical capacitances, compare the effective value of the capacitance to that of a single one if three identical capacitances are connected (1) in series and (2) in parallel.

Answer

(1) $C_{\mathrm{eff}} = C/3$; (2) $C_{\mathrm{eff}} = 3C$

17.18. In a particular circuit, two capacitors $C_1 = 5\,\mu\mathrm{F}$ and $C_2 = 2\,\mu\mathrm{F}$ are connected in parallel to a voltage source of $20\,\mathrm{V}$. Determine the charge on each capacitor.

Answer

$q_1 = 100\,\mu\mathrm{C}$; $q_2 = 40\,\mu\mathrm{C}$

17.19. A circuit consists of two capacitors $C_1 = 5\,\mu\mathrm{F}$ and $C_2 = 3\,\mu\mathrm{F}$ powered by a voltage source of $120\,\mathrm{V}$. Determine the effective capacitance, charge, and voltage if the capacitors are connected (1) in series and (2) in parallel.

Answer

(1) $C_{\mathrm{eff}} = 15/8\,\mu\mathrm{F}$; $q_1 = q_2 = 225\,\mathrm{C}$; $V_1 = 4.5 \times 10^7\,\mathrm{V}$, $V_2 = 7.5 \times 10^7\,\mathrm{V}$. (2) $C_{\mathrm{eff}} = 8\,\mu\mathrm{F}$; $q_1 = 6.0 \times 10^{-4}\,\mathrm{C}$, $q_2 = 3.6 \times 10^{-4}\,\mathrm{C}$; $V_1 = V_2 = 120\,\mathrm{V}$

17.20. For the electric circuit in Figure 17-9, given that $E = 7.0\,\mathrm{V}$, $R_1 = 9\,\Omega$, and $R_2 = 12\,\Omega$, determine the current through R_1 and R_2.

Answer

$i_1 = i_2 = 0.33\,\mathrm{A}$

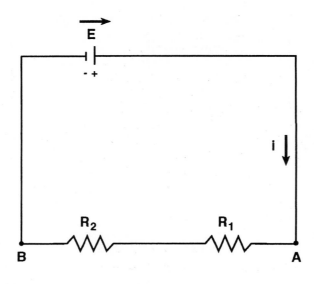

Fig. 17-9

17.21. Consider a single tube of cross-sectional area A and assemblies of three smaller tubes connected in parallel and in series, each with a cross-sectional area $A/3$. How does the resistance of the single larger tube compare with that of the three smaller tubes connected in series and in parallel?

Solution

Series: $R_T = R$; parallel: $R_T = R/9$

17.22. In an attempt to regain the normal beating of the heart, a cardiac defibrillator drives 10 A of current through the patient's chest over 5 ms. Determine the (1) charge passed through the patient's chest and (2) voltage required if 750 J of energy was dissipated by the current.

Solution

(1) 0.05 C; (2) 1.5×10^4 V

Chapter 18

Magnetism

This chapter is devoted to the study of magnets and magnetic fields or magnetism. In Chapter 16, we learned that for two static electric charges or electric charges at rest, there exists an associated force field that is either attractive or repulsive depending on the charge assignments. Similarly, in Chapter 5, we learned that two masses, separated by a given distance, exert a gravitational attractive force between the two masses. With regard to magnetism, although there is no known entity corresponding to charge or mass, magnetic fields can be generated by a permanent magnet or by a moving electric charge. A magnetic field can also be created by current (flowing charge) through a wire and permanent magnets. Magnetic fields also exert a force on the moving charge and resultant current. In this chapter, magnetic fields and their interaction with charged particles are discussed.

18.1 DEFINITIONS OF MAGNETISM

Magnetic field B is an attractive or repulsive force field generated by a moving charged particle. For a standard bar magnet with a north pole where the lines of force begin and a south pole where the lines of force end, the magnetic field is attractive for opposite poles and repulsive for like poles. Magnetic field is a vector quantity. Units of magnetic field are teslas (T) or newton/(ampere·meter) ($1 \text{ N A}^{-1} \text{ m}^{-1}$). An equivalent expression of 1 T is 1 weber/meter^2 (Wb m^{-2}). Smaller magnetic fields are measured in gauss (G), where $1 \text{ T} = 10^4 \text{ G}$.

Solved Problem 18.1. The earth can produce magnetic fields as high as 600 mG. Express this value of magnetic field in teslas.

Solution

As given above, units of tesla and gauss are related according to

$$1 \text{ T} = 10^4 \text{ G}$$

Therefore,

$$600 \text{ mG} = (600 \times 10^{-3} \text{ G})\left(\frac{1 \text{ T}}{10^4 \text{ G}}\right) = 6 \times 10^{-5} \text{ T}$$

Magnetic Force on a Charged Particle in Motion

Magnetic force F, or the force exerted on a charged particle q moving with a velocity **v** in a uniform magnetic field **B** is defined as

$$\mathbf{F} = q\mathbf{v} \times \mathbf{B} = qvB \sin \theta$$

where θ is the angle between the lines of the magnetic field **B** and the direction of the velocity of the charged particle **v**. The direction of the force can be determined by implementation of a version of the right-hand rule, as presented in Chapter 3 in the discussion on vectors. Given a charged particle moving with a velocity $q\mathbf{v}$ perpendicular to a uniform magnetic field **B**, the right hand is positioned such that the thumb points in the direction of $q\mathbf{v}$ and the remaining four fingers are aligned in the direction of **B**. The direction of the magnetic force **F** is perpendicular to the palm, as illustrated in Figure 18-1.

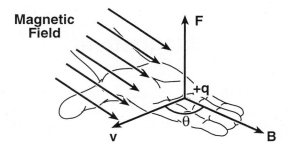

Fig. 18-1

Solved Problem 18.2. An electron moves with a velocity $0.7c\,\mathbf{i}$ in a uniform magnetic field $10\,\mathrm{W\,m^{-2}}\,\mathbf{j}$. Determine the magnitude and direction of the magnetic force exerted on the electron.

   Solution

   The magnetic force $\mathbf{F}$ on a charged particle moving with a velocity $\mathbf{v}$ in a uniform magnetic field $\mathbf{B}$ is

$$\mathbf{F} = q\mathbf{v} \times \mathbf{B} = (-1.6 \times 10^{-19}\,\mathrm{C})(0.7 \times 3 \times 10^{8}\,\mathrm{m\,s^{-1}})\mathbf{i} \times 10\,\mathrm{W\,m^{-2}}\,\mathbf{j}$$
$$= 3.36 \times 10^{-10}\,\mathrm{N}\,(\mathbf{i} \times \mathbf{j}) = 3.36 \times 10^{-10}\,\mathrm{N}\,\mathbf{k}$$

Solved Problem 18.3. Determine the force exerted on an electrically charged microsphere of $3.2 \times 10^{-19}\,\mathrm{C}$ that moves with a speed of $5 \times 10^{2}\,\mathrm{m\,s^{-1}}$ across a 2-T magnetic field.

   Solution

$$\mathbf{F} = q\mathbf{v} \times \mathbf{B} = (3.2 \times 10^{-19}\,\mathrm{C})(5 \times 10^{2}\,\mathrm{m\,s^{-1}})(2\,\mathrm{T}) = 1.28 \times 10^{-16}\,\mathrm{N}$$

Solved Problem 18.4. A proton assumes a counterclockwise circular orbit with a speed of $50\,\mathrm{m\,s^{-1}}$ in a uniform magnetic field of 2.0 T acting perpendicular to proton motion. Given the mass of the proton is $1.67 \times 10^{-27}\,\mathrm{kg}$, determine the radius of the circular orbit.

   Solution

   Equating the centripetal force F_c to the magnetic force F_m responsible for maintaining the circular orbit, we have

$$F_m = F_c$$
$$qvB = \frac{mv^2}{r}$$
$$r = \frac{mv}{qB}$$

Making the appropriate substitutions yields

$$r = \frac{mv}{qB} = \frac{(1.67 \times 10^{-27}\,\mathrm{kg})(50\,\mathrm{m\,s^{-1}})}{(1.6 \times 10^{-19}\,\mathrm{C})(2.0\,\mathrm{T})} = 2.61 \times 10^{-7}\,\mathrm{m}$$

Solved Problem 18.5. From Solved Problem 18.4, how would the results change if the particle were an electron rather than a proton?

Solution

The difference between the electron and proton is the mass and charge. For the electron, the mass is 9.11×10^{-31} kg, and the charge is $-e$. For the proton, the mass is 1.67×10^{-27} kg, and the charge is $+e$. Since the radius is dependent on particle mass, the numerical value of the radius will change according to

$$r = \frac{mv}{qB} = \frac{(9.11 \times 10^{-31} \text{ kg})(50 \text{ m s}^{-1})}{(1.6 \times 10^{-19} \text{ C})(2.0 \text{ T})} = 1.42 \times 10^{-10} \text{ m}$$

The negative charge of the electron will result in a reversal of direction, that is, the electron will traverse in a clockwise direction in the magnetic field.

Solved Problem 18.6.   Derive an expression for the momentum of a charged particle moving perpendicular to a uniform magnetic field.

Solution

The path of a charged particle moving perpendicular to a uniform magnetic field is circular. Thus, the two forces involved in maintaining circular motion are the magnetic force $F_m = qvB$ and the centripetal force $F_c = mv^2/r$. Equating these two forces gives

$$F_m = F_c$$
$$qvB = \frac{mv^2}{r}$$

This expression reduces to

$$qB = \frac{mv}{r}$$

The momentum is the product of mass and velocity and can be expressed as

$$p = mv = qBr$$

Magnetic Force on a Current-Carrying Wire

The magnetic force exerted on a current-carrying wire of current i of length L placed in a uniform magnetic field **B** is defined as

$$\mathbf{F} = i\mathbf{L} \times \mathbf{B} = iLB \sin \theta$$

where L is the length of the wire (conductor) and θ is the angle between the current and the magnetic field. The direction of the magnetic force on the wire can be found by orienting the thumb of the right hand along the axis of the wire with the remaining fingers in the direction of the magnetic field. The magnetic force is directed upward from the aligned palm.

Solved Problem 18.7.   A wire of length 40 cm carrying a current of 30 A is positioned at an angle of 50° to a uniform magnetic field of flux density 10.0×10^{-4} W m^{-2}. Determine the magnitude and direction of the force exerted on this wire.

Solution

The magnetic force F exerted on the wire of length L with current i placed in a magnetic field B is given by

$$F = iLB \sin \theta = (30 \text{ A})(40 \times 10^{-2} \text{ m})(10.0 \times 10^{-4} \text{ W m}^{-2})(\sin 50°) = 9.2 \times 10^{-5} \text{ N}$$

The wire is pushed from the stronger toward the weaker magnetic field.

Solved Problem 18.8.   A 3.5-m wire with a current of 7.0 A directed toward the $+y$ axis is subjected to a perpendicular uniform magnetic field of 0.5 T directed toward the $+x$ axis. Determine the magnitude and direction of the magnetic force exerted on the wire.

Solution

The magnetic force exerted on the wire can be determined by

$$F = iLB \sin \theta = (7.0 \text{ A})(3.5 \text{ m})(0.5 \text{ T})(\sin 90°) = 12.3 \text{ N}$$

The direction of the force can be determined by applying the right-hand rule. By orienting the thumb of the right hand upward (along the $+y$ axis) and the remaining fingers of the right hand forward (along the $+x$ axis), the magnetic force exerted on the wire is directed along the direction of the palm, which is to the left.

Magnetic Torque on a Current-Carrying Coil

The magnetic torque τ exerted on a current-carrying coil consisting of N loops, each carrying a current i, placed in a uniform magnetic field B is

$$\tau = NiAB \sin \theta$$

where A is the area of the coil and θ is the angle between the magnetic field lines and a plane perpendicular to that of the coil. The product iA is termed the *magnetic moment* μ of the loop, or

$$\mu = iA$$

The magnetic moment is expressed in units of A m².

Solved Problem 18.9.   Consider a wire of cross-sectional area $A = 0.25 \text{ m}^2$ with 50 loops carrying a current of 4 A in a perpendicular uniform magnetic field of $B = 1.0$ T. Determine the torque exerted on the loop.

Solution

The torque on the current-carrying loop can be determined from

$$\tau = NiAB \sin \theta = (50 \text{ loops})(4 \text{ A})(0.25 \text{ m}^2)(1.0 \text{ T})(\sin 90°) = 50 \text{ N m}$$

ILLUSTRATIVE EXAMPLE 18.1.   Mass Spectrometry and Quantitation of Ionic Mass

Mass spectrometry, a technique for the measurement and quantitation of ionic mass, is based on the fact that ions with different charge and mass will deviate to different degrees while propagating through electric and magnetic fields. Ions generated by electron bombardment in an ionization chamber such that they have only one positive charge are accelerated through a slit, forming a narrow beam, as shown in Figure 18-2. The ions then pass through a velocity selector when the electric force ($F_e = qE$) is equal to the magnetic force ($F_m = qvB$), yielding a velocity v:

$$v = \frac{E}{B}$$

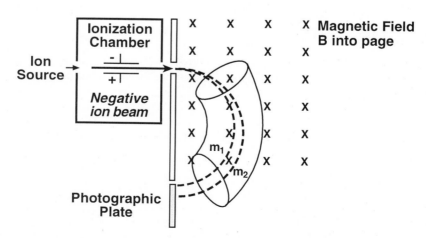

Fig. 18-2

The velocity selector prohibits all ions from moving farther through the spectrograph except those ions with speed v. The ions then pass through a uniform magnetic field arranged perpendicular to the path of the ions in a circular arc with a radius r:

$$r = \frac{mv}{qB}$$

Particles of varying mass move in a different circular arc, striking the photographic plate at different points. From the above relation, the mass m of the ion can be determined:

$$m = \frac{qBr}{v}$$

Solved Problem 18.10.   Mass spectrometric analysis of a chemical compound was performed using an electric field $E = 1.0 \times 10^5 \, \text{V m}^{-1}$ and a magnetic field $B = 0.1 \, \text{T}$. Determine the speed of the ions.

####### Solution

The speed of the ions is related to the electric and magnetic fields by

$$v = \frac{E}{B} = \frac{1 \times 10^5 \, \text{V m}^{-1}}{0.1 \, \text{T}} = 1.0 \times 10^6 \, \text{m s}^{-1}$$

Solved Problem 18.11.   In a mass spectrometry experiment, a positively charged ion travels in a circular arc of radius $0.4 \, \text{m}$ with a velocity of $2.0 \times 10^5 \, \text{m s}^{-1}$ in a 0.15-T magnetic field. Determine the mass of these ions.

####### Solution

By equating the magnetic and centripetal forces required for the ion to maintain a circular orbit, the following expression for the charge-to-mass (q/m) ratio can be derived:

$$\frac{q}{m} = \frac{v}{Br} = \frac{2.0 \times 10^5 \, \text{m s}^{-1}}{(1.5 \times 10^{-1} \, \text{T})(4 \times 10^{-1} \, \text{m})} = 3.3 \times 10^6 \, \text{C kg}^{-1}$$

The charge q for a positively charged ion is $1.6 \times 10^{-19} \, \text{C}$. Therefore, solving for m, we have

$$m = \frac{1.6 \times 10^{-19} \, \text{C}}{3.3 \times 10^6 \, \text{C kg}^{-1}} = 4.8 \times 10^{-26} \, \text{kg}$$

ILLUSTRATIVE EXAMPLE 18.2.   Magnetic Resonance Imaging

Magnetic resonance imaging (MRI), rapidly becoming the most widely used imaging modality for the diagnosis of many human injuries, diseases, and disorders, is based on the nuclear phenomenon of magnetic resonance. The structure of a typical atomic nucleus consists of a nucleus composed of neutrally charged neutrons and positively charged protons while negatively charged electrons are in continual orbital motion about the nucleus. Atoms possessing a nucleus with an even number of neutrons or protons are in nuclear equilibrium while those atoms possessing a nucleus with an odd number induce a charge distribution within the atoms which, in turn, causes the nuclei to spin or precess about a random axis. More importantly, the charge distribution between the intimately spaced entities (protons and neutrons) results in the nucleus possessing spin angular momentum and thus creates a magnetic moment. Examples of such nuclei are ^{1}H, ^{13}C, ^{19}F, and ^{31}P. Since the human body is composed primarily of water (H_2O), hydrogen atoms are very abundant, making MRI particularly applicable to imaging of the body. When subjected to an external magnetic field, the spinning nuclei tend to align along an axis parallel to the magnetic field.

In an MRI scan, the patient is positioned within the bore of a large magnet, in essence immersing the patient in a strong magnetic field ($\approx 2.5 \, \text{T}$) which is approximately 40,000 times larger than the earth's magnetic field ($\approx 0.6 \, \text{G}$). If a pulsating beam of radio-frequency (rf) energy is beamed onto the nucleus, energy is transfered to the spinning nuclei, causing these axially aligned nuclei or magnetic moments to flip from their lower energy state to their higher energy state. At the completion of an rf pulse, the magnetic moments return to their lower energy state. The frequency or energy released in the nuclear transitions of the magnetic moments between the higher and lower energy states is detected by an rf receiver. The signals recorded from the rf receiver are then mathematically processed and reconstructed to yield a visual image of the regions of tissue with characteristic frequencies.

18.2 MAGNETIC INDUCTION

Magnetic induction is the generation of a magnetic field due to the motion of charged particles either in free space or along a conductor.

(1) Point charge

The magnetic field B generated by a particle of charge q moving at a speed v is

$$B = \frac{\mu}{4\pi} \frac{qv \sin \theta}{r^2}$$

where r is the distance from the moving charge to the point where the magnetic field is to be calculated and $\mu =$ permeability of the medium in which the magnetic field exists. In free space, $\mu_o = 4\pi \times 10^{-7} \, \mathrm{T\,m\,A^{-1}} = 1.257 \times 10^{-6} \, \mathrm{T\,m\,A^{-1}}$.

Solved Problem 18.12. An alpha particle of $q = +2e$ is projected into a perpendicular uniform magnetic field of $0.01 \, \mathrm{T} \, \mathbf{i}$ with a velocity of $5 \times 10^6 \, \mathrm{m\,s^{-1}} \, \mathbf{j}$. Determine the magnitude and direction of the magnetic force exerted on the alpha particle.

Solution

$$\mathbf{F} = q\mathbf{v} \times \mathbf{B} = (3.2 \times 10^{-19} \, \mathrm{C})(5 \times 10^6 \, \mathrm{m\,s^{-1}} \, \mathbf{j}) \times 0.01 \, \mathrm{T} \, \mathbf{i} \sin 90° = 1.6 \times 10^{-14} \, \mathrm{N}$$

The direction can be found by applying the right-hand rule. A simpler way to determine direction of the magnetic force is to take the vector cross-product: $\mathbf{j} \times \mathbf{i} = -\mathbf{k}$, or in the negative z direction.

(2) Long, straight wire

The magnetic field B at any point A a distance r from the perpendicular axis of a wire carrying a current i is given by

$$B = \frac{\mu_o i}{2\pi r}$$

In a plane perpendicular to the axis of the wire, the direction of the magnetic field can be determined from the right-hand rule where the thumb is oriented along the direction of current and the remaining fingers of the right hand are wrapped around the wire, as shown in Figure 18-3. The orientation of the fingers represents the direction of the magnetic field which appears in concentric circles about the axis of the wire.

Solved Problem 18.13. A copper wire of length $0.60 \, \mathrm{m}$ carries a current of $12 \, \mathrm{A}$. Determine the magnetic field at a distance of (1) $2 \, \mathrm{cm}$, (2) $20 \, \mathrm{cm}$, and (3) $2 \, \mathrm{m}$.

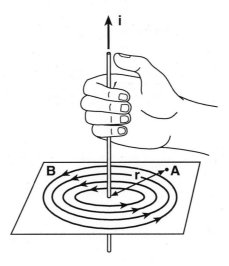

Fig. 18-3

Solution

The magnetic field at a distance r from a current-carrying wire is given by

$$B = \frac{\mu_o i}{2\pi r}$$

(1) $B = \dfrac{(4\pi \times 10^{-7}\,\text{T m A}^{-1})(12\,\text{A})}{2 \times 3.14 \times 0.02\,\text{m}} = 1.2 \times 10^{-4}\,\text{T}$

(2) $B = \dfrac{(4\pi \times 10^{-7}\,\text{T m A}^{-1})(12\,\text{A})}{2 \times 3.14 \times 0.2\,\text{m}} = 1.2 \times 10^{-5}\,\text{T}$

(3) $B = \dfrac{(4\pi \times 10^{-7}\,\text{T m A}^{-1})(12\,\text{A})}{2 \times 3.14 \times 2\,\text{m}} = 1.2 \times 10^{-6}\,\text{T}$

Solved Problem 18.14.   A straight wire of length 5 m carries a current of 10 A. Determine the distance of the point where the magnetic field is equivalent to the earth's magnetic field of $B = 5 \times 10^{-5}$ T.

Solution

The magnetic field B at a distance r from a current-carrying wire is

$$B = \frac{\mu_o i}{2\pi r}$$

Solving for r gives

$$r = \frac{\mu_o i}{2\pi B} = \frac{(4\pi \times 10^{-7}\,\text{T m A}^{-1})(10\,\text{A})}{2\pi(5 \times 10^{-5}\,\text{T})} = 0.04\,\text{m} = 4\,\text{cm}$$

(3)   Circular wire loop

At the center of a wire loop of radius r carrying a current i, the magnetic field B is

$$B = \frac{\mu_o i}{2r}$$

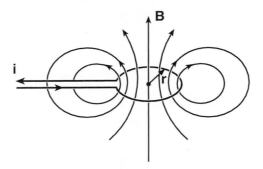

Fig. 18-4

as illustrated in Figure 18-4. If the loop contains N turns, the magnetic field B at the center of the loop becomes

$$B = \frac{N\mu_o i}{2r}$$

Solved Problem 18.15.   A circular conducting coil of diameter 0.4 m has 50 loops of wire and a current of 3 A flowing through it. Determine the magnetic field generated by the coil.

   Solution

$$B = \frac{N\mu_o i}{2r} = \frac{(50)(4\pi \times 10^{-7}\,\text{T m A}^{-1})(3\,\text{A})}{2(0.2\,\text{m})} = 4.7 \times 10^{-4}\,\text{T}$$

(4)   Solenoid

   The magnetic field at any point within the interior of a solenoid of length L with N loops carrying current i is depicted in Figure 18-5 and is given mathematically by

$$B = \frac{\mu_o N i}{L}$$

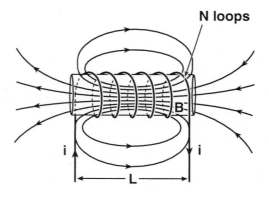

Fig. 18-5

Solved Problem 18.16. The effects of a magnetic field on a living organism can be investigated by placing the organism in a solenoid. Given a solenoid of diameter 2.0 cm, 10.0 cm in length, and constructed from 400 turns of a wire, determine the magnitude of the magnetic field exerted on the organism, given a current of 5 A. Assume the permeability of the organism is equal to that of air or free space.

Solution

The magnetic field B in the interior of a solenoid is given by

$$B = \frac{\mu_o Ni}{L} = \frac{(4\pi \times 10^{-7}\,\text{T m A}^{-1})(400)(5\,\text{A})}{0.1\,\text{m}} = 2.5 \times 10^{-2}\,\text{T}$$

18.3 INDUCED EMF

The movement of a conductor through a magnetic field produces current which is dependent on the rate at which the conductor moves through the magnetic field. More current can be generated by increasing the speed of the conductor through the field. If the conductor is a coil of N loops and it is exposed to a changing magnetic flux Φ or a magnetic field B exerted over a surface area A ($\Phi = BA$), an emf is induced in a conductor, as shown in a landmark demonstration by Michael Faraday in 1831. This phenomenon is illustrated in Figure 18-6.

The emf E induced by motion of a conductor causing a change in magnetic flux Φ during a time t is known as *Faraday's law for induced emf* and is given by

$$E = -N\frac{\Delta\Phi}{\Delta t}$$

where E is expressed in volts and $\Delta\Phi/\Delta t$ is expressed in Wb s^{-1}. The minus sign implies that the induced emf opposes the change in magnetic flux that generated it. This is expressed below as Lenz's law.

The emf induced by a straight conductor of length L moving with a velocity **v** in a magnetic field **B** perpendicular to its motion is given by

$$E = \mathbf{B}L\mathbf{v}$$

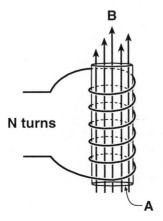

Fig. 18-6

18.4 LENZ'S LAW

Lenz's law states that upon induction of an emf, the direction of induced current opposes the change inducing the current.

Solved Problem 18.17.   A conducting coil with 100 loops of wire and a diameter of 0.30 m, originally subjected to a uniform magnetic flux of 5.5×10^{-4} Wb, is temporarily exposed to a uniform magnetic flux of 1.5×10^{-4} Wb for 60 ms. Determine the induced emf induced in the conducting coil.

Solution

The equation needed to determine the induced emf is

$$E = -N\frac{\Delta\Phi}{\Delta t}$$

Since we are only interested in obtaining the magnitude of the emf, the minus sign can be ignored and the above equation can be slightly rewritten to reflect the average value of the emf, or

$$|E| = N\left|\frac{\Delta\Phi}{\Delta t}\right| = (100 \text{ loops})\frac{5.5 \times 10^{-4} \text{ Wb} - 1.5 \times 10^{-4} \text{ Wb}}{0.06 \text{ s}} = 0.7 \text{ V}$$

Solved Problem 18.18.   An aluminum wire 75 cm in length is oriented perpendicular to a magnetic flux $\Phi = 1.4$ Wb m^{-2} and moves with a velocity of 0.2 m s^{-1} at right angles to the magnetic field. What is the induced emf?

Solution

The induced emf in a straight conductor is given by

$$E = \mathbf{B}L\mathbf{v} = (1.4 \text{ Wb m}^{-2})(0.75 \text{ m})(0.2 \text{ m s}^{-1}) = 0.21 \text{ V}$$

ILLUSTRATIVE EXAMPLE 18.3.   Electromagnetic Flowmeters and Blood Flow Measurements

An electromagnetic flowmeter is an instrument used to quantitatively measure blood flow velocity through a blood vessel. The flowmeter, shown in Figure 18-7, consists of a magnetic core or ring clamped around a vessel segment which produces a magnetic field. The flowmeter is based on the principle that an electromotive force is induced in a conductor moving so as to cut the lines of force in a magnetic field. The induced emf is detected by signal electrodes and subsequently is amplified. As the conductor (or, in this case, blood) moves through the field in a direction perpendicular to its own axis and to the lines of force, the potential difference in volts at the ends of the vessel, in terms of the blood flow velocity, is given as

$$V = B\, dv \times 10^{-8} \text{ volts}$$

where B is the strength of the magnetic field (G), d is the internal diameter of the vessel (cm), and v is the average velocity of the flowing blood (cm s^{-1}). The volumetric blood flow rate Q is determined from

$$Q = \frac{\pi d^2}{4}\frac{V}{Bd}$$

This equation for flow is true provided three criteria are met: (1) The magnetic field is uniform, (2) the conductor moves in a plane at right angles to the magnetic field, and (3) the length of the conductor extends at right angles to both the magnetic field and the direction of motion. Electromagnetic flowmeters reveal accurate measurements of blood flow, but require direct exposure of a vessel, limiting clinical applications to measurements during surgery.

Solved Problem 18.19.   An electromagnetic flowmeter applied to an artery of radius 0.25 cm yields a blood flow velocity of 150 mm s^{-1} using a magnetic field density of 500 G. What is the expected measured voltage?

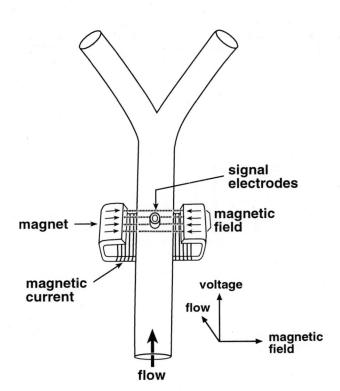

Fig. 18-7

Solution

From
$$V = B\,dv \times 10^{-8}\,\text{V}$$

where R = radius of the artery = 0.25 cm corresponding to a diameter $d = 2 \times 0.25$ cm = 0.5 cm; $B = 500$ G, and $v = 150$ mm s^{-1} = 15 cm s^{-1}, the voltage is

$$V = (0.5\,\text{cm})(500\,\text{G})(15\,\text{cm s}^{-1}) \times 10^{-8} = 3750 \times 10^{-8}\,\text{V} = 37.5\,\text{mV}$$

18.5 SELF-INDUCED EMF

In the previous section, the concept of induced emf was introduced and described as a phenomenon that occurs as a result of a moving conductor's cutting across magnetic lines of force or flux. However, an emf can be induced as a result of changes in current through the conductor. Changes in current cause a change in the magnetic field generated by the conductor, ultimately leading to a change in the magnetic flux and a self-induced emf. A self-induced emf E is given by

$$E = -L\frac{\Delta i}{\Delta t}$$

where L is known as the inductance, $\Delta i/\Delta t$ is the rate of change of current, and the minus sign represents opposing directions between the induced emf and the current Δi that generated the emf.

Inductance L is expressed in units of henry (H). From the equation for the self-induced emf, any conductor or circuit element of given inductance ($L = 1$ H) will generate a corresponding self-induced emf ($E = 1$ V) given a proportional rate of current change (1 A s^{-1}).

Energy of a Current-Carrying Inductor

The work done against the self-induced emf to generate current flow in the inductor is given by

$$W = \frac{1}{2}Li^2$$

where W is expressed in units of joules, given that L is in units of henrys and i is in units of amperes.

Solved Problem 18.20.   Consider a conducting coil of 500 loops with a current of 4 A flowing through it. If the flowing current generates a magnetic flux of 5×10^{-4} Wb over 0.1 s through the coil, determine the (1) inductance of the coil and (2) energy stored in the coil.

Solution

(1)   We must first determine the induced emf in the coil, which can be calculated from

$$|E| = N \left| \frac{\Delta \Phi}{\Delta t} \right| = 500 \left| \frac{5 \times 10^{-4}\,\text{Wb} - 0}{0.1\,\text{s}} \right| = 2.5\,\text{V}$$

The self-induced emf, given by E, generated by the conducting coil is

$$|E| = L \left| \frac{\Delta i}{\Delta t} \right|$$

Solving for L gives

$$L = \left| \frac{E\,\Delta t}{\Delta i} \right| = \left| \frac{(2.5\,\text{V})(0.1\,\text{s})}{4\,\text{A} - 0} \right| = 0.0625\,\text{H} = 62.5\,\text{mH}$$

(2)   The energy stored in an inductor can be calculated by

$$W = \frac{1}{2}Li^2 = \frac{1}{2}(0.0625\,\text{H})(4\,\text{A})^2 = 0.5\,\text{J}$$

Supplementary Problems

18.1.   In the cathode-ray tube of a television set, an electron is projected at a velocity of 2×10^7 m s^{-1} and travels perpendicular to a 0.6-T magnetic field. Determine the magnetic force exerted on the electron.

Solution

1.92×10^{-12} N

18.2.   Consider a square of side length a defined by the origin at A $(0, 0)$ and electrons at the remaining three corners: B $(a, 0)$, C (a, a), and D $(0, a)$ moving with a speed of 5×10^7 m s^{-1}. Given a uniform magnetic field of 1.0 T in the $+z$ direction, determine the magnetic force exerted on the electron at (1) B moving in the $+y$ direction, (2) C moving in the $+z$ direction, and (3) D moving diagonally toward B.

Solution

(1) 8×10^{-12} N; (2) 0; (3) 5.65×10^{-12} N

18.3.   Consider an electron moving in a circular orbit perpendicular to the magnetic field of the earth $B = 3.5 \times 10^{-5}$ T. If the radius of the circular orbit is 0.70 m, determine the speed of the electron. The mass of the electron is 9.1×10^{-31} kg.

Answer

$4.3 \times 10^6 \, \text{m s}^{-1}$

18.4.   In a mass spectrometer, positively charged ions of charge $+e$ exit the velocity selector at a speed of $2.5 \times 10^5 \, \text{m s}^{-1}$ before entering a 0.45-T magnetic field. Determine the electric field required to provide the electron speed.

Answer

$1.1 \times 10^5 \, \text{V m}^{-1}$

18.5.   A cyclotron is an instrument that accelerates protons moving in a circular path perpendicular to a uniform magnetic field. In a typical experiment, protons attain a speed of $2.0 \times 10^7 \, \text{m s}^{-1}$ while moving in a circular orbit of radius 0.15 m. Determine the (1) magnetic field and (2) the number of complete rotations made by a proton in 1 s.

Answer

(1)   13.9 T; (2) 2.1×10^7 rev

18.6.   In Bohr's theory of the hydrogen atom, the electron revolves around a proton in a circular orbit of radius 0.53 Å. Given that the electron completes one orbit in 2.0×10^{-16} s, calculate (1) the current generated by the motion of the electron and (2) the magnetic field exerted on the proton.

Answer

(1)   0.8 mA; (2) 0.48 T

18.7.   A proton assumes a circular orbit of radius r in a perpendicular uniform magnetic field B. Given the mass of the proton m_p, derive an expression for the frequency of the proton in terms of the protons mass, and magnetic field B.

Answer

$$f = \frac{qB}{2\pi m_p}$$

18.8   From Supplementary Problem 18.7, if $B = 1.0$ T, determine the frequency of the proton, given that the proton mass is $m_p = 1.67 \times 10^{-27}$ kg.

Answer

$1.5 \times 10^7 \, \text{s}^{-1}$

18.9.   An alpha particle of charge $+2e$ travels with a velocity of $3.0 \times 10^6 \, \text{m s}^{-1}$ in a magnetic field of flux density $2.0 \, \text{Wb m}^{-2}$ at an angle of $45°$ with the magnetic field. Determine the force exerted on the alpha particle.

Answer

$1.35 \times 10^{-12} \, \text{N}$

18.10. An electron in a cathode-ray tube is deflected in a circular arc of radius 1.5 cm by a uniform magnetic field of $B = 0.045$ T. Given the mass of the electron $m = 9.11 \times 10^{-31}$ kg, determine the speed of the electron.

Answer

$1.2 \times 10^8 \, \text{m s}^{-1}$

18.11. Suppose two straight conducting wires, each of length $L = 10$ m and carrying a current $i = 5$ A, are separated by a distance $r = 2.0$ cm. Determine the magnitude and direction of the magnetic force per unit length F exerted by wire 1 on wire 2.

Solution

2.5×10^{-4} N m^{-1}, directed toward wire 2 and is attractive

18.12. In Supplementary Problem 18.11, how would the magnitude and direction of the magnetic force change if the current in the two wires flowed in opposite directions?

Solution

Magnitude is the same, direction is repulsive.

18.13. A solenoid of length 0.8 m and cross-sectional area 4.5 cm^2 is wound with 200 coils of conducting wire that carries a current of 2.5 A. Determine the (1) magnetic field at any point in the interior of the solenoid and (2) magnetic flux.

Solution

(1) 7.85×10^{-4} T; (2) 3.53×10^{-7} T m^2

18.14. A circular cable of radius 40 cm consists of 300 turns of wire with a current of 12 A. Determine the magnetic induction at the center of the cable.

Solution

5.65×10^{-3} W m^{-2} directed perpendicular to plane of orbit

18.15. A circular conducting coil of radius 80 cm consists of 15 turns of wire carrying a current of 25 A. Determine the magnetic induction at the center of the coil.

Solution

2.9×10^{-4} W m^{-2} directed perpendicular to plane of orbit

18.16. A solenoid of length 0.85 m and radius 0.2 m consists of 2000 turns of wire carrying a current of 3 A. Determine the magnetic induction along the axis of the solenoid.

Solution

8.86×10^{-3} W m^{-2}

18.17. An electromagnetic flowmeter measures flow in a blood vessel according to

$$V = B\,dv \times 10^{-8}$$

where V is the voltage (V), B is the magnetic field strength (G), d is the internal diameter of the vessel (cm), and v is the velocity of the flowing blood (cm s^{-1}). Express the potential difference V in terms of the volumetric flow rate Q.

Solution

$$\frac{2BQ}{\pi R} \times 10^{-8} \text{ volts}$$

18.18. Given the influence of conductivity of the arterial wall, how would the voltage measured by the electromagnetic flowmeter be altered?

Solution

$V < B\, dv \times 10^{-8}$

18.19. An electromagnetic flowmeter applied to an artery of radius 0.75 cm yields a blood flow velocity of 100 cm s^{-1} using a magnetic field density of 300 G. What is the expected measured voltage?

Solution

450 μV

18.20. Given a conducting coil, determine its self-inductance if the rate of change of current of 15 A s^{-1} generates an emf of 3 V.

Solution

0.2 H

18.21. The self-inductance of a conducting coil is 2.4 mH. Given that the current changes steadily from 0 to 2.5 A in 0.3 s, determine the magnitude and direction of the self-induced emf.

Solution

-20 mV, opposite in direction to current

18.22. A magnetic flux of 7×10^{-5} Wb exists within a conducting coil with 500 turns. If 0.04 s is required to reduce the flux to 2×10^{-5} Wb, determine the average induced emf.

Solution

-0.63 V

Chapter 19

Alternating-Current Circuits

Alternating-current (ac) circuits are similar to direct-current (dc) circuits with one exception: the type of voltage source driving current through the circuit. In dc circuits, the voltage source is continuous and constant, resulting in current flow in only one direction, while in ac circuits the voltage is oscillatory or sinusoidal, resulting in periodic changes in direction of current flow. AC electricity is used in most households (typically at an oscillatory frequency of 60 Hz) because its voltage can easily be changed by a transformer. This chapter addresses electric circuits driven by oscillating or alternating voltage sources.

19.1 DEFINITIONS OF AC CIRCUITS

Voltage in ac circuits is produced by an electric power generator which generates an oscillating potential across its terminals. As a result, the voltage potential is sinusoidal, as shown in Figure 19-1, changing direction and consequently magnitude over time, and can be expressed mathematically by

$$V = V_o \sin \omega t = V_o \sin 2\pi f t$$

where V is the instantaneous voltage, in volts; V_o is the amplitude or maximum value of the voltage, in volts; ω is the angular velocity, in rad s^{-1}; f is the frequency, in hertz; t is time, in seconds. Because it continuously changes over time, ac voltage can be expressed as an effective or root mean square (r ms) voltage, given by

$$V = V_{\text{r ms}} = \frac{V_{\text{max}}}{\sqrt{2}}$$

Current in ac circuits, similar to voltage, is sinusoidal and is defined as

$$i = i_o \sin \omega t = i_o \sin 2\pi f t$$

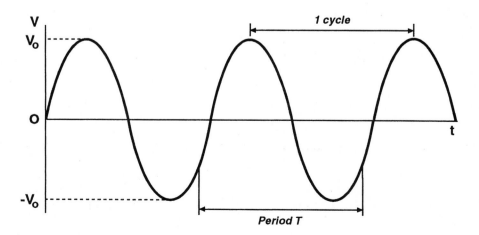

Fig. 19-1

282

Solved Problem 19.1. Determine the maximum or peak voltage from an ac rms voltage of 120 V.

Solution

The maximum voltage is related to the rms voltage by

$$V_{max} = \sqrt{2}\ V_{rms} = \sqrt{2} \times 120\ V = 169.7\ V$$

Solved Problem 19.2. Determine the maximum value of an 8.0-A alternating current.

Solution

It can be assumed that the rms current is given in the problem as 8.0 A. The rms current is related to the maximum current by

$$i_{max} = \sqrt{2}\ i_{rms} = \sqrt{2} \times 8.0\ A = 11.3\ A$$

Solved Problem 19.3. An ac circuit with an 80-Ω resistor is connected to an oscillating (sinusoidal) voltage with a maximum voltage of 190 V. Determine the instantaneous voltage when the voltage has reached (1) 0°, (2) 45°, and (3) 90° of its cycle.

Solution

The instantaneous voltage is given by

$$V = V_{max} \sin \theta$$

(1) $V = 190\ V \sin 0° = 0$

(2) $V = 190\ V \sin 45° = 134.3\ V$

(3) $V = 190\ V \sin 90° = 190\ V$

19.2   OHM'S LAW

Ohm's law for an ac circuit exists in four distinct forms, corresponding to the type and arrangement of circuit elements:

Pure Resistance

Given an ac circuit with a resistor connected to an oscillating power source, the rms voltage V is related to the rms or effective current i according to

$$V = iR$$

Pure Inductance

Inductance represents another circuit element that impedes voltage and is caused by a reversed emf, induced by changing magnetic fields as the voltage rises and falls in a wire. Given an ac circuit with an inductor connected to an oscillating voltage source, the voltage is given by

$$V = iX_L$$

where X_L is the inductance reactance. The inductance reactance represents the ability of the inductor to resist the flow of an ac circuit and is defined by

$$X_L = 2\pi f L = \omega L$$

Here X_L is expressed in units of ohms, given that L is in henrys, ω is in rad s^{-1}, and f is in hertz.

Pure Capacitance

As stated in Chapter 17 for dc circuits, capacitance represents the ability to store charge. Given an ac circuit with a capacitor connected to an oscillating voltage source, the voltage is given by

$$V = iX_C$$

where X_C is the capacitance reactance, defined by

$$X_C = \frac{1}{2\pi fC}$$

Here X_C is expressed in units of ohms, given that C is in farads and f is in hertz.

Resistance, Inductance, and Capacitance in Combination

Given an ac circuit with a resistor, capacitor, and inductor connected in series to an oscillating voltage source, as shown in Figure 19-2, the voltage is given by

$$V = iZ$$

where Z is the impedance, in ohms, defined by

$$Z = \sqrt{R^2 - (X_C - X_L)^2}$$

Impedance represents the total resistance or opposition to current flow in ac circuits.

Since the voltage and current are oscillatory, their values continually change with respect to phase and are dependent on the circuit element.

Pure resistance: Voltage and current are in phase.
Pure inductance: Voltage leads the current by $\frac{1}{4}$ cycle, or 90°.
Pure capacitance: Voltage lags behind the current by $\frac{1}{4}$ cycle, or 90°.

The phase angle ϕ which describes the phase between the voltage V and current i is

$$\tan \phi = \frac{X_L - X_C}{R}$$

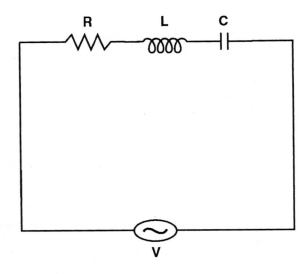

Fig. 19-2

19.3   POWER DISSIPATED IN AN AC CIRCUIT

Voltage V from an ac generator through an ac circuit with any type of impedance gives rise to a current i through the impedance, distinguished by the phase angle ϕ between V and i. The power P dissipated by the impedance is given by

$$P = iV \cos \phi$$

For an ac circuit with pure resistance $P = 1$, and $P = 0$ for an ac circuit with either pure capacitance or pure inductance.

19.4   RESONANCE IN AN AC CIRCUIT

In an ac circuit with a resistor, inductor, and capacitor connected in series, resonance occurs when $X_L = X_C$. When this condition holds, the total impedance is dependent on only the resistance and is thus a minimum, implying that current through the circuit is a maximum. The resonant or natural frequency f is given by

$$f = \frac{1}{2\pi\sqrt{LC}}$$

Solved Problem 19.4.   Given the ac circuit described in Solved Problem 19.3, determine the resultant current for each of the voltage cycles.

Solution

The current can be determined by Ohm's law with a resistance connected to an oscillating voltage source.

(1)   When $\phi = 0°$, $V = 0$. Thus, $i = 0$.

(2)   When $\phi = 45°$, $V = 134.3$ V. Thus,

$$i = \frac{V}{R} = \frac{134.3 \text{ V}}{80 \text{ }\Omega} = 1.7 \text{ A}$$

(3)   When $\phi = 90°$, $V = 190$ V. Thus,

$$i = \frac{V}{R} = \frac{190 \text{ V}}{80 \text{ }\Omega} = 2.4 \text{ A}$$

Solved Problem 19.5.   Determine the inductive reactance of a 10-mH inductor at angular frequencies of (1) 10 rad s^{-1}, (2) 100 rad s^{-1}, (3) 1000 rad s^{-1}, and (4) 10,000 rad s^{-1}.

Solution

The inductive reactance of an inductor is given as

$$X_L = \omega L$$

(1) $X_L = (10 \text{ rad s}^{-1})(10 \times 10^{-3} \text{ H}) = 0.1 \text{ }\Omega$

(2) $X_L = (100 \text{ rad s}^{-1})(10 \times 10^{-3} \text{ H}) = 1 \text{ }\Omega$

(3) $X_L = (1000 \text{ rad s}^{-1})(10 \times 10^{-3} \text{ H}) = 10 \text{ }\Omega$

(4) $X_L = (10,000 \text{ rad s}^{-1})(10 \times 10^{-3} \text{ H}) = 100 \text{ }\Omega$

Solved Problem 19.6.   Determine the capacitive reactance of a 3-μF capacitor at angular frequencies of (1) 1000 rad s^{-1} and (2) 10,000 rad s^{-1}.

Solution

The capacitive reactance of a capacitor is given as

$$X_C = \frac{1}{\omega C}$$

(1) $X_C = \dfrac{1}{(1000 \text{ rad s}^{-1})(3 \times 10^{-6} \text{ F})} = 333 \ \Omega$

(2) $X_C = \dfrac{1}{(10,000 \text{ rad s}^{-1})(3 \times 10^{-6} \text{ F})} = 33 \ \Omega$

Solved Problem 19.7.   In an ac circuit powered by an alternating frequency of 500 Hz, a capacitor exhibits a capacitance reactance of 40 Ω. Determine the capacitance of the capacitor.

Solution

By definition, the capacitive reactance is given by

$$X_C = \frac{1}{2\pi f C}$$

Solving for C yields

$$C = \frac{1}{2\pi f X_C} = \frac{1}{6.28 \times 500 \text{ Hz} \times 40 \ \Omega} = 7.9 \ \mu\text{F}$$

Solved Problem 19.8.   In an ac circuit driven by 120 V, a coil with an inductive reactance of 75 Ω is connected in series to a 30-Ω resistor. Determine the current through the circuit.

Solution

We first must calculate the impedance of the ac circuit.

$$Z = \sqrt{R^2 + (X_L - X_C)^2} = \sqrt{(30 \ \Omega)^2 + (75 \ \Omega - 0)^2} = 80.7 \ \Omega$$

From Ohm's law for an ac circuit,

$$i = \frac{V}{Z} = \frac{120 \text{ V}}{80.7 \ \Omega} = 1.5 \text{ A}$$

Solved Problem 19.9.   For an ac circuit with a resistor, inductor, and capacitor connected in series, as shown in Figure 19-3, determine the impedance.

Solution

From Figure 19-3, it can be seen that $C = 2 \ \mu\text{F}$, $L = 50 \text{ mH}$, $R = 70 \ \Omega$, $E = 120 \text{ V}$, and $f = 60 \text{ Hz}$. The impedance for an ac circuit is given by

$$Z = \sqrt{R^2 + (X_L - X_C)^2}$$

We must next calculate X_L and X_C:

$$X_L = 2\pi f L = (6.28)(60 \text{ Hz})(0.05 \text{ H}) = 18.8 \ \Omega$$

$$X_C = \frac{1}{2\pi f C} = \frac{1}{(6.28)(60 \text{ Hz})(2 \ \mu\text{F})} = 1326 \ \Omega$$

The impedance can now be calculated:

$$Z = \sqrt{(70 \ \Omega)^2 + (18.8 \ \Omega - 1326 \ \Omega)^2} = 1309 \ \Omega$$

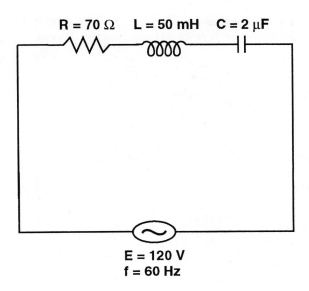

$R = 70 \ \Omega \quad L = 50 \ mH \quad C = 2 \ \mu F$

$E = 120 \ V$
$f = 60 \ Hz$

Fig. 19-3

Solved Problem 19.10.   For the circuit in Solved Problem 19.9, determine the phase angle.

Solution

The phase angle is related to the impedance by

$$\phi = \tan^{-1} \frac{X_L - X_C}{R} = \tan^{-1} \frac{18.8 \ \Omega - 1326 \ \Omega}{70 \ \Omega} = -86.9°$$

Solved Problem 19.11.   For the circuit in Solved Problem 19.9, determine the power dissipated by the impedance.

Solution

The power dissipated by the impedance is given by

$$P = iV \cos \phi$$

The voltage is 120 V, $\phi = -86.9°$, and the current can be determined by

$$i = \frac{V}{Z} = \frac{120 \ V}{1309 \ \Omega} = 0.09 \ A$$

Thus

$$P = (0.09 \ A)(120 \ V)[\cos (-86.9°)] = 0.58 \ W$$

Solved Problem 19.12.   Define the phase angle in terms of the impedance.

Solution

Elementary trigonometric relations can be used to solve this problem. The impedance, in effect, is the hypotenuse of a right triangle where R is the base and the quantity $X_L - X_C$ is the height. One can also visualize this relationship in terms of vectors, where R is a vector directed along the x axis, the quantity $X_L - X_C$ is a vector directed along the y axis, and Z is the resultant vector. Regardless of the approach, it can be seen that

$$\cos \phi = \frac{\text{adjacent}}{\text{hypotenuse}} = \frac{R}{Z}$$

$$\sin \phi = \frac{\text{opposite}}{\text{hypotenuse}} = \frac{X_L - X_C}{Z}$$

Solved Problem 19.13.   Given a 60-mH inductor that exhibits an inductive reactance of 2.4 kΩ, determine the resonant frequency of the ac circuit.

Solution

The inductive reactance of an inductor is given by

$$X_L = 2\pi f L$$
$$2.4 \times 10^3 \ \Omega = (6.28)(f)(60 \times 10^{-3} \text{ H})$$

Solving for f yields

$$f = 6.4 \text{ Hz}$$

19.5   TRANSFORMER

A *transformer* is an electrical device, shown in Figure 19-4, consisting of two coils wound around a ferromagnetic material, which can easily be magnetized and demagnetized. As current in the primary coil changes, the magnetic field changes everywhere, including at the secondary coil. The voltage generated in the secondary coil will be greater than that in the primary coil if the secondary coil contains more coils.

The power in the primary coil is equal to the power in the secondary coil, or

$$V_1 i_1 = V_2 i_2$$

The ratio of the voltages in the two coils is related to the number of turns in each coil by

$$\frac{V_1}{V_2} = \frac{N_1}{N_2}$$

The ratio of current in the two coils is related to the number of turns in each coil by

$$\frac{i_1}{i_2} = \frac{N_2}{N_1}$$

Solved Problem 19.14.   The primary coil of a transformer contains 1000 turns and is connected to an ac generator with an effective voltage of 120 V. If the secondary coil contains 600 turns, determine the effective voltage across the secondary coil.

Solution

The voltage in a transformer is related to the number of coils by

$$\left(\frac{N}{V}\right)_{\text{primary coil}} = \left(\frac{N}{V}\right)_{\text{secondary coil}}$$
$$\left(\frac{1000 \text{ turns}}{120 \text{ V}}\right)_{\text{primary coil}} = \left(\frac{600 \text{ turns}}{x \text{ V}}\right)_{\text{secondary coil}}$$

Solving for x yields $V_{\text{secondary coil}} = 72 \text{ V}$.

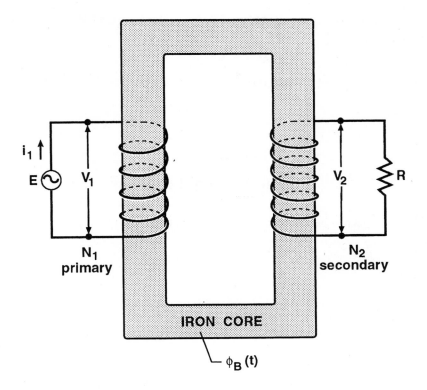

Step-up transformer: N$_2$ > N$_1$
$$V_2 > V_1$$

Step-down transformer: N$_2$ < N$_1$
$$V_2 < V_1$$

Fig. 19-4

ILLUSTRATIVE EXAMPLE 19.1. Electrical Analogs and Blood Flow Revisited

In Illustrative Example 17.1, electrical entities were introduced and discussed in analogous terms to fluid flow, particularly as they pertain to the human circulatory system. In brief, it was stated that blood flow through a vessel could be approximated by steady fluid flow Q through a rigid pipe of resistance R, as produced by a steady pressure gradient ΔP according to the general principles behind Ohm's law:

$$\Delta P = QR$$

Although this is admittedly a simplistic representation of the human circulatory system, one physiological factor which cannot be easily accounted for by the description above is the type of fluid flow. Steady fluid flow was assumed in the previous demonstration, which tends to hold for the smaller blood vessels. However, the oscillatory nature of the heartbeat introduces pulsatile flow through the circulatory system, and thus blood flow can be approximated in a more accurate manner by an ac circuit. All the parameters displayed in Table 17-1 hold with the exception of inductance. Inductance L represents the inertance or fluid inertia of the flowing blood. Thus, the oscillatory pressure gradient is rewritten as

$$\Delta P = L\frac{\Delta Q}{\Delta t} + QR$$

In terms of the geometry of a blood vessel, the inductance and resistance are defined as

$$R = \frac{8\pi\mu L}{A^2} \qquad \text{and} \qquad L = \frac{\rho L}{A}$$

where μ is blood viscosity, L is vessel length, A is cross-sectional area of the vessel, and ρ is blood density.

Supplementary Problems

19.1. Determine the maximum or peak voltage from an ac rms voltage of 240 V.

Solution

339.4 V

19.2. From Supplementary Problem 19.1, express the oscillating voltage in equation form, given a generator with an operating frequency of 60 Hz.

Solution

339.4 V sin $120\pi t$

19.3. In an ac circuit, a voltage $V = 80$ V sin $120\pi t$ is applied across a resistor of 45 Ω. Determine the current through the resistor.

Solution

1.25 A

19.4. Given an rms current of 15 A, determine its maximum value.

Solution

21.2 A

19.5. An ac voltage source of 120 V with a frequency of 60 Hz is connected in series to a 4-μF capacitor. Determine the (1) capacitive reactance and (2) current through the capacitor.

Solution

(1) 663.5 Ω; (2) 0.18 A

19.6. In Supplementary Problem 19.5, determine the power dissipated in the capacitor.

Solution

0

19.7. An ac circuit consists of a coil of 30-mH inductance connected to a 60-Hz voltage source. Determine the inductance reactance.

Solution

11.3 Ω

19.8. In an ac circuit, an inductor with an inductance of 0.75 H is connected to an ac generator with a maximum voltage of 240 V and a frequency of 50 Hz. Determine the effective current through the inductor.

Answer

0.72 A

19.9. Given an angular frequency of 100 rad s^{-1}, determine the inductive reactance of a (1) 50-mH and (2) 500-mH inductor.

Answer

(1) 5 Ω; (2) 50 Ω

19.10. An ac circuit consists of a 2-mH inductor connected in series to a 5-μF capacitor. Determine the resonance frequency of the ac circuit.

Answer

1.6 kHz

19.11. For an ac circuit with an inductor, capacitor, and resistor connected in series, where $C = 60\,\mu$F, $L = 0.3$ H, $R = 50\,\Omega$, $E = 120$ V, and $f = 60$ Hz, determine the impedance.

Answer

85.0 Ω

19.12. Given an ac series circuit with $R = 1500\,\Omega$, $C = 3\,\mu$F, $L = 1.5$ H, $\omega = 500$ rad s^{-1}, and $E = 100\sin\omega t$ V, determine (1) Z, (2) i_{max}, and (3) i_{rms}.

Answer

(1) 1502 Ω; (2) 0.067 A; (3) 0.047 A

19.13. From Supplementary Problem 19.12, determine the rms voltage through the (1) inductor, (2) resistor, and (3) capacitor.

Answer

(1) 35.3 V; (2) 70.5 V; (3) 31.5 V

19.14. From Supplementary Problem 19.12, determine the (1) phase angle and (2) power dissipated by the impedance.

Answer

(1) 3.6°; (2) 3.3 W

19.15. The primary coil of a transformer is connected to a 120-V ac generator with a frequency of 50 Hz. Given that the effective current through the primary coil is 1.2 A, determine the inductance of the primary coil of the transformer.

Answer

0.32 H

19.16. The primary coil of a transformer consists of 600 turns and is connected to a 120-V source. If the secondary coil carries a voltage of 10 V, determine the number of turns of the secondary coil.

Answer

50 turns

Chapter 20

Light

We are accustomed to thinking about light in terms of the output of a lamp or lightbulb radiated as brightness and/or heat. This type of light, that is, visible light, represents only a small subset of a much broader physical entity: electromagnetic waves. Although distinct similarities exist between electromagnetic waves and physical waves, as described in Chapter 15, there also exist differences, particularly in their interactions between various surfaces or media. In this chapter, we introduce the physics involved in the behavior of light.

20.1 DEFINITIONS OF LIGHT

Electromagnetic waves are not generated by the motion of matter, as is the case with physical waves, but consist of coupled oscillating electric and magnetic fields. These waves are, however, transverse because the vibrating electric and magnetic fields are at right angles to the direction of propagation, as shown in Figure 20-1. Electromagnetic waves transport energy and vary only in frequency. Examples of electromagnetic waves are radio waves, microwaves, infrared waves, visible light waves, ultraviolet waves, X-rays, and γ rays which range in frequency from 3×10^6 Hz for radio waves to 3×10^{20} Hz for γ rays. Regardless of the type, all electromagnetic waves move with the velocity of light which, in vacuum or free space, is given by $c = 3 \times 10^8$ m s^{-1}. The frequency f of electromagnetic waves is related to their wavelength λ by

$$f = \frac{c}{\lambda}$$

Solved Problem 20.1.   The frequency of ultraviolet waves is 2×10^{16} Hz. Determine its wavelength.

Solution

By definition, the frequency of electromagnetic waves is related to the wavelength by

$$f = \frac{c}{\lambda}$$

$$\lambda = \frac{c}{f} = \frac{3 \times 10^8 \text{ m s}^{-1}}{2 \times 10^{16} \text{ Hz}} = 1.5 \times 10^{-8} \text{ m}$$

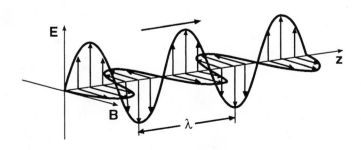

Fig. 20-1

Solved Problem 20.2.   Radio waves have a typical wavelength of 10^2 m. Determine their frequency.

   Solution

$$f = \frac{c}{\lambda} = \frac{3 \times 10^8 \text{ m s}^{-1}}{1 \times 10^2 \text{ m}} = 3 \times 10^6 \text{ Hz}$$

Solved Problem 20.3.   What is the speed of waves from radio station KQRS that broadcasts at a frequency of 710 kHz?

   Solution

   All electromagnetic waves, including radio waves, move with the same speed, the speed of light. Thus, the speed of these waves is 3×10^8 m s^{-1}.

ILLUSTRATIVE EXAMPLE 20.1.   Marine Organisms and Bioluminescence

   Several types of marine organisms such as squid, jellyfish, and sea anemone, exhibit the unique ability of glowing in the dark, known as *bioluminescence*. On a chemical level, bioluminescence occurs as the result of oxygen added to luciferin, a substance capable of emitting light. An enzyme—luciferase—is required to promote this reaction mechanism. Energy generated from the reaction is released as visible light, primarily in the blue-green range. The conversion of chemical energy to light energy can be nearly 100 percent efficient, producing very little heat. In contrast, a common household lightbulb is only 10 percent efficient, wasting 90 percent of its energy as heat.

20.2   REFLECTION AND REFRACTION OF LIGHT

   As light rays propagate through one medium and approach a second medium, they can interact with the boundary between the two media in two separate ways: reflection and refraction. Consider a light ray that propagates through one medium and strikes the boundary at an incidence angle of θ_i with the normal, or perpendicular, to the plane of the surface boundary, as shown in Figure 20-2. The light ray can reflect or bounce off the surface boundary at an angle of reflection θ_r, also measured with respect to the normal, which is equal to the angle of incidence, or

Law of reflection: $\theta_i = \theta_r$

   In addition to reflection, the light ray can bend or refract as it enters the second medium. The angle at which the light ray refracts is dependent on the ratio of the indices of refraction between the two media. The index of refraction n for a given material is defined as

$$n = \frac{\text{speed of light in vacuum}}{\text{speed of light in material}} = \frac{c}{v}$$

The index of refraction for common substances is as follows:

Air	$n = 1.00$	Ethyl alcohol	$n = 1.36$
Glass	$n = 1.52$	Polystyrene	$n = 1.55$
Water	$n = 1.33$	Sodium chloride	$n = 1.54$
Diamond	$n = 2.42$	Acetone	$n = 1.36$

The angle of refraction θ_r with respect to the normal can be determined from the relation known as *Snell's law*:

Law of refraction: $n_1 \sin \theta_i = n_2 \sin \theta_r$

Solved Problem 20.4.   A ray of light strikes the surface of a swimming pool at an angle of 45° from the normal. Determine the angle of reflection.

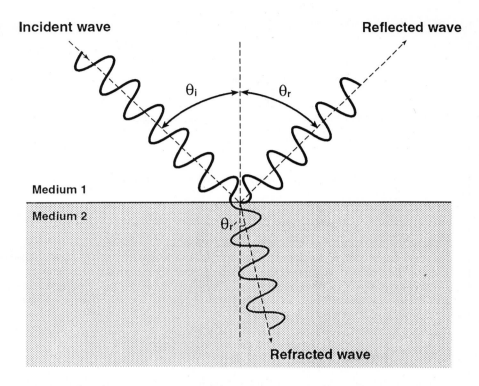

Fig. 20-2

Solution

From the law of reflection, the angle of incidence is equal to the angle of reflection. Thus, $\theta_i = \theta_r = 45°$ from the normal.

Solved Problem 20.5.   Calculate the speed of a light ray traveling through a sample of glass ($n = 1.52$).

Solution

By definition, the index of refraction is given by

$$n = \frac{c}{v}$$

$$v = \frac{c}{n} = \frac{3 \times 10^8 \text{ m s}^{-1}}{1.52} = 1.97 \times 10^8 \text{ m s}^{-1}$$

Solved Problem 20.6.   A light ray in air ($n = 1.00$) strikes the surface of a glass pane ($n = 1.52$) at an angle of incidence of 35°. Determine the angle of refraction.

Solution

From Snell's law,

$$n_1 \sin \theta_i = n_2 \sin \theta_{r'}$$
$$1.00 \sin 35° = 1.52 \sin \theta_{r'}$$
$$\sin \theta_{r'} = \frac{1.00 \sin 35°}{1.52} = 0.377$$
$$\theta_{r'} = \sin^{-1} 0.377 = 22.1°$$

Solved Problem 20.7.   Light rays propagating through air strike a rectangular block of ice 3.5 in thick at an angle of 50° from the surface. The light rays are then refracted at an angle of 25° with the normal. Determine the (1) index of refraction of ice and (2) velocity of the light rays in the ice.

Solution

(1)   The index of refraction of the block of ice can be determined from Snell's law:

$$n_{air} = \sin \theta_i = n_{ice} \sin \theta_{r'}$$

Since the angle of incidence is measured from the normal and the angle presented in the text of the problem is with respect to the surface, the required angle is $90° - 50° = 40°$.

$$1.00 \sin 40° = n_{ice} \sin 25°$$
$$n_{ice} = 1.52$$

(2)   The velocity of the light rays through the block of ice is

$$v = \frac{c}{n} = \frac{3 \times 10^8 \text{ m s}^{-1}}{1.52} = 1.97 \times 10^8 \text{ m s}^{-1}$$

Solved Problem 20.8.   For Solved Problem 20.7, determine the angle at which the light rays exit the block of ice.

Solution

We now want to apply Snell's law for light in the block of ice striking the interface between the ice and air. Since the light, upon incidence from air into the ice, was refracted at 25°, the angle of incidence for light at the interface between the ice and air is also 25°, since the geometry of a line bisecting two parallel lines yields identical angles. We can now reapply Snell's law to yield

$$n_{ice} \sin \theta_i = n_{air} \sin \theta_{r'}$$
$$1.52 \sin 25° = 1.0 \sin \theta_{r'}$$
$$\theta_{r'} = 39.9°$$

Solved Problem 20.9.   Image formation in the human eye occurs on the retina, as shown in Figure 20-3, by the refraction of light at three individual interfaces:

(1)   Surface of the cornea where light passes from air ($n = 1.00$) into the cornea (filled with aqueous humor) ($n = 1.33$)

(2)   Anterior surface of the lens where light passes from the aqueous humor ($n = 1.33$) into the lens ($n = 1.41$)

(3)   Posterior surface of the lens where light passes from the lens ($n = 1.41$) through the vitreous humor ($n = 1.33$) until it strikes the retina

For a light ray that impinges on the cornea at an angle of incidence of 35° from the normal, determine the angle at which the light ray strikes the retina.

Solution

This is a problem that requires repeated use of Snell's law. At the first interface,

$$n_{air} \sin \theta_i = n_{aq\,humor} \sin \theta_{r'}$$
$$1.00 \sin 35° = 1.33 \sin \theta_{r'}$$
$$\theta_{r'} = 25.5°$$

This angle now becomes the angle of incidence for the light rays at the second interface. At the second interface,

$$n_{aq\,humor} \sin \theta_i = n_{lens} \sin \theta_{r'}$$
$$1.33 \sin 25.5° = 1.41 \sin \theta_{r'}$$
$$\theta_{r'} = 23.9°$$

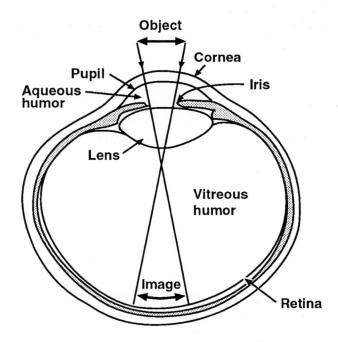

Fig. 20-3

This angle now becomes the angle of incidence for the light rays at the third interface. At the third interface,

$$n_{\text{lens}} \sin \theta_i = n_{\text{vi humor}} \sin \theta_{r'}$$
$$1.41 \sin 23.9° = 1.33 \sin \theta_{r'}$$
$$\theta_{r'} = 25.4°$$

This is the angle at which the light rays will strike the retina to form the image.

Solved Problem 20.10.   From Solved Problem 20.9, what is the speed of the light ray as it strikes the retina to form the image?

Solution

The medium through which the light propagates prior to striking the retina is the vitreous humor whose index of refraction n is 1.33. Thus, by definition of the index of refraction

$$n = \frac{c}{v}$$

The velocity of the light ray can be found by

$$v = \frac{c}{n} = \frac{3 \times 10^8 \text{ m s}^{-1}}{1.33} = 2.25 \times 10^8 \text{ m s}^{-1}$$

20.3   TOTAL INTERNAL REFLECTION

Total internal reflection is a phenomenon that occurs when light passes from a medium or material with a high index of refraction to a medium with a low index of refraction. For example, consider a light ray passing from glass ($n = 1.52$) to air ($n = 1.00$) at an angle of incidence θ_i, as shown in Figure 20-4. In this case, the

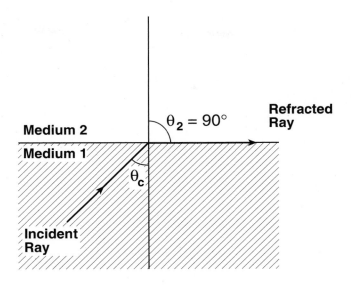

Fig. 20-4

angle of refraction is greater than the angle of incidence. In fact, as one increases the angle of incidence, the light ray has an angle of refraction of 90° and is subsequently refracted along the interface between the two media. The angle of incidence that causes light to be refracted at 90° from the normal is called the *critical angle* θ_c, given by

$$\sin \theta_c = \frac{n_2}{n_1}$$

ILLUSTRATIVE EXAMPLE 20.2. Endoscopy: Imaging inside the Body

Endoscopy is a technique used to obtain images of internal structures inside the human body. This technique is accomplished by advancing long, thin, and flexible bundles of optical fibers through the major vessels of the human body to the region or organ of interest. The tips of these cables are equipped with miniature light sources. The fiber-optic assembly, in essence, acts as a camera, propagating light rays reflected from the region of interest. The light ray from one point of the structure is channeled through one fiber, which is the case for every point and its corresponding fiber. A light source is used to eliminate the fiber-optic bundle which is emitted in all directions and subsequently reflected from surrounding structures. Each optical fiber consists of a core of index of refraction n_{core} encased in a cladding of index of refraction n_{clad} such that $n_{\text{core}} > n_{\text{clad}}$. When the reflected light strikes the surface of the fiber, as shown in Figure 20-5, total internal reflection occurs for angles of incidence greater than the critical angle θ_c, given by

$$\sin \theta_c = \frac{n_{\text{clad}}}{n_{\text{core}}}$$

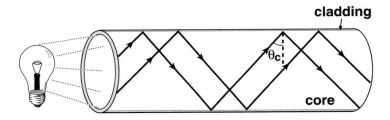

Fig. 20-5

where n_{clad} and n_{core} represent the index of refraction for the medium outside and inside, respectively, through which the light is traveling. The higher the value of n_{core} for a material, the smaller the critical angle and the greater the range of incident angles that undergo total internal reflection.

Solved Problem 20.11.   Assuming air as the external medium, determine the critical angle for total internal reflection for diamond ($n = 2.42$).

Solution

Using Snell's law, we have

$$n_{\text{diamond}} \sin \theta_c = n_{\text{air}} \sin 90°$$
$$\sin \theta_c = \frac{n_{\text{air}}}{n_{\text{diamond}}} = \frac{1.00}{2.42} = 0.413$$
$$\theta_c = \sin^{-1} 0.413 = 24.4°$$

Solved Problem 20.12.   Determine the critical angle for light rays passing from glass ($n = 1.52$) to water ($n = 1.33$).

Solution

$$\sin \theta_c = \frac{n_{\text{water}}}{n_{\text{glass}}} = \frac{1.33}{1.52} = 0.875$$
$$\theta_c = \sin^{-1} 0.875 = 61.0°$$

Solved Problem 20.13.   The inner core of a scintillating optical fiber is composed of polystyrene ($n_{\text{core}} = 1.60$) and is clad with poly methyl methacrylate ($n_{\text{clad}} = 1.48$). Determine the critical angle of the scintillating fiber.

Solution

$$\sin \theta_c = \frac{n_{\text{clad}}}{n_{\text{core}}} = \frac{1.48}{1.60} = 0.925$$
$$\theta_c = \sin^{-1} 0.925 = 67.6°$$

Supplementary Problems

20.1.   Microwaves oscillate at a frequency of 3×10^{10} Hz. Determine their wavelength.

Answer

1.0×10^{-2} m

20.2.   The wavelength of infrared waves is 10^{-4} m. Determine their frequency.

Answer

3×10^{12} Hz

20.3.   Radio station KTUV-AM broadcasts at a frequency of 1100 kHz while WXYZ-FM broadcasts at a frequency of 95 MHz. Which radio station will reach a car radio 3 mi away in the shortest time?

Solution

Both will reach at the same time.

20.4. Determine the time required for X-rays of frequency $f = 3 \times 10^{18}$ Hz to travel (1) 1 cm and (2) 1 m.

Solution

(1) 3.3×10^{-11} s; (2) 3.3×10^{-9} s

20.5. Determine the frequency of an electromagnetic wave with a wavelength of (1) 1.0 km, (2) 1.0 m, and (3) 1.0 mm.

Solution

(1) 3×10^5 Hz; (2) 3×10^8 Hz; (3) 3×10^{11} Hz

20.6. Light strikes the surface of a diamond ($n = 2.42$) at an angle of 45° from the normal. Determine the angle of reflection.

Solution

45°

20.7. For Supplementary Problem 20.6, determine the speed of light through the diamond.

Solution

1.24×10^8 m s^{-1}

20.8. Light rays are incident on a plastic object at an angle of 30° from the surface. Determine the angle of reflection.

Solution

60°

20.9. The incident angle of light rays on a plastic substance is 35° from the normal. If the angle of refraction within the substance is 22° from the normal, determine the index of refraction of the substance.

Solution

1.53

20.10. Light rays, originally traveling in air, strike a glass bottle filled with water at an incident angle of 20° from the normal. Given that the indices of refraction for glass and water are 1.52 and 1.33, respectively, determine the angle of refraction in the water.

Solution

14.9°

20.11. Light, originally traveling in air ($n = 1.00$), enters a pane of glass ($n = 1.52$) at the side of an aquarium and then passes into water ($n = 1.33$). If the angle of incidence of light at the air-glass interface is 25° with the normal, determine the angle at which the light wave travels (1) in the glass and (2) in the water.

Solution

(1) 15.5°; (2) 17.8°

20.12. Determine the critical angle for total internal reflection to occur in sodium chloride ($n = 1.54$), assuming air is the external medium.

Solution

40.5°

20.13. Repeat Supplementary Problem 20.12, assuming that water ($n = 1.33$) is the external medium.

Solution

59.7°

Chapter 21

Geometric Optics

In a continuation of the discussion on light in Chapter 20, we now become interested in using light for the formation of images. Light or, more generally, electromagnetic waves travel in a straight line unless an obstacle is placed in the path, causing the light rays to alter their path, resulting in convergence or divergence of the light rays, ultimately resulting in the formation of an image. Two specific types of obstacles or optical components which are integral in image formation are spherical reflecting surfaces (mirrors) and spherical refracting surfaces (thin lens), and these are discussed in detail in this chapter.

21.1 DEFINITIONS OF GEOMETRIC OPTICS

Geometric optics is a branch of physics concerned with the propagation of light utilized in the formation of images. In typical problems of geometric optics, a light source is allowed to impinge upon one of two basic types of optical components: mirrors and thin lenses. *Mirrors* are optical components that reflect light rays from a light source to form an image. Two types of mirror are plane mirrors and spherical mirrors. Spherical mirrors can further be subdivided into concave (curved-inward) mirrors and convex (curved-outward) mirrors. *Thin lenses* are optical components that refract light rays from a light source to form an image. Two general forms of spherical thin lenses are concave (curved inward) and convex (curved outward). Regardless of type, the optical component is positioned at an arbitrary point along an optical axis.

Several terms describing parameters critical to such problems in geometric optics include these:

Object distance o is the distance between the object and the optical component.

Image distance i is the distance between the image and the optical component.

Focal length f is the distance between the image and the optical component observed when the source is imaged from an infinite distance. The point at which the incoming light rays converge to form the image is known as the *focal point F*.

Radius of curvature R is the radius of a circle which most closely approximates the curvature of the spherical optical component. The point representing the center of this circle is termed the *center of curvature C*.

A *real image* is formed by the convergence of light rays and is characterized by a *positive* image distance *i*. A *virtual image* is formed by the divergence of light rays and is characterized by a *negative* image distance *i*.

Magnification m of an optical component (thin lens or mirror) is defined as

$$\text{Magnification } m = -\frac{\text{image distance } i}{\text{object distance } o} = \frac{\text{image size } h_i}{\text{object size } h_o}$$

A *negative* magnification value indicates that the image is inverted, while a *positive* magnification value indicates that the image is upright or erect.

In solving problems in geometric optics, presented with an optical component and an object (light source) positioned a given distance from the component, we would like to know the image distance from the component where the image is formed and the type of image formed. This objective can be accomplished utilizing two techniques, analytical and graphical. The analytical technique consists of a mathematical relation between the image distance *i*, object distance *o*, focal length *f*, and radius of curvature *r*. The graphical technique involves ray tracing for the location of the image formed by the optical component. Ray tracing graphically follows the path of three principal light rays emitted from the object and interacting with the optical component.

21.2 MIRRORS

Plane Mirrors

Parallel light rays emanating from an object toward a plane mirror remain parallel after reflection. The images created by plane mirrors are always virtual and erect; that is, the image appears to be positioned the same distance from the mirror as the distance from the object to the mirror, as shown in Figure 21-1.

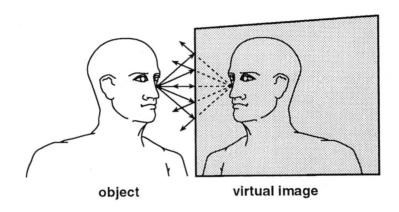

object   **virtual image**

Fig. 21-1

Spherical Mirrors

Spherical mirrors are mirrors that are curved and exist primarily in two forms, dependent on their curvature. Concave mirrors are curved inward, and convex mirrors are curved outward. The spherical mirrors are represented by their radius of curvature R, which is related to their focal length f by

Concave mirror: $R = +2f$
Convex mirror: $R = -2f$

In Figure 21-2a, parallel light rays emanating from an object toward a concave mirror reflect and ultimately converge at the real focal point F. Similarly, in Figure 21-2b parallel light rays emanating from an object toward a convex mirror reflect and ultimately diverge, such that the reflected rays appear to converge at a virtual focal point F' behind the mirror.

Mirror equation represents the analytical approach in the determination of the image type and location. The mirror equation relates the object distance o, the image distance i, and the focal length f by

$$\frac{1}{o} + \frac{1}{i} = \frac{1}{f}$$

where o, i, and f are expressed in units of length. The mirror equation and associated parameters are displayed in Figure 21-2c and d.

Ray tracing represents a graphical approach by which the image position and image size formed by a mirror can be determined. Ray tracing follows the path of three different principal light rays emanating from an object (typically represented by an arrow), interacting with the mirror, and either converging or diverging at the real or virtual focal point, depending on whether the lens is convex or concave, as shown in Figure 21-2e and f.

- One principal light ray (ray 1) travels parallel to the optical axis of the mirror, reflects off the mirror, and passes through the real focal point F of a concave mirror or appears to originate from the virtual focal point F' of a convex mirror.

Spherical Mirrors

General Properties

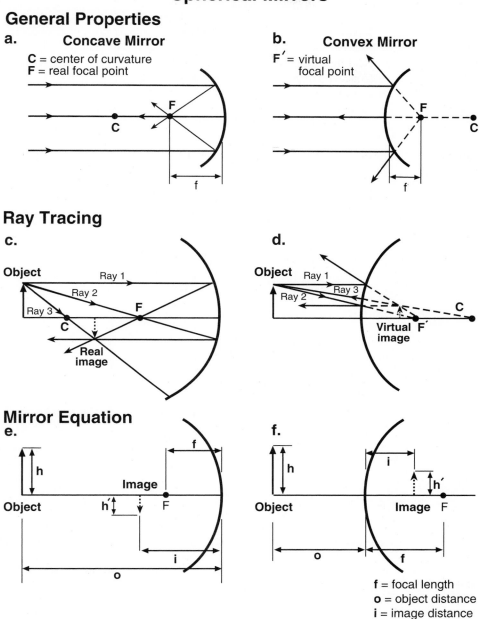

a.   Concave Mirror

C = center of curvature
F = real focal point

b.   Convex Mirror

F' = virtual
focal point

Ray Tracing

c.

Object   Ray 1
Ray 2
Ray 3
C   F
Real
image

d.

Object   Ray 1
Ray 3
Ray 2
Virtual **F'**   C
image

Mirror Equation

e.

f
h
Image
Object   h'   F
i
o

f.

i
h
h'
Object   Image   F
o   f

f = focal length
o = object distance
i = image distance

Fig. 21-2

- One principal light ray (ray 2) travels along the optical axis, directed toward the real focal point F and reflected in a direction parallel to the mirror axis.

- One principal light ray (ray 3) travels along a radius of the mirror toward the center of curvature C, striking the mirror perpendicular to its surface. After reflection, the reflected ray travels back along its original path.

Sign Conventions for Spherical Mirrors

Parameter	Positive	Negative
Object distance o	Real object	Virtual object
Image distance i	Real image	Virtual image
Focal length f	Concave mirror	Convex mirror
Radius of curvature R	Concave mirror	Convex mirror
Magnification m	Erect image	Inverted image

Solved Problem 21.1.   Express the mirror equation in terms of the radius of curvature.

   Solution

   The focal length f is related to the radius of curvature R by

$$f = \pm \frac{R}{2}$$

   Substituting this expression for f into the mirror equation yields

$$\frac{1}{o} + \frac{1}{i} = \frac{1}{R}$$

Solved Problem 21.2.   From the mirror equation derived in Solved Problem 21.1, derive an expression for (1) object distance o, (2) image distance i, and (3) radius of curvature R.

   Solution

   The mirror equation is written as

$$\frac{1}{o} + \frac{1}{i} = \frac{2}{R}$$

   (1)   Solving for o yields

$$o = \frac{iR}{2i - R}$$

   (2)   Solving for i yields

$$i = \frac{oR}{2o - R}$$

   (3)   Solving for R yields

$$R = \frac{2oi}{i + o}$$

Solved Problem 21.3.   An object is placed in front of a spherical mirror of radius 50 cm. Determine the image distance if the object distance is (1) 10 cm and (2) 100 cm.

   Solution

   The image distance is related to the radius of curvature by

$$\frac{1}{o} + \frac{1}{i} = \frac{2}{r}$$

(1) $$\frac{1}{10\text{ cm}} + \frac{1}{i} = \frac{2}{50\text{ cm}}$$
$$i = -16.6\text{ cm}$$

(2) $$\frac{1}{100\text{ cm}} + \frac{1}{i} = \frac{2}{50\text{ cm}}$$
$$i = 33.3\text{ cm}$$

Solved Problem 21.4.   For the mirrors in Solved Problem 21.3, determine the magnification.

> **Solution**
>
> Magnification m is defined as
>
> $$m = -\frac{i}{o}$$
>
> (1) $m = -\dfrac{-16.6\text{ cm}}{10\text{ cm}} = 1.7$
>
> (2) $m = -\dfrac{33.3\text{ cm}}{10\text{ cm}} = -3.3$

Solved Problem 21.5.   An object is placed 0.05 m in front of a convex mirror with a radius of curvature of 0.15 m. Determine the (1) image distance and (2) magnification.

> **Solution**
>
> (1)   The image distance can be determined from
>
> $$\frac{1}{o} + \frac{1}{i} = \frac{2}{r}$$
> $$\frac{1}{5\text{ cm}} + \frac{1}{i} = \frac{2}{15\text{ cm}}$$
> $$i = -15\text{ cm}$$
>
> The minus sign implies that the image is behind the mirror.
>
> (2)   The magnification is given by
>
> $$m = -\frac{i}{o} = -\frac{-15\text{ cm}}{5\text{ cm}} = 3.0$$

Solved Problem 21.6.   The focal length of a concave mirror is 15 cm. For an object placed 20 cm in front of the mirror, determine the (1) image distance and (2) magnification.

> **Solution**
>
> (1)   The image distance is related to the focal length by
>
> $$\frac{1}{o} + \frac{1}{i} = \frac{1}{f}$$
> $$\frac{1}{20\text{ cm}} + \frac{1}{i} = \frac{1}{15\text{ cm}}$$
> $$i = 60\text{ cm}$$
>
> (2)   The magnification is given by
>
> $$m = -\frac{i}{o} = -\frac{-60\text{ cm}}{20\text{ cm}} = 3.0$$

Solved Problem 21.7. While trying on hats, Kelly stands in front of a concave mirror with a focal length of 45 cm. How far should Kelly be from the mirror in order to produce an image magnified twofold?

Solution

From the definition of magnification

$$m = -\frac{i}{o} = 2$$

$$i = -2o$$

Substituting into the mirror equation gives

$$\frac{1}{i} + \frac{1}{o} = \frac{1}{f}$$

$$\frac{1}{-2o} + \frac{1}{o} = \frac{1}{45 \text{ cm}}$$

$$\frac{2-1}{2o} = \frac{1}{45 \text{ cm}}$$

Solving for o yields $o = 22.5$ cm.

21.3 THIN LENSES

Thin lenses, like mirrors, exist as convex or concave lenses. However, lenses form images by refracting light rays, as opposed to mirrors that form images by reflecting light rays. Thin lenses have two surfaces and are thus characterized by two radii of curvature R_1 (side of lens closest to the object) and R_2 (side of lens opposite to object). A convex lens has a positive R_1 and a negative R_2 while a concave lens has a negative R_1 and a positive R_2. For a convex or converging lens (Figure 21-3a), all light rays emanating from an object refract through the lens and converge at the real focal point F on the side of the lens opposing the object. For a concave or diverging lens (Figure 21-3b), all light rays emanating from an object refract through the lens and diverge but generate an image at the virtual focal point F' on the same side of the lens as the object.

The *lens equation* relates the object distance o, image distance i, and the focal length f of a lens according to

$$\frac{1}{f} = \frac{1}{i} + \frac{1}{o}$$

The lens equation and associated parameters are illustrated in Figure 21-3c and d.

The *lens maker's equation* can be expressed as

$$\frac{1}{f} = (n-1)\left(\frac{1}{R_1} + \frac{1}{R_2}\right)$$

where f is the focal length, n is the index of refraction for the lens material, R_1 is the radius of the lens closest to the object, and R_2 is the radius of the lens farthest from object. If the lens, characterized by an index of refraction n_1, is placed within a medium with an index of refraction n_2, then n in the lens maker's equation is replaced by n_1/n_2.

Power P of a lens is defined by

$$P = \frac{1}{f}$$

where P is expressed in units of diopters or inverse length (L^{-1}). The sign convention for P is similar to that for f: *positive* for a converging lens and *negative* for a diverging lens.

Ray tracing can also be used to determine image position and image size formed by a thin lens. As is the

Thin Lenses

General Properties

a.
Converging Lens
F = real focal point

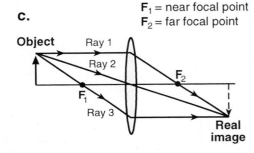

focal length

b.
Diverging Lens
F′ = virtual focal point

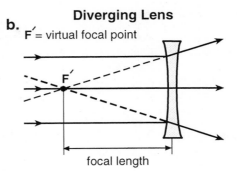

focal length

Ray Tracing

F_1 = near focal point
F_2 = far focal point

c.

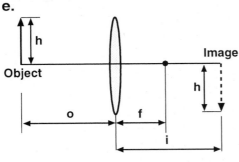

d.

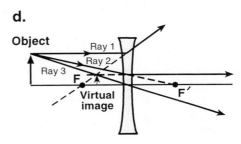

Lens Equation

e.

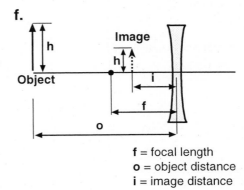

f.

f = focal length
o = object distance
i = image distance

Fig. 21-3

case for mirrors, three particular principal rays of light are followed upon emission from an object moving toward and ultimately interacting with the lens, as shown in Figure 21-3*e* and *f*.

- One principal light ray (ray 1) travels parallel to the optical axis of the lens, refracts through the lens, and passes through a second real focal point F_2 for a convex lens, or appears to originate from a second virtual focal point F_2' for a concave lens.

- One principal light ray (ray 2) travels along the optical axis, directed toward the center of the lens, and reflects through the lens in a direction parallel to the lens axis.

- One principal light ray (ray 3) travels through the first real focal point F of the lens and emerges from the lens parallel to the optical axis.

Sign Conventions for Spherical Lenses

Parameter	Positive	Negative
Object distance o	Real object	Virtual object
Image distance i	Real image	Virtual image
Focal length f	Converging lens	Diverging lens
Magnification m	Erect image	Inverted image

Solved Problem 21.8.   An object is placed 15 cm in front of a concave lens with a focal length of -20 cm. Determine the image type and distance.

Solution

The image distance can be determined from

$$\frac{1}{15\text{ cm}} + \frac{1}{i} = \frac{1}{-20\text{ cm}}$$
$$i = -8.55\text{ cm}$$

The negative value of the image distance indicates that the image is virtual. The image orientation, that is, inverted or erect, can be determined from the magnification:

$$m = -\frac{i}{o} = -\frac{-(-8.55\text{ cm})}{15\text{ cm}} = 0.57$$

A positive value for the magnification indicates that the image is erect.

Solved Problem 21.9.   An object is placed 25 in from a convex lens with a focal length of 8 in. Determine the distance of the image from the lens.

Solution

Using the lens equation, we have

$$\frac{1}{o} + \frac{1}{i} = \frac{1}{f}$$
$$\frac{1}{i} = \frac{1}{f} - \frac{1}{o} = \frac{1}{8\text{ in}} - \frac{1}{25\text{ in}} = 0.085\text{ in}^{-1}$$
$$i = 11.8\text{ in}$$

Solved Problem 21.10.   An object of height 5 cm is positioned 24 cm in front of a convex lens with a focal length of 15 cm. Determine the location and height of the produced image, using the (1) ray-tracing technique and (2) analytical technique.

Solution

(1) From Figure 21-4, three principal rays emanating from the object O are followed through the convex lens, ultimately forming the real image.

- Ray 1 strikes the lens toward the top and refracts through it. Following refraction, ray 1 passes through the principal focal point F.

- Ray 2 passes through the optical center of the convex lens and continues until it converges with ray 1.

- Ray 3 strikes the lens toward the bottom and refracts through it. Following refraction, ray 3 continues parallel to the optical axis until it converges with rays 1 and 2.

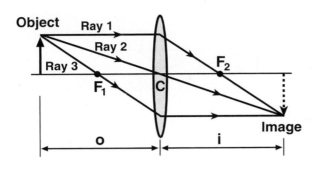

Fig. 21-4

The point where all three rays converge represents the size and location of the image. From the ray tracing in Figure 21-4, it can be concluded that the image is real, inverted, magnified, and located farther from the lens than the object.

(2) The analytical technique uses applicable equations to determine the desired information. Using the lens equation, we can find the image distance:

$$\frac{1}{i} = \frac{1}{f} - \frac{1}{o} = \frac{1}{15\,\text{cm}} - \frac{1}{24\,\text{cm}} = 0.025\,\text{cm}^{-1}$$
$$i = 40\,\text{cm}$$

A positive value of i indicates the image is real and on the side of the lens opposite the object.

The height of the image h_i can be determined from the magnification m:

$$m = -\frac{i}{o} = -\frac{40\,\text{cm}}{24\,\text{cm}} = -1.67$$

The magnification m is also related to the image height by

$$m = \frac{h_i}{h_o}$$
$$h_i = mh_o = (-1.67)(5\,\text{cm}) = -8.35\,\text{cm}$$

The negative value of h_i indicates that the image is inverted.

Solved Problem 21.11. An object 8 cm in height is positioned 24 cm in front of a concave lens with a focal length of -16 cm. Determine the location and size of the image, using the (1) ray-tracing technique and (2) analytical technique.

Solution

(1) From Figure 21-5, three principal rays emanating from object O are followed through the concave lens, ultimately forming the virtual image.

- Ray 1 strikes the lens toward the top and refracts through it. Ray 1 refracts outward but appears to have passed through the principal focal point F.
- Ray 2 passes through the optical center of the convex lens, refracts outward, and continues on.
- Ray 3 strikes the lens toward the bottom and refracts through it. Ray 3 refracts outward but appears to have passed through the principal focal point F.

The principal focal point F where all three rays converge prior to the outward refraction of the rays by the diverging lens represents the size and location of the image. From the ray tracing in Figure 21-5, it can be concluded that the image is virtual, erect, and minified.

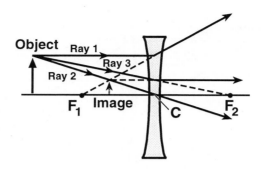

Fig. 21-5

(2) The analytical technique uses applicable equations to determine the desired information. Using the lens equation, we can find the image distance:

$$\frac{1}{i} = \frac{1}{f} - \frac{1}{o} = \frac{1}{-16 \, \text{cm}} - \frac{1}{24 \, \text{cm}} = -0.1045 \, \text{cm}^{-1}$$
$$i = -9.6 \, \text{cm}$$

A negative value of i indicates the image is virtual and on the same side of the lens as the object.

The height of the image h_i can be determined from the magnification m:

$$m = -\frac{i}{o} = -\frac{-9.6 \, \text{cm}}{24 \, \text{cm}} = 0.40$$

The magnification m is also related to the image height by

$$m = \frac{h_i}{h_o}$$
$$h_i = mh_o = (0.40)(8 \, \text{cm}) = 3.2 \, \text{cm}$$

The positive value of h_i indicates that the image is inverted.

ILLUSTRATIVE EXAMPLE 21.1. The Physics of Vision and Common Visual Defects: Myopia and Hyperopia

The human eye is a remarkable visual instrument capable of visualizing objects from as far away as infinity to as close as 20 cm in proximity into focus. The human eye can resolve particles of matter approximately 0.1 mm ($<\frac{1}{250}$ in). The ability to focus occurs as the result of ciliary muscles that control the curvature of the cornea.

Myopia, or nearsightedness, and hyperopia, or farsightedness, are the most common visual defects, both of which can be readily corrected with the proper choice of spectacle or contact lenses. In myopia, focused vision is achieved for objects that are near while vision becomes blurred for distance objects. The exact opposite is true in hyperopia, where distant objects can be seen clearly and close objects appear blurry.

The nature of these defects involves the point of convergence for parallel rays of light from the object. In the normal eye, parallel rays from an object enter through the lens and converge on the retina where the image is formed. A myopic person can easily see close objects clearly, but vision becomes blurry for distant objects. The myopic eye focuses light in front of the retina due in part to a lens with too short a focal length. The shortened focal length occurs as a result of either an elongated eyeball or excessive curvature of the cornea. The objective in treating myopia is to reduce the converging power of the lens, which can be accomplished with the placement of a diverging lens in front of the eye. A hyperopic person can easily see distant objects clearly, but vision becomes blurry for close objects. The hyperopic eye focuses light behind the retina due in part to inadequate curvature or too flat a lens or too short an eyeball. The objective in treating hyperopia is to increase the converging power of the lens, which can be accomplished with placement of a converging lens in front of the eye. These common vision defects and their related cures are illustrated in Figure 21-6.

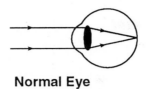

Normal Eye

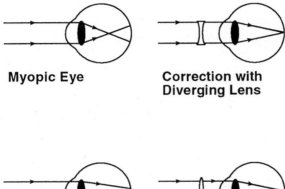

Myopic Eye   **Correction with
                                  Diverging Lens**

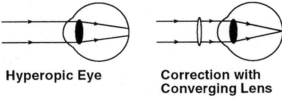

Hyperopic Eye   **Correction with
                                  Converging Lens**

Fig. 21-6

ILLUSTRATIVE EXAMPLE 21.2.   Radial Keratotomy

An alternative to eyeglasses in the treatment of myopia or nearsightedness is a surgical procedure termed *radial keratotomy* (RK). As stated previously, myopia is caused by an elongated eyeball or a lens that is too convex. RK involves 8 to 16 hairline incisions, 0.001 mm in depth, performed by a laser in the curved surface of the cornea. The objective of RK is to flatten the cornea, thereby reducing the converging power of the eye to focus distant objects on the retina, as opposed to before it. Therefore, instead of placing a lens in front of the eye to compensate for the abnormal curvature of the eyeball in hyperopia, this surgical procedure reduces the convexity of the eyeball to correct the myopic condition.

Solved Problem 21.12.   In a visit to her optometrist, Angela, who is nearsighted, is being fitted for glasses. During the examination, she discovers that the farthest distance of an object that she can properly focus on is 2 m from her eye. If this distance for a normal eye is infinity, determine the focal length and power of the diverging lens needed for her glasses.

Solution

To adjust the glasses such that normal vision is attained, the object distance is infinity, and the desired location for the image from the diverging lens is -200 cm. The focal length f of the diverging lens for Angela's eyeglass prescription can be determined by

$$\frac{1}{f} = \frac{1}{o} + \frac{1}{i} = \frac{1}{\infty} + \frac{1}{-200 \text{ cm}}$$

$$f = -200 \text{ cm}$$

The power of the diverging lens is

$$P = \frac{1}{f} = \frac{1}{-2.0 \text{ m}} = -0.5 \text{ diopter}$$

When light passes from a medium of index n_1 to a medium of index n_2 at a spherical refracting surface, the equation relating object distance o, image distance i, and radius of curvature R is

$$\frac{n_1}{o} + \frac{n_2}{i} = \frac{n_2 - n_1}{R}$$

Solved Problem 21.13.   Angela's cousin, Catherine, who is farsighted, is being fitted for glasses by her optometrist. The closest distance of an object that she can properly focus on is 150 cm from her eye. If this distance for a normal eye is 25 cm, determine the focal length and power of the converging lens needed for her glasses.

Solution

In this problem, the object distance o is 25 cm, and the image distance i is -150 cm. Using the lens equation, we have

$$\frac{1}{f} = \frac{1}{o} + \frac{1}{i} = \frac{1}{25 \text{ cm}} + \frac{1}{-150 \text{ cm}}$$
$$f = -30 \text{ cm}$$

The power P of the converging lens is then

$$P = \frac{1}{f} = \frac{1}{0.30 \text{ m}} = 3.33 \text{ diopters}$$

Lenses in Combination

For a series of lenses in contact, the equivalent focal length f_{eq} is given by

$$\frac{1}{f_{\text{eq}}} = \frac{1}{f_1} + \frac{1}{f_2} + \frac{1}{f_3} + \cdots + \frac{1}{f_n}$$
$$P_{\text{eq}} = P_1 + P_2 + P_3 + \cdots + P_n$$

An example of an optical instrument with lenses in contact is the eye.

For a series of lenses not in contact, the image created by one lens is used as the object for the next lens in succession and continues similarly for the number of lenses in the system. Thus, the image formed by the system is the image generated by the last lens. The magnification of the optical system m_t is equal to the product of the magnification from each optical component comprising the optical system, or $m_t = m_1 m_2 m_3 \cdots m_n$.

Examples of optical instruments with lenses not in contact are the microscope and telescope.

ILLUSTRATIVE EXAMPLE 21.3.   Compound Microscope

The compound microscope, shown in Figure 21-7, uses essentially two lenses, an objective lens and an eyepiece lens, separated by an optical tube of length 16.0 cm, to produce a magnified image of an object. The object of interest is placed beyond the first focal point F_1 of the objective lens. The objective lens produces an intermediate image I_1 that is real, inverted, and magnified. This intermediate image serves as the object for the eyepiece lens, which further magnifies the image I_1, to form the final virtual image I_2 seen by the observer.

Solved Problem 21.14.   A compound microscope has an objective lens with a focal length of 0.58 cm and an eyepiece with a focal length of 10.0 cm. The two lenses are separated by 25.0 cm. If an object 0.025 cm in size is placed 0.60 cm from the objective, determine the (1) location, (2) magnification, and (3) size of the image.

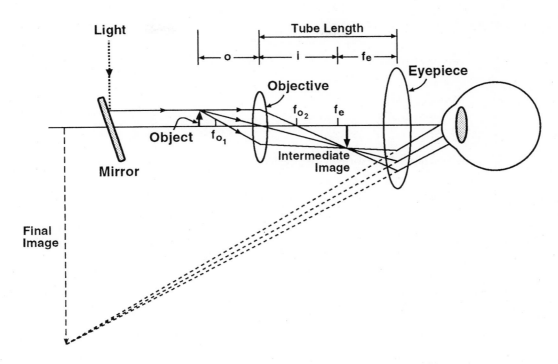

Fig. 21-7

Solution

(1)   The image distance i can be determined from the lens equation:

$$\frac{1}{i} = \frac{1}{f} - \frac{1}{o} = \frac{1}{0.58\text{ cm}} - \frac{1}{0.60\text{ cm}} = 0.06\text{ cm}^{-1}$$

$$i = 16.7\text{ cm}$$

(2)   The magnification m is given by

$$m = -\frac{i}{o} = -\frac{16.7\text{ cm}}{0.60\text{ cm}} = -27.8$$

(3)   The size of the image h_i is

$$m = \frac{h_i}{h_o}$$

$$h_i = mh_o = (-27.8)(0.025\text{ cm}) = -0.695\text{ cm}$$

Solved Problem 21.15.   From the microscope in Solved Problem 21.14, determine the (1) location, (2) magnification, and (3) size of the image generated by the eyepiece lens, using the image produced by the objective lens as the object.

Solution

The objective lens and the eyepiece are separated by 25.0 cm. The focal length of the eyepiece lens is 10.0 cm. The location of the image produced by the objective lens is 16.7 cm, or 25.0 cm − 16.7 cm = 8.3 cm from the eyepiece lens. Thus, the object distance for the eyepiece lens is 8.3 cm.

(1) The image distance for the eyepiece lens is

$$\frac{1}{i} = \frac{1}{f} - \frac{1}{o} = \frac{1}{10.0\,\text{cm}} - \frac{1}{8.3\,\text{cm}} = -0.02\,\text{cm}^{-1}$$
$$i = -50\,\text{cm}$$

(2) The magnification m is defined by

$$m = -\frac{i}{o} = -\frac{-50\,\text{cm}}{8.3\,\text{cm}} = 6.02$$

(3) The size of the final image h_i is related to the magnification m by

$$h_i = mh_o = (6.02)(-0.695\,\text{cm}) = -4.18\,\text{cm}$$

Supplementary Problems

21.1. An object is placed 1.3 m from a concave mirror whose focal length is 0.5 m. Determine the image distance from the mirror.

Solution

0.81 m

21.2. An object is placed 18 cm from a convex mirror whose focal length is 9 cm. Determine the image distance from the mirror.

Solution

−6.1 cm

21.3. An object placed in front of a concave mirror with a radius of curvature of 20 cm produces a virtual image 20 cm behind the mirror. Where was the object placed in front of the mirror?

Solution

6.7 cm

21.4. An object placed 3.0 in in front of a convex mirror generates a virtual image −12.0 in from the mirror. Determine the focal length of the convex mirror.

Solution

4 in

21.5. From Supplementary Problem 21.4, determine the radius of curvature of the convex mirror.

Solution

−8 in

21.6. An object is positioned 10 cm from a concave mirror with a radius of curvature of 25 cm. Determine the position and magnification of the image.

Solution

−50 cm; 5X

21.7. Fran uses a concave mirror with a focal length of 25 in to apply her makeup. Determine the distance she should sit in front of the mirror to yield a 1.5-fold magnification.

Solution

8.3 in

21.8. An object is placed 5 cm in front of a diverging lens with a focal length of -10 cm. Determine the image type and distance.

Solution

-3.33 cm; virtual and erect

21.9. An object is placed 23 in from a convex lens with a focal length of 15 in. Determine the distance of the image from the lens.

Solution

41.7 in

21.10. An object is placed at the focal point of a concave mirror. Describe the formed image.

Solution

No image is produced.

21.11. Upon optometric examination, a person with myopia (or nearsightedness) was unable to focus on objects beyond 90 cm from the eye. Determine the power of the diverging lens required to correct the person's vision to normal standards.

Solution

-1.11 diopters

21.12. A person with hyperopia (or farsightedness) was found to be unable to clearly focus on objects closer than 80 cm from the eye. Determine the power of the converging lens required to allow the person to see the same object at a distance of 25 cm.

Solution

2.75 diopters

21.13. A compound microscope consists of an objective lens with a focal length of 10 mm and an eyepiece lens with a focal length of 25 mm. Determine the distance between the lenses when it is completely focused on a sample 10.5 mm from the objective lens.

Solution

23.3 cm

21.14. For the compound microscope described in Supplementary Problem 21.13, determine the magnification by the (1) objective lens, (2) eyepiece lens, and (3) entire microscope.

Solution

(1) 20.0; (2) -11.0; (3) -220

Chapter 22

Nuclear Physics and Radioactivity

The structure of an atom and the physics of the electron were encountered in Chapter 16 in the discussion of electricity. In this chapter, we focus primarily not on the electrons, but on the particles comprising the nucleus, that is, protons and neutrons. These particles dictate the energy and stability of the atom and whether the atom is radioactive.

22.1   DEFINITIONS OF NUCLEAR PHYSICS AND RADIOACTIVITY

Atomic Structure

The smallest building block or unit of all matter is the atom. All atoms consist of three basic particles: proton, neutron, and electron. The atom can be further subdivided into two distinct entities: (1) a positively charged nucleus or densely packed core of protons and neutrons situated at the center of the atom and (2) negatively charged electrons in continual orbit about the nucleus.

Atomic Nucleus

The nucleus is a compact, positively charged core composed of protons and neutrons, collectively known as *nucleons*, and it accounts for the majority of the atomic mass. The proton charge is identical in magnitude to that of the electron but is positive while the electron is negatively charged. The neutron is neutrally charged (uncharged) and is slightly heavier than the proton. The mass and charge of the proton and neutron are as follows:

	Mass	Charge
Proton	1.67×10^{-27} kg	1.6×10^{-19} C
Neutron	1.68×10^{-27} kg	0

Several terms used to describe the identity and structure of an atom are:

Mass number A (number of protons and neutrons)

Atomic number Z (number of protons)

Neutron number $N = A - Z$ (number of neutrons)

Also note that Z is equal to the number of electrons. The number of protons must always equal the number of electrons so that charge is conserved. In symbolic form, the atom can be represented by

$$_Z^A X$$

where X symbolizes the chemical element. For example, hydrogen, which has one proton and one electron, can be expressed as

$$_1^1 \text{H}$$

Solved Problem 22.1.   Helium has two protons and two neutrons. What is its chemical symbol?

Solution

Since helium has two protons and two neutrons, the atomic mass A, which is the sum of protons and neutrons, is 4. The atomic number Z is the number of protons, or 2. Thus, the chemical symbol is

$$_2^4 \text{He}$$

Solved Problem 22.2. Uranium has a mass number of 235 and an atomic number of 92. Determine the (1) number of protons, (2) number of neutrons, and (3) number of electrons.

Solution

(1) Uranium has an atomic number $Z = 92$, which indicates the number of protons. Thus, uranium has 92 protons.

(2) The number of neutrons is the mass number A minus the atomic number Z. Thus, for uranium, the number of neutrons is

$$N = A - Z = 235 - 92 = 143 \text{ neutrons}$$

(3) Since the normal atom is electrically neutral, the atom has Z electrons. Thus, uranium has 92 electrons.

Solved Problem 22.3. From Solved Problem 22.2, what is the chemical symbol of uranium?

Solution

The symbol for a chemical element is

$$^A_Z X$$

Thus, for uranium, the symbol is

$$^{235}_{92} U$$

Solved Problem 22.4. A certain element has 10 neutrons and 9 electrons. Determine the (1) atomic number Z and (2) mass number A.

Solution

(1) The atomic number Z is equal to the number of protons, which is equal to the number of electrons. Thus, $Z = 9$.

(2) The mass number A is equal to the number of neutrons and protons. The element has nine electrons, which is the number of protons. Thus,

$$A = 10 \text{ neutrons} + 9 \text{ protons} = 19$$

22.2 ISOTOPES

Isotopes represent the class of nuclei with the same number of protons (Z) but different numbers of nucleons (A). They are usually stable and do not decay into different nuclei. As an example, the nuclides ^1_1H, ^2_1H (deuterium), and ^3_1H (tritium) are all isotopes of hydrogen; and $^{10}_6C$, $^{11}_6C$, $^{12}_6C$, $^{13}_6C$, $^{14}_6C$, and $^{15}_6C$ are all isotopes of carbon.

Solved Problem 22.5. An isotope of argon has 22 neutrons and 18 electrons. What is the chemical element of this isotope?

Solution

We must determine the mass number A and the atomic number Z. Keeping in mind that the number of electrons is equal to the number of protons,

$$A = 22 \text{ neutrons} + 18 \text{ protons} = 40$$
$$Z = 18 \text{ electrons} = 18$$

Thus, the element can be written as

$$^{40}_{18} Ar$$

22.3 QUANTUM THEORY OF RADIATION

In Chapter 19, light (which is equivalent to radiation) was discussed in terms of electromagnetic waves propagating through space at a speed of $c = 3.0 \times 10^8$ m s^{-1}. However, certain aspects of light can be explained by representing light as individual packets or quanta of energy, known as *photons*. The photon has no mass but does possess energy and momentum. The energy of a photon is defined as

$$E = h\nu = h\frac{c}{\lambda}$$

where ν is the photon frequency, λ is the photon wavelength, c is the speed of light ($= 3 \times 10^8$ m s^{-1}), and h is Planck's constant ($= 6.63 \times 10^{-34}$ J s $= 4.136 \times 10^{-15}$ eV s). Another related factor of convenience is $hc = 12.4$ keV Å.

Solved Problem 22.6. Given a 140-keV X-ray, determine its (1) frequency and (2) wavelength.

Solution

The energy of the X-ray is related to the frequency and wavelength by

$$E = h\nu = h\frac{c}{\lambda}$$

where h = Planck's constant $= 4.15 \times 10^{-15}$ eV s.

(1) Rearranging the equation for the energy gives

$$\nu = \frac{E}{h} = \frac{140 \times 10^3 \text{ eV}}{4.15 \times 10^{-15} \text{ eV s}} = 3.37 \times 10^{19} \text{ Hz}$$

(2) The wavelength can be determined by rearranging the equation for energy:

$$\lambda = \frac{hc}{E} = \frac{(4.15 \times 10^{-15} \text{ eV s})(3 \times 10^8 \text{ m s}^{-1})}{140 \times 10^3 \text{ eV}} = 8.9 \times 10^{-12} \text{ m}$$

Solved Problem 22.7. Given a photon whose frequency is 1.7×10^{19} Hz, determine its (1) wavelength and (2) energy.

Solution

(1) $\lambda = \dfrac{c}{\nu} = \dfrac{3 \times 10^8 \text{ m s}^{-1}}{1.7 \times 10^{19} \text{ s}^{-1}} = 1.76 \times 10^{-11} \text{ m}$

(2) $E = h\nu = (4.15 \times 10^{-15} \text{ eV s})(1.7 \times 10^{19} \text{ Hz}) = 7.1 \times 10^4 \text{ eV}$

22.4 RADIOACTIVITY

Radioactive decay is a nuclear phenomenon exhibited by radioisotopes or elements with an atomic number Z greater than that of lead ($Z = 82$). In these elements which contain generally more Z than A, the repulsive electric forces in the nucleus become greater than the attractive nuclear forces, making the nuclei unstable. The radioactive element, in nature, strives toward a stabilized state of existence and, in the process, spontaneously emits particles (photons and charged and uncharged particles) in the transformation to a different nucleus and hence a different element. This process is referred to as *radioactive decay* and is dependent on the amount and identity of the radioactive element.

Given a radioactive element originally with N_o number of atoms, the number of atoms present at any time t is

$$N(t) = N_o e^{-\lambda t}$$

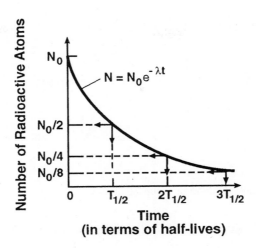

Fig. 22-1

where λ is a decay constant, defined by

$$\lambda = \frac{0.693}{T_{1/2}}$$

Here $T_{1/2}$ is the half-life of the radioactive element and represents the time required for one-half of the radioactive atoms to remain unchanged. Half-lives for radioactive elements range from fractions of seconds (e.g., polonium, $T_{1/2} = 3 \times 10^{-7}$ s) to millions of years (e.g., thorium, $T_{1/2} = 1.39 \times 10^{10}$ yr). Radioactive decay is an exponential curve and is illustrated in Figure 22-1.

An alternative method of expressing radioactivity is by the activity or disintegration rate

$$A = \lambda N$$

which is expressed in units of curies (Ci), where

$$1 \text{ Ci} = 3.7 \times 10^{10} \text{ disintegrations per second (dps)}$$

Natural radioactivity in the human body is 10 nCi, nuclear medicine scanning techniques use 10 μCi to 10 mCi, and γ ray sources for radiotherapy are 1000 Ci.

Solved Problem 22.8.   Derive the equation for radioactive decay.

Solution

Assume we have a sample of a radioactive element where the following variables are identified: N, number of radioactive atoms in the sample; λ, decay constant or constant describing the rate of decay unique to the radioactive element. Given this, the rate of change of the number of radioactive atoms (dN/dt) is equal to the rate of decay or decay constant multiplied by the number of radioactive atoms, as shown in Figure 22-1, written in equation form as

$$\frac{dN}{dt} = -\lambda N$$

This equation can be solved by first collecting like terms on the same side of the equation:

$$\frac{dN}{N} = -\lambda t$$

Integrating the above equation over the limits of N and N_o yields

$$\ln \frac{N}{N_o} = -\lambda t$$

This equation can be simplified by taking the exponential of both sides:

$$\frac{N}{N_o} = e^{-\lambda t}$$

where N_o is the original population of radioactive atoms. Further simplification of the equation results in the familiar equation of radioactivity:

$$N = N_o e^{-\lambda t}$$

As can be seen, the radioactive decay is an exponential process.

Solved Problem 22.9.   Radioactive elements are typically characterized by a decay constant in terms of their half-life ($k_{1/2}$). Derive the decay constant, assuming a quarter-life or $k_{1/4}$.

Solution

From the radioactive decay equation,

$$N = N_o e^{-kt}$$

We substitute $N = N_o/4$ to yield

$$\frac{N_o}{4} = N_o e^{-kt}$$

$$\frac{1}{4} = e^{-kt}$$

Taking the natural logarithm of both sides reveals

$$\ln \tfrac{1}{4} = -kt$$

Solving for k gives

$$k = \frac{1.38}{t}$$

Solved Problem 22.10.   If the photon energy of a 1-mCi sample of Tc-99m is 140 keV at $t = 0$, determine the energy of a photon emitted from the sample at its half-life of $t = 6$ h.

Solution

The energy of the emitted particle never changes—only the number of particles in a radioactive sample does. Thus, regardless of the time elapsed, the energy of the emitted photon is, and will always be, 140 keV.

Solved Problem 22.11.   Oxygen-15 is a radioisotope with a half-life of 2.1 min. (1) Determine its decay constant. (2) Determine the number of radioactive atoms present in a source of activity 4 mCi. (3) Determine the time elapsed before the activity in this sample is reduced by a factor of 8.

Solution

(1) $\lambda = \dfrac{0.693}{2.1 \text{ min} \times 60 \text{ s min}^{-1}} = 5.5 \times 10^{-3} \text{ s}^{-1}$

(2) $A = \lambda N$

$N = \dfrac{A}{\lambda} = \dfrac{(4.0 \times 10^{-3} \text{ Ci})(3.7 \times 10^{10} \text{ s}^{-1})}{5.5 \times 10^{-3} \text{ s}^{-1}} = 2.7 \times 10^{10} \text{ atoms}$

(3) $t = \dfrac{\ln 8}{\lambda} = \dfrac{2.079}{5.5 \times 10^{-3} \text{ s}^{-1}} = 3.8 \times 10^2 \text{ s}$

22.5   TYPES OF RADIOACTIVE DECAY

Radionuclides undergo radioactive decay typically according to three common types:

- *Alpha decay*, caused by the repulsive electric forces between the protons, involves the emission of an alpha particle (or a helium nucleus that consists of two protons and two neutrons) by nuclei with many protons. In alpha decay, the radioactive nucleus *decreases* in A by 4 and *decreases* in Z by 2.

- *Beta decay*, which occurs in nuclei that have too many neutrons, can occur by emission of a β particle. A β particle exists in two forms. A β^- particle is an electron, and a β^+ particle is a positively charged electron, known as a *positron*. In β^- decay, the radioactive nucleus *remains unchanged* in A and *increases* in Z by 1. In β^+ decay, the radioactive nucleus *remains unchanged* in A by 4 and *decreases* in Z by 1.

- *Gamma decay* occurs by the emission of highly energetic photons. In γ decay, the radioactive nucleus *remains unchanged* in A and *remains unchanged* in Z.

Solved Problem 22.12.   One way of identifying the radiation emitted from a radioactive sample is to analyze the behavior of the radioactive emissions through an electric field. As a result, one particle was attracted to the negative pole, one was not affected at all, and a third was attracted to the positive pole. Identify the particles.

Solution

The beta particle is negatively charged and was thus attracted toward the positive pole. The gamma rays are neutrally charged and thus were not affected by the electric field. The alpha particle is positively charged and was thus attracted to the negative pole of the electric field.

Solved Problem 22.13.   An isotope of hydrogen, tritium ($_1^3$H) undergoes β^- decay. Determine the resultant nuclide following its decay.

Solution

In the β^- decay of a radioactive nucleus, one of its neutrons changes to a proton, causing an increase in Z by 1 with A remaining unchanged. Thus, the resultant nuclide following decay is

$$_2^3\text{He}$$

ILLUSTRATIVE EXAMPLE 22.1.   Nuclear Medicine and Radioactive Tracers

Nuclear medicine is a branch of medical imaging which uses radioactive tracers to obtain diagnostic information of the human body. Nuclear medicine uses radioactive tracers, injected into the circulation, that emit energetic particles upon decay. Two specific types of nuclear medicine procedures are *single photon emission computed tomography (SPECT)* and *positron emission tomography (PET)*.

Single Photon Emission Computed Tomography

One type of medical imaging procedure that utilizes radioactive agents is single photon emission computed tomography. In a SPECT procedure, a specific type of radioisotope chemically attached to a drug or pharmaceutical known to specifically target a particular organ, called a *radiopharmaceutical*, is first administered into the bloodstream of the patient. The radioisotope is a unique type of radioactive source that is a single-photon emitter or emits photons as a primary emission upon decay. As the source distributes itself and localizes within the body, it is continually emitting photons which have sufficient energy to penetrate surrounding tissues and escape the human body. The photons that escape the body are then registered and collected by a detector assembly. The detector assembly consists of a collimator which acts to restrict unwanted scattered photons from reaching the detector and therefore increases image quality and sharpness. Once they pass the collimator, the photons impinge upon a gamma ray detector which is coupled to photomultiplier tubes. The two-dimensional position and intensity of the detected gamma ray are registered to form a "snapshot" image of the object or organ. Snapshots are acquired at angular increments within a circular or elliptical orbit about the organ of interest. These snapshots are then mathematically reconstructed to provide the physician with a three-dimensional distribution of the activity within the organ.

Positron Emission Tomography

Positron emission tomography is similar in principle to SPECT in that both are forms of nuclear medicine involving the detection of radioactive emissions from a circulating radiotracer. The PET radionuclide [^{13}N, ^{15}O, ^{18}F] has an extremely short half-life and decays by emitting a positron which travels a short distance, approximately 1 to 2 mm, before a collision with an electron and subsequent annihilation. Annihilation results in two photons of equal energy, 511 keV, moving in opposite directions from each other. Because of the 180° ejection of annihilation photon pairs, two detectors situated opposite each other define a line along which the radioactive decay process occurred and facilitate accurate positioning of the decay location. The superposition of a very large number of these lines is evaluated by an image reconstruction program to generate a three-dimensional image of the organ or region being investigated. Major disadvantages of PET are that (1) radionuclides are extremely short-lived and (2) it requires the use of an on-site cyclotron.

ILLUSTRATIVE EXAMPLE 22.2.   Radiotherapy of Tumors

X-ray and γ ray radiation emitted by radioactive nuclei has energies far greater than the energy required to bind electrons to atoms and molecules. As an example, the energy contained within a carbon–carbon single bond is 4.9 eV. Therefore, when radiation of this type penetrates human tissue, it collides with an atom or molecule, prompting the localized release of large amounts of energy sufficient to break the chemical bonds between electrons and the atom or molecule, ultimately resulting in the ejection of electrons from the tissue molecules, which substantially alters their structure.

In radiotherapy, the diseased region of tissue, that is, the tumor, is exposed to high-energy radiation. Although all cells within the exposed area are affected, the healthy cells recover quickly while the diseased cells are severely damaged or completely destroyed, causing a substantial reduction or cessation of their ability to reproduce. Continual radiation therapy over an extended time permits total recovery of the healthy cells and destruction and subsequent elimination of the tumor.

Supplementary Problems

22.1.   Determine the number of (1) protons and (2) neutrons in $^{209}_{83}$Bi.

Solution

(1) 83; (2) 126

22.2.   Determine the (1) mass number and (2) atomic number of $^{32}_{15}$P.

Solution

(1) 32; (2) 15

22.3.   Determine the number of electrons in $^{36}_{16}$S.

Solution

16

22.4.   Silicon-30 has a mass number of 30 and 16 neutrons. Write the chemical element of silicon 30.

Solution

$^{30}_{14}$Si

22.5.   Determine the (1) frequency and (2) energy of 0.1-Å X-rays.

Solution

(1) 3.0×10^{19} Hz; (2) 1.25×10^5 eV

22.6.   Determine the (1) frequency and (2) wavelength of 511-keV γ rays.

Solution

(1) 1.23×10^{20} Hz; (2) 0.024 Å

22.7.   Find the (1) wavelength and (2) frequency of a 1-keV photon.

Solution

(1) 12.4 Å; (2) 2.42×10^{17} Hz

22.8.   An isotope of oxygen contains nine neutrons and eight electrons. Write the chemical element.

Solution

$^{17}_{8}$O

22.9.   Given that the half-life of ^{99m}Tc is 6 h, determine the decay constant after (1) 1 s and (2) 1 h.

Solution

(1) 3.2×10^{-5} s^{-1}; (2) 3.2×10^{-5} s^{-1}

22.10.   Given that the half-life of ^{131}I is 8.05 days, determine the decay constant.

Solution

9.96×10^{-7} s^{-1}

22.11.   If the activity of a radioactive sample decreases to one-eighth its original activity within 12 h, determine the half-life of the radioactive sample.

Solution

3.99 h

22.12.   Nitrogen 13, which decays by emission of a positron, has a half-life of 10 min. Determine the decay constant.

Solution

1.15×10^{-3} s^{-1}

22.13.   A radioactive isotope of carbon, $^{15}_{6}$C, undergoes β^- decay. Identify the new nuclide following radioactive decay.

Solution

$^{15}_{7}$N

22.14.   The half-life of tritium ($^{3}_{1}$H), an isotope of hydrogen, is 12.3 yr. Given that there are 5×10^{20} nuclei of $^{3}_{1}$H at time t, how many nuclei will remain 2.0 min after t?

Solution

4.9999×10^{20} nuclei

22.15.   The decay constant of a certain radioactive nuclide is 3.5×10^{-3} s^{-1}. Determine the half-life of this nuclide.

Solution

3.3 min

Index